Uranium prospecting handbook

Uranium prospecting handbook

Proceedings of a NATO-sponsored Advanced Study Institute on methods of prospecting for uranium minerals, London, 21 September–2 October, 1971

Edited by
S. H. U. Bowie
Michael Davis
Dennis Ostle

London: The Institution of Mining and Metallurgy

First published 1972

Reprinted 1974

ISBN 0 900488 15 8

Library of Congress Catalogue Card Number 72–87929

Contents

Organizing Committee vii

Foreword viii

List of participants xi

Opening address *Lord Penney* xiii

The status of uranium prospecting *S. H. U. Bowie* 1
  Discussion 12

Uranium supply and demand *Michael Davis* 17
  Discussion 30

Some environments of formation of uranium deposits *R. G. Dodson* .. 33
  Discussion 45

Prospecting criteria for sandstone-type uranium deposits *E. W. Grutt Jr.* 47
  Discussion 77

Uranium exploration costs *F. Q. Barnes* 79
  Discussion 93

Neutron activation analysis as an aid to geochemical prospecting for uranium *D. Ostle, R. F. Coleman and T. K. Ball* 95
  Discussion 107

Use of geochemical techniques in uranium prospecting *A. Grimbert* .. 110
  Discussion 119

Planning and interpretation criteria in hydrogeochemical prospecting for uranium *M. Dall'Aglio* 121
  Discussion 133

Instrumental techniques for uranium prospecting *J. M. Miller and W. R. Loosemore* 135
  Discussion 147

Portable equipment for uranium prospecting *Y. Puibaraud* 149
  Discussion 155

Assessment of uranium by gamma-ray spectrometry *Leif Løvborg* 157
  Discussion 171

Airborne gamma-ray survey techniques *A. G. Darnley* 174
Discussion 208

Radon methods of prospecting in Canada *Willy Dyck* 212
Discussion 241

Borehole logging techniques for uranium exploration and evaluation *P. H. Dodd and D. H. Eschliman* 244
Discussion 275

Application of computers to the assessment of uranium deposits *R. Coulomb* 277
Discussion 287

Uranium search in Australia, 1968: a case history *J. L. Montgomery* .. 289
Discussion 295

Ranger 1: a case history *G. R. Ryan* 296
Discussion 300

The Nabarlek area, Arnhemland, Australia: a case history *D. B. Tipper and G. Lawrence* 301
Discussion 304

Experimental survey with a portable gamma-ray spectrometer, Blind River area, Ontario: a case history *P. G. Killeen and C. M. Carmichael* 306
Discussion 311

Exploring for uranium in the Northern Highlands of Scotland: a case history *M. J. Gallagher* 313
Discussion 319

Final discussion session 320

Name index 333

Subject index 337

Organizing Committee

Dr. Michael Davis (*Chairman*)

Dr. S. H. U. Bowie

C. F. Bradley

G. R. Clarke

Dr. M. J. Gallagher

D. F. Kell

H. J. Millen

D. Ostle

Miss E. Ward

Foreword

Uranium prospecting handbook is based on the proceedings of an Advanced Study Institute entitled 'Methods of prospecting for uranium minerals', which was sponsored by the Science Council of the North Atlantic Treaty Organisation (NATO) and held in London from 21 September to 2 October, 1971. The meeting was organized jointly by the United Kingdom Atomic Energy Authority and the Institute of Geological Sciences.

Held at Imperial College, University of London, the Study Institute comprised a series of lectures, discussions, case histories and prospecting instrument demonstrations, followed by a field excursion to southwest England. It was gratifying that not only did it benefit from the services of a distinguished international panel of lecturers, each of whom is an expert in his particular field, but also to find among the participants (more than 100 from 22 countries) so many whose experience brought much to the formal as well as to the informal proceedings.

Although the uranium market is currently depressed by conditions of oversupply, the demands for the future to satisfy the rapidly growing installation of nuclear power reactors for electricity generation make it important that more uranium reserves be discovered. New discoveries will depend to an increasing extent on the successful application of prospecting techniques capable of identifying the most tenuous surface indications, or of detecting deposits which are hidden by barren overburden.

Each geological environment in which uranium is sought is in some way special, but it is becoming less and less likely that significant uranium deposits will be stumbled upon by the lone prospector with a Geiger or scintillation counter. The search requires shrewd geological insight, as well as knowledge and application of the most appropriate prospecting methods selected from the wide range available, if it is to have reasonable chances of economic success.

This *Handbook* describes the principles and use of virtually all the latest methods of uranium prospecting, and illustrates their use in different environments. As such, it is hoped that it will provide a convenient and authoritative reference source for all those concerned in uranium discovery.

Acknowledgment is made to the NATO Science Council, the U.K. Atomic Energy Authority and the Institute of Geological Sciences, whose Director, Dr. K. C. Dunham (now Sir Kingsley Dunham) acted as Chairman of the opening session, at which Lord Penney, Rector of Imperial College and Past Chairman of the U.K.A.E.A., gave the opening address.

The Director of the Advanced Study Institute, Dr. Michael Davis, and the Deputy Director, Dr. S. H. U. Bowie, also wish to acknowledge the help and advice they received from Mr. R. F. St. G. Lethbridge, Senior Mining Adviser to the U.K.A.E.A., members of the Organizing Committee, and other colleagues.

Thanks are due to the Institution of Mining and Metallurgy for its agreement to publish *Uranium prospecting handbook* and to Mr. M. J. Jones and his staff for their conscientious editorial work.

S. H. U. Bowie
Michael Davis
Dennis Ostle
Editors
July, 1972

List of participants*

AUSTRALIA
Armstrong C. W.
Cameron E.
Dodson R. G.
Dunlop A. C.
Kitto P. L.
McGregor P. W.
Mackenzie D. H.
Meyertons C. T.
Pratten R. D.
Robinson P.
Ryan G. R.

AUSTRIA
Cameron J.

BELGIUM
Dejonghe L.

CANADA
Aitken R. N.
Anzalone S. A.
Barnes F. Q.
Darnley A. G.
Dyck W.
Killeen P. G.
Kvendbo B. O.
Little H. W.
Meyer W.
Montgomery J. L.
Newson N. R.
Phillipson A. D.
Robertson J.
Roscoe S.
Smith E. E. N.
Williams R. M.

DENMARK
Kunzendorf H.
Løvborg L.
Nielsen B. L.

FRANCE
Coulomb M. R.
Gangloff A.
Grimbert A.
Puibaraud M. Y.
Schiltz J. C.

GERMANY
Bundrock G.
Dahlkamp F.
Fuchs H. D.
Ghosh A. K.
Hauerstein G.
Klinge U.
Müller H.
Nottmeyer D.
Ziehr H.

GREECE
Anastopoulos J.
Stavropodis J. D.

ISRAEL
Weissbrod T.

ITALY
Dall'Aglio M.

NETHERLANDS
Harsveldt H. M.

NIGER
Diallo O.

NORWAY
Nixon F.

PORTUGAL
Martins Vicente C. A. C. M.

SOUTH AFRICA
Antrobus E. S. A.
Backström J. W. von
Bagnall P. S.
Biesheuvel K.
Herzberg W.

SPAIN
Ruesgas F. C.

SWEDEN
Adamek P. M.
Lundberg B.

*Countries given are those in which participants reside.

TAIWAN
Ching Chao W.

TURKEY
Cetincelik M.

UNITED KINGDOM
Ball T. K.
Beckinsale R. D.
Berg G. W.
Bowie S. H. U.
Branscombe K.
Butler J. R.
Cole H. A.
Coleman R. F.
Davis Michael
Domzalski W.
Dowie D. L.
Eccles R. G.
Fenning P. J.
Ford I. H.
Gallagher M. J.
Huggett M. G.
Johnson P.
Kingston G. A.
Lawrence G.
Lethbridge R. F. St. G.
Loosemore W. R.
Miller J. M.
Morgan D. A. O.
Morris D. B.
Napier R. S.
Ostle D.
Plant J. A.
Riddler G. P.
Sprague R. B.
Taylor R. J.
Thompson P. J.
Tweedie K. A. M.
Waldron O. C.
Wallace R. J.
Ward E. H.
Watson J. F.
Weller R. K.
Wilson I. R.
Wilson R. A. M.

U.S.A.
Chico R. J.
Dodd P. H.
Grutt E. W. Jr.
Koch L. W.
Schwarzer T. F.

ZAMBIA
Rodger T. H.

Opening address

Lord Penney Dr. Dunham, Ladies and Gentlemen: May I first offer you a very cordial welcome to this Study Institute. To those of you from overseas, and especially any who are visiting this country for the first time, I extend a particularly warm greeting. I hope you will enjoy your stay here and find it useful.

Those of you who, like me, first became associated with the exploitation of nuclear energy many years ago, will know how the interplay of science, technological development, economics and politics has swayed the fortunes of nuclear power. Energy, especially electric energy, is vital to human society as we know it. Nuclear energy, in spite of the ups and downs of the last twenty years—and that is a short time, even compared with one man's lifetime—is the only certain source of the vast quantities of electric power needed to continue living as we do into the next century and beyond.

The early days of nuclear energy, the mid-1950s, did not get the perspective right. The popular press exaggerated wildly, and there was a general misunderstanding about the cost of nuclear electricity. Suez and oil politics seemed to have an easy solution. There was a rush to find uranium—a rush reminiscent, in some ways, of the gold rushes of earlier days. But, unfortunately, uranium did not prove as universal a currency as gold; and when the time—and cost—of nuclear power development proved greater than expected, and both defence and national nuclear power programmes were cut back, the actual and potential supply of uranium was quickly seen to exceed the likely demand for many years. The consequences were that mining operations had to be stretched out and interest in prospecting waned rapidly.

There was some revival in prospecting activity in the second half of the 1960s, significant discoveries leading, in recent years, to known reserves at $10 per lb or less being increased to almost 1000000 short tons of U_3O_8. These relatively recent discoveries have coincided with uncertainty about the size of nuclear power programmes, which has, at least for the moment, dulled the incentive for mining companies to undertake, or sponsor, major uranium prospecting programmes.

Our experts believe, however, that we may look forward to a dramatic change in the situation during the course of the 1970s. On current nuclear power plans the known uranium reserves at a price not exceeding $10 per lb of ore will be exhausted by about 1985, so to do no more than maintain the reserves at the current level in absolute terms will require the discovery of a further million tons of ore during the next 10–15 years.

This becomes the more urgent because of the position of oil. The recent round of price increases and the likelihood of more to come should cause utilities to think seriously about building too many oil-fired power stations. There must also be doubts about how long the world's oil resources are going to last. On present indications they will be exhausted within the lifetime of today's children. Moreover, oil, like coal, has pollution problems, which either have to be endured or cured at great cost.

So I agree that the nuclear energy experts are right in thinking that uranium is a key and precious material. Before long, and here I mean perhaps a couple of decades, we will have to integrate the uranium into the plutonium fast-breeder fuel cycle, and then we really do have a large world reservoir of energy.

I think that we have to assume that the intensive uranium prospecting work of earlier years will have found most of the more readily detectable uranium deposits, and that the task of uncovering further deposits of good yield and of sufficient size will depend on the use of sophisticated techniques capable of identifying the most tenuous surface indications and of detecting deposits covered by entirely barren overburden.

This Study Institute has been planned to introduce you to these techniques and has brought together experts of world standing in their fields to lecture and share with you their experiences—and yours with them. Next week a number of you will be going down to Cornwall—a very delightful county at this time of year—to see some demonstrations of equipment in the field and something of an area which is one of the cradles of the world's mining industry.

I hope that all of you will enjoy the course and find it valuable. You are all very welcome.

The status of uranium prospecting

S. H. U. Bowie D.Sc., F.R.S.E., F.I.M.M.

Institute of Geological Sciences, London, England

550.8: 553.495 2

Synopsis

Geological knowledge is considered to be a prerequisite of successful uranium reconnaissance: hence, the prospector should obtain as much information as possible on the different types of deposit presently known and on the environments in which they occur. Understanding of the genesis of deposits is also vitally important, particularly when attempts are being made to predict the whereabouts of a new uranium province. The main features of known orebodies are therefore described and some thoughts on their genesis are conveyed, reference being made to important publications which stem from information gathered during the intense period of search from 1945 to 1955.

No single prospecting technique can be regarded as a panacea: rather, each of the methods described is regarded as having specific applications in particular environments and for different types of deposit. If one approach could be singled out, it would probably be aeroradiometric surveying with a sensitive total-count instrument with strict geological control during and after flight. Recent information, however, shows that geochemical techniques based on neutron activation analysis could prove to be equally effective and appreciably cheaper.

Emphasis is placed on the need for continued search for uranium orebodies—particularly those hidden by barren overburden. And, to make this feasible, suggestions are made concerning relatively untried techniques, such as radon monitoring of ground air and isotopic analysis for radiogenic lead in non-radioactive minerals. In addition, reference is made to the novel techniques of measuring radiogenic heat, the determination of radon and its daughter products in the atmosphere and the possible value of ^{222}Rn:^{4}He ratio measurements.

It seems reasonable that anyone who has been closely associated with most aspects of the uranium industry for 25 years should be permitted to attempt to put the various techniques of prospecting for uranium into perspective—it being realized, of course, that the most appropriate method for one type of terrain may be entirely unsuited to another. In 1955, at the Second United Nations International Conference on the Peaceful Uses of Atomic Energy, it was stated that *in no phase of mineral exploration is there a wider range of technical methods at the disposal of the prospector than in the search for radioactive ores. Indeed, so many are the techniques available, and so vociferous are their respective advocates, that it tends to be forgotten that successful prospecting is almost always based upon the application of fundamental geological knowledge.*[1] The same applies today, despite the recent development of new and important methods of detecting uranium orebodies.

Some geological concepts

Several authors attempted to assimilate the immense amount of information obtained during the decade of activity in the search for uranium from 1945 to 1955. Of particular interest is the work of Page, Stocking and Smith[2] and Kerr.[3] Later, Nininger and co-workers[4] and Page[5] discussed the genesis of uranium ore deposits. Previously, Davidson[6] had discussed the occurrence of uranium in ancient conglomerates, and Nel[7] followed this by a paper on the conglomerates of the Witwatersrand system. Numerous other contributions to uranium deposit genesis have since been published, including a review of uranium in Precambrian shield areas by van Wambeke[8] and a detailed account of Huronian rocks and uraniferous conglomerates in the Canadian Shield.[9] More recently, worldwide favourability for uranium deposition has been discussed by a panel of uranium geologists.[10]

Perhaps the most significant feature of presently known uranium deposits is their association with Precambrian rocks. More than 90% of reserves occur in Precambrian conglomerates or in Phanerozoic rocks closely underlain by Precambrian rocks. Uranium appears to have been concentrated in metallogenic provinces in early Precambrian times during an era of partial or complete melting of the outer crust of the earth when lithophile elements were concentrated in acidic rocks. Subsequent redistribution of uranium in provinces formed at that time most probably resulted from orogenic processes and associated anatexis.

Nearly all uranium geologists accept the association of uranium with acid rather than basic rocks, though uranium vein deposits are frequently found in close proximity to basic derivatives of magmatic differentiation—for example, diabase dykes. The reason why uranium occurs in Precambrian oligomictic conglomerates is still speculative, though a straightforward placer origin is losing favour.[6,7,10] Likewise, although there is no widely accepted mode of origin for peneconcordant uranium deposits in fluviatile sandstones, age determination data indicate episodic crystallization of uranium minerals, and there is growing evidence that the uranium was not simply derived from the erosion of granitic rocks and redeposited in a closed basin. Some geologists consider that uranium is mainly contained in the resistate accessory minerals of granites, and hence survives in detrital minerals; others believe that the uranium is dissolved during weathering processes and precipitated in favourable horizons in sediments. Neither of these views appears to be valid. They are negated by new information on uranium ore genesis—and this probably applies to most metalliferous deposits—which shows that the time parameter in deposition was extensive.[11] For example, the evolution of the uranium deposits of the Beaverlodge area took place during six discrete periods covering a time-span of more than 2200 m.y.;[12] mineralization in the Blind River–Elliot Lake field ranges from 2500 to 600 m.y. ago; and primary mineralization took place in sandstones in the Colorado Plateau over at least two main periods 210 and 110 m.y. ago.[13] As was indicated above, phases of mineralization appear to be associated with orogenic events and the uranium was either introduced into favourable sediments by way of deep-seated faults or, alternatively, uranium in existing minerals was mobilized and redistributed.

Whatever the facts are, enough is known about uranium and its association for useful geological guides to be given to assist in the choice of areas in which there is a good chance of uranium being discovered. Some uranium will, no doubt, be found where it is not expected, and regions not yet well enough known will, of course, become regarded as favourable as geological mapping advances, particularly in the less well developed countries.

Types of uranium deposit and the environments in which they occur

OLIGOMICTIC CONGLOMERATES

Uranium-bearing quartz-pebble conglomerates occur in Precambrian intracratonic basins in the relatively stable Shield areas of the world, where further deposits can be expected to be found. Mineralization is peneconcordant and usually of oxide minerals such as uraninite and brannerite. Some hydrocarbon material is generally present and abundant pyrite is almost always developed. Pay-grade mineralization is sometimes confined to one or two horizons—which is essentially so at Blind River—or occurs in several different widely separated horizons—as in the Witwatersrand. Grades are in the range of 0·025–0·15% U_3O_8, but mineralization covers large areas and is related to sedimentary features such as proximity to the basement, unconformities, troughs and ancient stream channels.

SANDSTONES

Uranium deposits occur mainly in fluviatile quartzose sandstones and mudstones of continental origin and tend to be relatively high-grade (0·2% U_3O_8) and to be of two different types, exemplified by those of the Colorado Plateau and of the Wyoming basins. The former are thin tabular peneconcordant layers in lithified moderately arkosic sandstone. The latter are long, arcuate bodies that occur sporadically in unlithified highly arkosic sandstone units.

Ore minerals in the Colorado-type deposits are of uraninite and coffinite, with associated vanadium and copper minerals, whereas in the Wyoming deposits uraninite and coffinite are the only ore minerals.

The most favourable environments for uranium deposits in sandstones are intermontane basins in relatively stable regions subjected to regional uplift and with associated acid igneous rocks, abundant interbedded organic matter and the presence of tectonic and stratigraphic features that acted as traps for mineralization. Subtle changes in Eh and pH and in confining pressures are believed to have caused precipitation of uranium from solution. The Colorado-type deposits apparently formed *in situ* under highly reducing conditions and were probably of earlier origin than those of Wyoming, which are of the so-called 'roll' type. The latter clearly formed by the movement of oxygenated groundwater down-dip until free oxygen was used up mainly in oxidizing sulphides and other minerals in the sandstone. At the point when reducing conditions prevailed, and some way beyond, uranium precipitated. This process is possibly still taking place.

In the U.S.A. experienced prospectors can recognize the red or drab colouring in sandstone due to the oxidation of pyrite-bearing carbonaceous facies, which is particularly favourable to uranium mineralization. Oxidation–reduction boundaries in roll-type deposits can also be recognized in this way.[14]

VEIN-TYPE DEPOSITS

Vein-type deposits tend to occur in metallogenic provinces in mobile belts—though not all deposits in mobile belts are vein-type—and are associated with leucocratic rather than melanocratic rocks. Provinces commonly exhibit a zoning of elements, which is essentially temperature-dependent and has developed both in space and time. Uranium occurs with hypothermal and mesothermal zone minerals, such as tin, copper, cobalt, vanadium and arsenic, rather than with typical epithermal assemblages. Examples are the vein deposits of France, Spain, Portugal and Czechoslovakia, in which the main uranium mineral is pitchblende.

PEGMATITES

So far, pegmatites have only contributed a small portion of the world's output of uranium. The most important to have been worked are the complex granitic dykes and lenses in highly metamorphosed Precambrian sediments in the

Bancroft area of Ontario. The main ore minerals are uraninite and uranothorite, but other radioactive species, such as betafite, fergusonite and allanite, are also present.

Somewhat similar pegmatites, also in highly metamorphosed Precambrian sediments, occur at Rössing near Swakopmund, South West Africa.[15] Uraninite is the dominant uranium mineral, with betafite as an accessory, and there are several uranium secondaries, such as uranophane, in the oxidized zone. The uraninite tends to occur in narrow stringers a few millimetres across, and mining is therefore likely to be by open-pit methods only.

MAGMATIC URANIUM

Uranium in acid igneous rocks rarely exceeds 40 ppm. Some peralkaline rocks, however, such as those of Ilímaussaq, Greenland, and Seal Lake, Labrador, contain up to 1500 ppm U_3O_8. Uranium is mainly concentrated in refractory minerals, which makes recovery uneconomic unless circumstances exist where associated elements, such as thorium, beryllium or niobium, could be co-products.

PYROMETASOMATIC DEPOSITS

The only important deposit of this type is that of Mary Kathleen, Queensland. The uranium mineral, uraninite, occurs along with rare-earth silicates and sulphide minerals in a garnetized zone of calc–silicate rocks of early Precambrian age. The mineralization is closely associated with the Mount Burstall granite, and is localized by faulting and shearing.

SYNGENETIC URANIUM IN SEDIMENTS

Uranium is frequently enriched by sorption or chemical combination in shales, lignites and phosphorites. Such occurrences are usually of low grade, but contain vast resources of uranium. Best examples are the alum shales of the Västergotland and Närke districts of Sweden, where uranium forms an organo–uranium complex associated with pyrite, quartz and feldspar and the clay minerals illite and kaolinite. Reserves are more than 1000000 tons U_3O_8 at a grade of the order of 300 ppm. Similar or higher concentrations of uranium occur in lignites, such as those of North and South Dakota and Saskatchewan. Syngenetic uranium—almost certainly derived from sea water—also occurs in amounts as high as 100–200 ppm U_3O_8 in beds several feet thick in the phosphorite deposits of the Phosphoria and Bone Valley Formations of the U.S.A.; and even richer phosphates, with between 0·05 and 0·20% U_3O_8, occur in belts of phosphates in Cabinda and southwest Angola. Extraction of uranium from the West African localities seems economically possible if the pilot-plant studies which employ a bisulphate of ammonia attack prove successful.[16]

Prospecting techniques and their application

AERORADIOMETRIC SURVEYS

The choice of method of approach to uranium reconnaissance depends to a large extent on finances available and the time-scale involved. Thus, if funds are plentiful and quick results are required, there is little doubt that the best method to follow would be to obtain such geological data as were available and choose favourable regions to be covered by a sensitive total-count airborne scintillometer. Costs of total-count surveys are less than one-third of those for a high-sensitivity gamma-spectrometry survey, and in many circumstances they are likely to be equally effective in discovering uranium orebodies. No single prospecting technique, however, can be regarded as a panacea. The aeroradiometric method suffers, like others, in particular conditions of terrain. For example, relief must not be so excessive that the aircraft cannot maintain the specified ground clearance with an accuracy of better than $\pm 20\%$. Flying should be possible at or near the optimum ground clearance of approximately 150 m; and to obtain virtually complete cover, line spacing should be at 1000-m intervals. Climatic conditions must be suitable for low flying and the terrain not covered by dense vegetation, muskeg, aeolian sand or other superficial material, which would seriously reduce or stop gamma radiation.

Instrument sensitivity is important in aeroradiometric surveying and should be such that anomalies with a concentration factor of 10 or less can be detected. Concentration factor in this case is simply area of anomaly (m^2) times eU_3O_8 (%). Of even greater significance is the need for geological interpretation of all anomalies, preferably by a trained geologist in the aircraft. This can eliminate most of the anomalies due to potassium or to detrital thorium minerals and lead to immense reductions in the cost of follow-up work.

For gamma spectrometry from the air to be successful even more accurate ground clearance must be maintained, preferably by linear radioaltimeter, and, generally, flying has to be at about 100 m. In order to achieve statistically significant count rates in the respective channels representative of U, Th and K, a NaI(Tl) crystal volume of the order of 50000 cm^3 is necessary. Furthermore, a high degree of stabilization is essential to avoid spectral drift outside the selected channels.[17,18] A serious weakness of most currently available gamma spectrometers is the overaccentuation of potassium due to the degradation of high-energy radiation by Compton scattering and to this lower energy radiation being recorded in the potassium channel. This gives the false impression that uranium or thorium occurrences are potassium-rich. An important use of gamma spectrometers from the air is in follow-up studies of anomalous regions by use of either light fixed-wing aircraft or helicopters. Gamma spectrometers, however, probably have their greatest use in geological mapping,[18] but it should be remembered that total gamma activity also gives an excellent correlation with geology[19] since U, Th and K are often closely associated in nature.

Although it is accepted that additional information is rarely useless, there is an overriding reason why gamma spectrometry in uranium reconnaissance must be treated with particular care. The reason is that more often than not uranium at the surface is not in equilibrium with its daughter products. Thus, in circumstances where ^{214}Bi has been depleted by surface leaching there could well be a tendency to ignore an anomaly caused by combined uranium and thorium deposits which showed up as being appreciable only in the ^{208}Tl channel. All anomalies that cannot readily be explained on geological grounds require follow-up, because radiation intensity measured by an instrument at 100–150 m above ground is no guide to the economic significance of an occurrence.

VEHICLE-BORNE RADIOMETRIC SURVEYS

Vehicle-borne Geiger–Müller and scintillation equipment has been used successfully in the discovery of uranium. The method has the advantage of low cost per mile of traverse and can be used either systematically or simply installed in a vehicle used for other purposes and likely to cover a considerable non-repetitive mileage. The method suffers because of the limited area accessible and because radiation is only detected from a relatively narrow strip adjacent to the track of the vehicle. This can be overcome to some extent by mounting the detector on a mast high above the vehicle.[17]

Fully stabilized gamma spectrometers with a probe that can be carried a few hundred metres from the path of the vehicle offer good prospects for follow-up studies since samples can be taken and tested without the complications of air attenuation or Compton scattering experienced in airborne quantitative measurements.

ON-FOOT RADIOMETRIC SURVEYS

On-foot surveying is of paramount importance in secondary prospecting—that is, prospecting in known uraniferous provinces—and in the detailed mapping of known anomalies. Both Geiger–Müller and scintillation counters are of value, but preference is for a light-weight scintillation instrument with a NaI(Tl) crystal of about 30 cm^3. Such a counter permits changes of less than 2 ppm eU_3O_8 in bedrock to be recorded and can be used at normal walking pace. Geiger–Müller instruments have the advantage of being more readily adaptable for beta-radiation surveys, which can be valuable both in the examination of pits and trenches and in the delineation of ore underground. Beta–gamma analysis is also possible and the degree of secular disequilibrium of sample material can readily be established by this method—though not with a high degree of accuracy. Either type of instrument also has an important use in detecting uranium under overburden if fitted with a probe unit that can be lowered into holes penetrating to bedrock.

Portable gamma spectrometers are valuable for the analysis of mixed uranium–thorium ores, such as those in the Elliot Lake–Agnew Lake area, where there are regional changes in uranium to thorium ratios of from 4:1 to 1:4. Similar conditions can be expected in other conglomerate deposits—for example, the Dominion Reef, South Africa.

There is the same need for stability in portable gamma spectrometers as there is with air- and vehicle-borne instruments. This has been achieved in few instruments to date, but it is clearly only a matter of time before stable two-, three- or four-channel instruments become available.[21] It is doubtful, however, if the more complicated spectrometer instruments will ever replace a good scintillometer for the geological assessment of most radiometric anomalies.

MEASUREMENT OF RADON

The quantitative determination of the concentration of radon in 'ground air' has been applied successfully in recent years as a method of detecting uranium deposits obscured by sand, peat, glacial drift or other superficial material. The method is particularly suited to secondary surveys in areas known to be favourable to uranium mineralization. Two types of apparatus are available: one measures radon in ground air in the immediate vicinity of a shallow hole in which the probe is inserted; the other employs a pump to extract ground air from the hole.[17,20,21] The former method tends to be less sensitive, but is more specific and is particularly suited to defining the trace of veins, pegmatites and other relatively narrow orebodies. It is also effective in locating low-grade disseminated uranium, provided that the overburden is not too thick, and has a considerable potential in tracing faults hidden by overburden. The latter method is more appropriate for the location of more deeply buried deposits of large area.[21] The measurement of radon in water has been successfully used by Dyck[22] in the study of lake and stream samples in Canada, and a modified radon monitor has been used for water samples in the United Kingdom.[21]

HYDROGEOCHEMICAL, STREAM-SEDIMENT AND SOIL SAMPLING METHODS

Geochemical techniques have not yet been widely applied in the search for uranium because of the ease with which radiation detectors can be used semi-quantitatively in the field. It has long been realized, however, that hydrogeochemical sampling is a valuable method of search in heavily forested mountain terrain or other regions, such as lakelands, that are difficult of access.[17,20] A variety of analytical methods exists capable of detecting uranium in water down to 0·2 μg/l or less. Also, a colorimetric method has been used to determine uranium down to 1·0 μg/l at the water side.[23]

In recent years the availability of neutron activation techniques has revitalized geochemical prospecting by use of water, stream-sediment and soil samples, particularly in the location of hidden uranium orebodies. The method is specific, rapid and relatively cheap,[24] though it has the disadvantage that access to a reactor is necessary. The analysis of stream sediments for uranium is useful both in providing evidence of uranium provinces and in secondary prospecting. It is strongly recommended that stream-sediment samples collected in the course of base-metal exploration programmes be analysed for uranium—particularly when associated lithophile elements are being sought.

Because radium (^{226}Ra) is less readily removed from soils and weathered rock than uranium, it has been suggested that it might be the preferred element to measure.[29] There is little evidence, however, as to why radium might be expected to give better results than could be obtained by the precise measurement of the element sought. A correlation between uranium and radium could only be expected on the basis of chemical similarity in the behaviour of the elements. It therefore seems likely that radon is a better indicator of the general whereabouts of a uranium deposit, neutron activation or chemical analysis for uranium being used as a follow-up. Radium recombines readily with ions of similar radius, such as Ba^{2+} and Sr^{2+}, and can be fixed (for example, as radian baryte) some distance from its original source. An overriding reason for not measuring radium chemically is that the radium: uranium ratio in uraniferous minerals in equilibrium is 1:3 500 000: hence, it can scarcely be regarded as a sensitive method of locating uranium ore.

GEOBOTANICAL OR BIOGEOCHEMICAL METHODS

No direct ecological aid to uranium prospecting is known, but in semi-desert regions where soil cover is thick and plants are deep-rooted (down to 20 m) the analysis of plant leaves may give an indication of uranium in depth. Two prospecting methods have been used. In one the foliage of identical species of trees or shrubs is collected on a grid system, ashed and analysed; in the second the distribution of indicator plants, which flourish on soils containing elements associated with uranium, is studied.

Common values for uranium in ashed foliage in barren ground is $\approx$0·5 ppm U, whereas in trees or shrubs rooted in or near uranium ore deposits values of 1–100 ppm U or more have been recorded. The depth to which ore can be detected depends on the root habits of the plant and on groundwater circulation. In the Colorado Plateau the deep-rooted vetch, *Astragalus pattersoni*, requires considerable amounts of selenium for its healthy growth and, since selenium is commonly associated with uranium, the presence of abundant *Astragalus* can be regarded as indicative of the presence of uranium.[25]

INDICATOR-ELEMENT TECHNIQUES

Indicator elements used in the search for uranium are those which are either more mobile or occur in greater abundance. Thus, they either give a much wider dispersion halo or are easier to detect chemically than uranium. Selenium, mentioned above, is a good example of an associated element. Other elements, such as copper, lead, cobalt, vanadium, molybdenum or arsenic, may in certain terrain be regarded as indicator elements. On the Colorado Plateau, for example, regional metal zoning related to orogeny[26] shows a halo of uranium and vanadium with vanadium from 5 to 40 times more abundant than uranium, and, hence, an excellent indicator element.

Relatively untried techniques

RADIOGENIC LEAD IN NON-RADIOACTIVE MINERALS

The presence of abnormal amounts of radiogenic lead (^{206}Pb and ^{207}Pb) in non-radioactive minerals, such as galena or pyrite, was first suggested by Cannon and co-workers[27] in 1958 as being indicative of close proximity to uranium occurrences. Subsequently, Tugarinov *et al.*[28] confirmed the usefulness of the method and drew specific attention to $^{206}Pb:^{208}Pb$ ratios in the total lead of a mineral as a direct indicator of uranium. They recognized, however, that in a thorium province ^{208}Pb would present complications. It is clear that at least in some environments the method has considerable potential and there is a need for further research to be carried out on a worldwide scale to define its potentialities as well as its limitations. An obvious restraint in the use of isotope ratios results from the need to employ mass spectrometers, which are readily available only to the larger geological institutes and mining houses.

GEOPHYSICAL METHODS

Orthodox geophysical techniques have rarely been used in the search for uranium orebodies and probably have little application at the prospecting stage, except in detecting associated elements. In proving uranium deposits, however, several methods have been used with success in defining structural controls and other features. Resistivity, induced polarization and electromagnetic methods have important applications and could well be used with other more direct methods, such as radon concentration, in the location of buried orebodies.

Novel techniques

MEASUREMENT OF RADIOGENIC HEAT

It is well known that considerable radiogenic heat is produced by uranium and its decay products. 1 g of uranium in secular equilibrium produces 0·73 cal/g/year. Heat flow to surface over rocks containing a uranium deposit must therefore be considerably greater than normal for comparable crustal rocks. For example, a column of granite 1 km long, with a cross-sectional area of 1 cm^2, produces $\approx$2·0 cal/year. It seems, therefore, that radiogenic heat, even from a fairly deep-seated uranium orebody, could be measured at surface by thermal sensors. Alternatively, high-resolution thermal infrared scanning techniques could be employed from an aircraft. Modern equipment of this type is capable of detecting thermal changes of $\pm$0·25°C at altitudes of up to 3000 m. If normal surface variations in thermal emissivity characteristics are such as to mask radiogenic heat variations, heat measurements by probing below the surface layers might prove successful.

AIRBORNE GEOCHEMISTRY

Although gamma spectrometry is, strictly speaking, a geochemical and not a geophysical technique, more positive elemental information could almost certainly be obtained by studying the distribution of radon and its daughter products in the lower levels of the atmosphere. Two methods are possible: one in which radon (^{222}Rn) is monitored by use of an ionization chamber or other detector system in an aircraft; the other in which one (or more) of the decay products ^{210}Pb, ^{210}Bi and ^{210}Po is measured.

Radon escapes from uranium deposits, whether exposed or buried, and mixes with atmospheric air. In turbulent conditions it tends to be dispersed and to decrease with altitude, but under an inversion—which under some meteorological conditions is a diurnal phenomenon—abnormal amounts of radon build up. In addition, there is an accumulation of the longer-lived decay products of radon. Decay products of thoron (^{220}Rn) are all short-lived and not likely to be a problem. In fact, the combined measurement of ^{212}Pb with ^{210}Pb, ^{210}Bi or ^{210}Po might well be used to evaluate atmospheric mixing and wash-out processes and so make airborne geochemistry more quantitative. The daughter nuclides of radon and thoron tend to become attached on to aerosols and could easily be collected by the method of Weiss.[30] Surface conditions will always be a hazard in quantitative work, as is the case with radon in ground air.

Assessment of uranium occurrences

It is not within the scope of this paper to cover the evaluation of uranium occurrences save in outline. Briefly, the techniques are those normally employed in assessing concentrations of other metallic ores. But, in addition, many of the methods used in primary prospecting can be used, sometimes in modified form, in mineral appraisal.

Investigations normally take the form of a preliminary examination, which involves geological mapping, gridding and the production of a map showing contours of gamma radiation or levels of concentration of radon. This, together with all available information on topography, drainage, primary and secondary mineral dispersal and equilibrium conditions, helps to locate the mineralization more precisely and provides useful information on the type of occurrences. Subsequent procedure will depend much on circumstances, but if further work is justified, it will normally involve more detailed mapping, pitting, trenching, sampling and exploration drilling.

Where core recovery from diamond drill holes is poor, or where extensive pattern drilling is justified, borehole logging techniques with a relatively small number of cored holes are used. Down-the-hole gamma spectrometry is invaluable in conditions where the U:Th ratio changes appreciably. Useful correlation of gamma-logging data with ore grade is only usually possible in the case of relatively homogeneous concordant or peneconcordant deposits. Secular disequilibrium of uranium can lead to erroneous results and in all such cases frequent checks should be made by normal chemical or neutron activation methods. Disequilibrium is usually marked only where secondary hydroxide, phosphate or vanadate minerals are developed. The percentage disequilibrium may vary markedly from place to place without the bulk of the orebody being seriously out of equilibrium.

A counter with a beta-probe attachment[17] is invaluable for *in situ* analysis and for the approximate determination of equilibrium conditions by beta to gamma ratio measurements.

Conclusions

It is no longer valid to say that *uranium is where you find it*, although occasional wildcat discoveries will continue to be made. Successful exploration programmes are likely to be based on geological knowledge, incomplete though this may be, and on the application of one or more of the many chemical and physical search techniques now available. Uranium is a lithophile element associated with other elements that have ions with an outermost eight-electron shell and is concentrated in the outermost part of the earth's crust. Primary chemical differentiation, which took place in the early history of the earth, almost certainly accounts for the preponderance of uranium in granitic rocks of the Precambrian and in early quartz-pebble conglomerates derived from acid igneous rocks. Uranium has been redistributed since Precambrian times by magmatic, metasomatic, pneumatolytic and hydrothermal processes, as well as in the sedimentary cycle and by biological processes—particularly associated with organic matter.

Once a search area has been selected geologically, a decision will have to be taken on the best method or methods of reconnaissance survey to employ to locate orebodies. The facile answer is usually aeroradiometric survey—preferably spectrometric—or a multi-method approach, but, as has been indicated earlier, the former approach has its limitations and the latter is likely to prove inordinately expensive. The exploration geologist should decide on the prospecting method or methods on known geology, or on such information as can be gleaned from aerial photography, together with physiographical and ecological conditions in the region under study. Hydrogeochemical methods, for example, are likely to prove more successful in the lake-covered areas of Canada or Finland than aeroradiometric surveying; and the analysis of tree leaves could well give more useful information in forested country than any other technique. In country where the mode of occurrence of uranium is little known, orientation surveys to establish the most suitable search techniques are recommended and any information available from stream-sediment analysis should be carefully studied.

Of the recently developed prospecting methods, the measurement of radon in ground air and in water seems to hold most promise as a means of detecting hidden uranium deposits. The possibility of combining the measurement of ^{222}Rn with ^{4}He could give valuable additional information on the location of uranium if a relatively simple method of helium isotope determination could be devised. The measurement of ^{222}Rn and one or more of its daughter products in the atmosphere is a relatively easy method to apply and holds considerable promise.

The presence of abnormal amounts of radiogenic lead in non-radioactive minerals has as yet been inadequately tested, but the method could well prove effective in defining broad areas of uranium favourability.

It is by no means certain that adequate reserves of low-cost uranium will be available to satisfy future needs unless research both into the geology of uranium and into methods of discovering workable deposits continues. The solution is not, as is often suggested, the extraction of uranium from lower and lower grade ores by opencast methods—which could create more problems than it would solve—or of extracting uranium from sea water. Rather, for the foreseeable future, effort should be concentrated on establishing resources of uranium in ore deposits of types that are currently economic. Such researches need not be excessive, but they must be sustained. In this age of remote sensing, sophisticated electronics and geochemistry it tends to be forgotten that all discoveries have to be assessed by experienced geologists on foot. Proving reserves and the amenability of ore to treatment is a highly skilled task in which shortcuts are dangerous. Franc R. Joubin's discovery of the Blind River field and the subsequent drilling of 700000 ft of strata between 1953 and 1958, which indicated much of Canada's uranium reserves, is a unique example of the application of available geological knowledge combined with sound common sense. The efforts that led to this addition of ore reserves are worthy of emulation, though they are unlikely to be repeated. Likewise, it is not to be expected that a major mine can be opened up after as few as seven diamond drill holes have been put down.[31] As a guide to the non-geologist, it is safe to say that deposits of high grade tend to be of small tonnage; those of large tonnage are more likely to be of low or sub-marginal grade.

Acknowledgment

This paper is published by permission of the Director, Institute of Geological Sciences, London.

References

1. Davidson C. F. and Bowie S. H. U. Methods of prospecting for uranium and thorium. In *Proc. Int. Conf. peaceful Uses atom. Energy, Geneva, 1955* (New York: U.N., 1956), vol. 6, 659–62.
2. Page L. R., Stocking H. E. and Smith H. B. Contributions to the geology of uranium and thorium by the United States Geological Survey and Atomic Energy Commission for the United Nations International Conference on Peaceful Uses of Atomic Energy, Geneva, Switzerland, 1955. *Prof. Pap. U.S. geol. Surv.* 300, 1956, 739 p.
3. Kerr P. F. The natural occurrence of uranium and thorium. In *Proc. Int. Conf. peaceful Uses atom. Energy, Geneva, 1955* (New York: U.N., 1956), vol. 6, 5–59.
4. Nininger R. D. *et al.* The genesis of uranium deposits. In *Rep. 21st Int. geol. Congr.* (Copenhagen: Berlingske Bogtrykkori, 1960), pt. 15, 40–50.
5. Page L. R. The source of uranium in ore deposits. In *Rep. 21st Int. geol. Congr.* (Copenhagen: Berlingske Bogtrykkori, 1960), pt. 15, 149–64.
6. Davidson C. F. On the occurrence of uranium in ancient conglomerates. *Econ. Geol.*, **52**, 1957, 668–93.
7. Nel L. T. The genetic problem of uraninite in the South African gold-bearing conglomerates. In *Rep. 21st Int. geol. Congr.* (Copenhagen: Berlingske Bogtrykkori, 1960), pt. 15, 15–25.
8. van Wambeke L. *Some geologic concepts as a guide for the search of uranium in the Precambrian shields.* EUR 3481.e, Euratom, Joint Nuclear Research Center, Ispra, 1967, 63 p.
9. Roscoe S. M. Huronian rocks and uraniferous conglomerates in the Canadian Shield. *Pap. geol. Surv. Can.* 68–40, 1969, 205 p.
10. *Uranium exploration geology* (Vienna: IAEA, 1970), 384 p.
11. Bowie S. H. U. Uranium: the present and the future. In *Mining and petroleum geology* (London: IMM, 1970), 405–20. (*Proc. 9th Commonw. Min. Metall. Congr. 1969, vol. 2*)
12. Koeppel V. Age and history of the uranium mineralization of the Beaverlodge area, Saskatchewan. *Pap. geol. Surv. Can.* 67–31, 1968, 111 p.
13. Miller D. S. and Kulp J. L. Isotopic evidence on the origin of the Colorado Plateau uranium ores. *Bull. geol. Soc. Am.*, **74**, 1963, 609–30.
14. Adler H. H. Interpretation of colour relations in sandstone as a guide to uranium exploration and ore genesis. Reference 10, 331–42.
15. von Backström J. W. The Rössing uranium deposit near Swakopmund, South West Africa: a preliminary report. Reference 10, 143–50.
16. Gautier R. Perspective nouvelles dans le traitement des minerais d'uranium et l'obtention de concentrés directement utilisables. In *Recovery of uranium* (Vienna: IAEA, 1971), 17–32. (STI/PUB 262)
17. Bowie S. H. U. and Bisby H. Methods of detecting and assessing low grade uranium deposits. *J. Br. nucl. Energy Soc.*, **7**, 1968, 169–77.
18. Darnley A. G. and Grasty R. L. Mapping from the air by gamma-ray spectrometry. In *Geochemical exploration* (Toronto: CIM, 1970) 485–500. (*Proc. 3rd Int. geochem. Explor. Symp. 1970*) (*CIM Spec. vol.* 11)
19. Bowie S. H. U. *et al.* Airborne radiometric survey of Cornwall. In *Proc. 2nd U.N. Int. Conf. peaceful Uses atom. Energy, Geneva, 1958* (Geneva: U.N., 1958), vol. 2, 787–98.
20. Peacock J. D. and Williamson R. Radon determination as a prospecting technique. *Trans. Instn Min. Metall.*, **71**, Nov. 1961, 75–85.
21. Miller J. M. and Loosemore W. R. Instrumental techniques for uranium prospecting. This volume, 135–46.
22. Dyck W. Field and laboratory methods used by the Geological Survey of Canada in geochemical surveys. No. 10. Radon determination apparatus for geochemical prospecting for uranium. *Pap. geol. Surv. Can.* 68–21, 1969, 30 p.

23. Smith G. H. and Chandler T. R. D. A field method for the determination of uranium in natural waters. In *Proc. 2nd U.N. Int. Conf. peaceful Uses atom. Energy Geneva, 1958* (Geneva: U.N., 1958), vol. 2, 148–52.
24. Ostle D., Coleman R. F. and Ball T. K. Neutron activation analysis as an aid to geochemical prospecting for uranium. This volume, 95–107.
25. Cannon H. L. and Kleinhampl F. J. Botanical methods of prospecting for uranium. In *Proc. Int. Conf. peaceful Uses atom. Energy, Geneva, 1955* (New York: U.N., 1956), vol. 6, 801–5.
26. Gabelman J. W. and Krusiewski S. V. Zonal distribution of uranium deposits in Wyoming, U.S.A. In *Proc. 22nd Int. geol. Congr. Delhi 1964*, in press.
27. Cannon R. S. Jr., Stieff L. R. and Stern T. W. Radiogenic lead in nonradioactive minerals: a clue in the search for uranium and thorium. In *Proc. 2nd U.N. Int. Conf. peaceful Uses atom. Energy, Geneva, 1958* (Geneva: U.N., 1958), vol. 2, 215–23.
28. Tugarinov A. I. *et al.* Using variations in the isotope composition of lead in survey work on uranium-bearing regions. *Atomn. Energ.*, **25**, 1968, 483–9; *Soviet atom. Energ.*, **25**, 1968, 1311–6.
29. Morse R. H. Comparison of geochemical prospecting methods using radium with those using radon and uranium. In *Geochemical exploration* (Toronto: CIM, 1970), 215–30. (*Proc. 3rd Int. geochem. Explor. Symp. 1970*) (*CIM Spec. vol.* 11)
30. Weiss O. Airborne geochemical prospecting. In *Geochemical exploration* (Toronto: CIM, 1971), 502–14. (*Proc. 3rd Int. geochem. Explor. Symp. 1970*) (*CIM Spec. vol.* 11)
31. Griffith J. W. The uranium industry—its history, technology, and prospects. *Dep. Mines Resour. Can., Miner. Rep.* 12, 1967, 335 p.

DISCUSSION

M. G. Huggett said that he thought that a 50000-cm^3 crystal volume would be almost prohibitively costly for most commercial operations. He asked if the author had considered the practical factors affecting the performance of a geologist on board an aircraft at 100–150 m flying height and, say, 250-m spacing in difficult terrain. Reference had been made to airborne and vehicle-borne radiometric surveys: he asked if any marine survey work had been carried out.

The author said that he entirely agreed with what had been said about the size of the crystal, and that was the point which he was making—in not always necessarily employing gamma spectrometry from the air. In many cases a good total-count instrument would give the information required. But, if reliable counts in the three channels were necessary, a crystal volume of the order of 50000 cm^3 was required. If such an equipment did not incorporate spectrum stripping, there would be accentuation of the potassium channel. Smaller crystal sizes could be used, particularly in light aircraft and helicopters.

The task of the geologist in the aircraft was to look at the regional geology and say whether anomalies were due to non-economic concentrations of uranium, such as granitic outcroppings and accumulations of alluvial heavy minerals, or to potentially economic mineralization.

At the present time he and his colleagues were engaged in developing a portable gamma spectrometer. They were also carrying out trials with another gamma-spectrometer equipment whose probe was to be towed on the sea-bed. That should be of great value in assisting geological mapping and detecting uranium orebodies on sea or lake bottoms.

In reply to a question by Dr. F. Q. Barnes, the author said that diabase dykes and uranium mineralization were associated both spatially and genetically with acid igneous rocks, which, in their turn, were derivatives of a larger magma chamber, the diabases being derived at a later stage than the granites. The age relations of the uranium mineralization and the diabases were somewhat similar. In many cases the uranium pre-dated the diabase, but in some cases it post-dated it.

R. J. Chico said that it seemed that they should include among the North American districts noted by Dr. Bowie the uranium occurrences of the eastern U.S.A. Although there was no U_3O_8 production in the Appalachians, that district contained several uranium occurrences of probable significance. Gabelman* had shown how the uranium occurrences were related to other metal/mineral distributions. The eastern U.S.A. sandstones were not necessarily arkosic. Many U.S.A. occurrences were of greywacke type, and that different composition pointed to less acid igneous, indeed basic, rocks as sources of the sandstones. That should not be a surprise, nor a disappointment. The reasons were that most of the Blind River sandstones were not of arkosic nature but sub-greywackes and even greywackes, according to the Twenhofel and Pettijohn classification of sandstones.

Another comment derived from his knowledge of the results of gamma-spectrometry surveys over known radiometric anomalies: 'However, the numerous 250 counts/2 sec anomalies and isorad patterns based on lower-order values turn out to have more significance than the local high-order anomalies'.† That conclusion, made after study of the gamma-spectrometer survey records, had proved to be excellent. In fact, two orebodies with 10 000 000 lb of U_3O_8 were found within the isorad areas of 200 counts/2 sec. Two areas with 500 counts/2 sec anomalies contained uranium mineralization but not one orebody. Geologists should, therefore, not rely only on geophysical advice and be careful not to overspend on geophysical surveys!

The author said that he thought that most of Gabelman's recent work had not been connected with Shield areas but, rather, with mobile belts. He agreed, however, with the statement made.

Dr. K. C. Dunham said that from the viewpoint of deposits other than uranium, they were very fortunate in having deposits which generated their own signals, and to a large extent that, so far, had been exploited. But he wondered if they were not now approaching the stage, reached long ago in the general world of metalliferous geology, that the deposits which generated their own signal had been found and the need now, as the author had said, was for geological deduction to be used as the basis for exploration—not merely the search of the ground surface. He was convinced that, as far as most metalliferous deposits were concerned, the real key was the existence of permeability channels, present-day or old, but chiefly old, and traps within those channels. The real question was to what extent geologists were able to forecast where those conditions were likely to occur. They would certainly come to a stage where the kind of drilling referred to by Dr. Bowie would be the sort of thing that had to be done. They should face the fact that although every means must be taken to narrow the target, they could not pattern drill the whole of the earth's crust. Nevertheless, the whole process of geological argument and deduction had not yet been used to the extent that it would have to be used in the future.

G. R. Ryan commented that in areas where geology was relatively unknown and where there might be government-imposed time-limits, airborne scintillometer surveys were not only quick but cheap. There was a tendency to regard airborne work as an expensive tool, but that was not so when it was compared with car-borne scintillometry. From the field geologist's point of view, the former was usually the most effective way to begin the evaluation of a large area.

The author said that the quickest possible way of covering an area effectively was by aeroradiometric methods, with total-count or gamma spectrometry. If one used the latter, however, one had to be absolutely sure that one was employing an instrument which had the requisite sensitivity, particularly in the uranium and thorium channels—otherwise one was hoodwinking oneself. Unless the instrument were fully stabilized, the channels would move outside their limits. Without spectrum stripping there would be enhancement of the potassium count due to Compton scattering. Examination of the records of many spectrometer surveys showed that wherever there was a uranium anomaly there was also a 'ghost' potassium one. One of the best ways of eliminating the potassium anomalies was to have a geologist in the aircraft.

In reply to the further comment by Mr. Ryan that, very often, aeroradiometric surveys were flown over areas of unknown geology, the author said that the use of a geologist

*Gabelman J. W. Uranium in the Appalachian mobile belt. *U.S. AEC Rep.* RME–4107, 1968, 41 p.

†Texas Instruments, Denver, Colorado, U.S.A., personal communication.

presupposed knowledge of the geology of an area; and the geologist in most areas of the world had some idea what was underneath him. That argument became less valid, however, the less one knew about the geology; but a great deal of pre-survey information could be obtained from photogeology.

A. Gangloff agreed that it was useful for a geologist to be on the aircraft, but the appreciation he was able to make of the anomalies encountered could only be very general and provisional. A more comprehensive and systematic interpretation was made after the flight by examining the traces (radiometric and altitude) and by comparison with the geological map (photogeology).

The author agreed that that was the task of a geologist, but said that he would like to go back to a point that he had not emphasized sufficiently. One of the reasons why one had to be very careful with gamma spectrometry was that the uranium channel was measuring a ^{214}Bi peak; and ^{214}Bi tended to be leached out of the surface layers very readily, along with the whole uranium family. That would leave only the thorium, so that in cases where there was a mixed uranium–thorium ore and the uranium had been leached out, the remaining anomaly would show up in the thorium channel only. Should one therefore ignore that anomaly? In those circumstances one ran the risk of ignoring a uranium deposit. No information was useless information, but one had to be very careful how one used gamma-spectrometric data.

R. B. Sprague asked the age range indicated by isotopic dating in uranium deposits classified as being syngenetic. Although the Proterozoic conglomeratic deposits were noted as peneconcordant, they were not listed as penecontemporaneous. Past reports had shown the age of contained uranium to be much younger than the host rocks.

The author said that he preferred to call the syngenetic deposits concordant, such as shales, phosphorites and so on. Generally speaking, their uranium content came from ion adsorption or chemical combination with the host rock. The most recent work on conglomerates indicated several periods of crystallization. That might, of course, be recrystallization, because in age dating one was measuring the date of the last crystallization of that particular mineral. But if one looked at the Blind River field, the age of the Basement was somewhere around 2500 m.y. old and the brannerite in the conglomerate was 1850 m.y., which was younger than the so-called detrital minerals. Concordant ages of the uraninite showed a range from 1740 to 600 m.y. That indicated either an introduction of those minerals or that they had been recrystallized or remobilized and redistributed. The answer to that was not known, but they were trying to do some work on Witwatersrand material which might well give the answer. That work would include dating the rocks very carefully by a number of methods—not only the uranium minerals but the whole rock as well—with a view to showing if remobilization had taken place and whether material had moved over considerable distances. The mineralogical evidence available indicated that remobilization had not taken place and that movement of material was over perhaps only centimetres. True syngenetic deposits did not show any appreciable range of mineralization dates.

Dr. W. Domzalski said that the author had discussed a vast subject in a very able and erudite manner. As it was not possible for him to dwell in much detail on various aspects of uranium prospecting, the contribution aimed at highlighting a situation frequently encountered when a large area had to be prospected with a view to blocking out zones of greater interest. Such circumstances arose when a vast, unmapped and geologically unknown or very poorly known region was investigated. Furthermore, the region might be difficult of access and logistics of any ground operation could be formidable if the whole area had to be covered with an adequate density of observations. Also, the mining permits or options might impose considerable time limitations. Under such circumstances the most feasible approach was an airborne survey. Such a survey would normally involve a magnetometer in addition to a spectrometer or a scintillation counter. The advantages of simultaneously flying a magnetometer arose because of the importance of interpretation of magnetic data in terms of structure. Structure was one of the very important factors in localizing possible uranium mineralization. In his experience that approach had been adopted on numerous occasions and had been proved to be a sound basis for further exploration. Even over densely forested areas (rain forest) the airborne radiometric survey produced results sufficiently differentiated for decisions to be made regarding more or less favourable areas. One might expect that dense vegetation and moisture would effectively mask the radiation:

that was not necessarily the case; and it was useful to remember that the range of conditions in which airborne radiometric survey was applicable was perhaps more extensive than purely theoretical considerations would suggest.

The author commented that, if he were introduced into a completely new area, what he would begin with was a photogeological interpretation to obtain information on structure and then, almost certainly, use an aeroradiometric method. The most important thing to know was the kind of deposit one was looking for. Was one seeking a Wyoming Basin type of deposit or vein-type mineralization? In the latter case, the line spacing had to be relatively narrow, whereas for larger disseminations the interval might be 2 or 3 miles. In addition, the survey method must be adapted to the funds available. Thus, if other things were equal, he would go for the cheapest method of aeroradiometric survey in the first instance: that was, a very sensitive total-count instrument. He would follow that up in the second stage with a gamma spectrometer mounted, perhaps, in a helicopter; but, in the end, a geologist had to go and look at the anomalies on the ground.

With regard to the measurement of radiogenic heat in exploration for uranium, Professor M. Dall'Aglio said that it was not easy to detect and localize a source of relatively low heat flow. Experience gathered through exploration for geothermal fields had shown that one met with many difficulties in localizing and measuring heat flow, even 100 times higher than that caused by an important uranium orebody. That was due to many local conditions—for example, water circulation.

S. A. Anzalone said that the author had stated that the 'Colorado-type' deposits were apparently formed *in situ*. Those had been contrasted with the 'Wyoming roll-type' occurrences, believed to be formed by the remobilization of uranium through the action of oxygenated groundwater on weak concentrations of uranium in a permeable host rock, with subsequent enrichment and redeposition in a suitable reducing environment. He had examined numerous uranium occurrences in the Colorado Plateau, including those at Ambrosia Lake, New Mexico, and had been struck by the fact that, in a broad sense, those deposits displayed many of the 'roll-front' characteristics noted in the Wyoming Basin. They did not appear to be formed *in situ*.

The author replied that the 'roll-type' deposits did not occur entirely in the Wyoming Basin but were more typical of that area than of the Colorado Plateau. He thought that those 'roll-front' deposits were formed in the way that had been suggested by U.S. geologists; and one must accept that they were formed later than the peneconcordant type characteristic of the Colorado Plateau. There had been movement of uranium in the 'roll-front' deposits, but some uranium had stayed in the original deposits: that was why the age determinations carried out showed a very wide range. The tendency was for the material formed very recently to be of the 'roll-front' type—some was probably forming at the present time.

W. Meyer said that he had the impression that the Ambrosia Lake deposit was, in fact, a very large 'roll front' in which there had been considerable movement.

R. D. Beckinsale said that several recent studies, for example, of the Climax deposit, the Marysvale deposit, Utah,* and Bingham† indicated that mineralization was a short event, possibly lasting only one million years or so. Thus, the long time-scale advocated by the present author might not be such a general feature of mineralization. Whether the range of U–Pb ages on pitchblende, etc., from particular deposits should be interpreted as indicating continued fresh introduction of uranium into the host rock from an external source or simply minor recrystallization of uranium minerals, more or less *in situ*, was a problem which might best be solved by other isotopic methods.

The author said that he would agree with that in reference to porphyry copper deposits. He thought that the mineralization was a single episode which could not be measured in time by normal age-dating techniques; but, if one looked at work in southwest England, it was evident that there there was a long spread of uranium

*Basset W. A. *et al.* Potassium–argon dating of the late Tertiary volcanic rocks and mineralization of Marysvale, Utah. *Bull. geol. Soc. Am.*, **74**, 1963, 213–20.

†Moore J. M. and Lanphere M. A. The age of porphyry-type copper mineralization in the Bingham mining district, Utah—a refined estimate. *Econ. Geol.*, **66**, 1971, 331–4.

ages. *The early mineralization was uraninite associated with tin, which was of the same age as the granite (280 m.y.), followed by a whole series of ages which agreed extremely closely with geological observation. Those indicated that the mineralization took place at a number of periods which could have been any length of time apart. Uranium mineralization started at 280 m.y., when the huge southwest England batholith was crystallizing, and continued in stages until 50 m.y. ago. The more dates were determinne, the more different locations were shown to have successive periods of mineralization; but he did not say that that was universal.

*Darnley A. G. *et al.* Ages of uraninite and coffinite from south-west England. *Mineralog. Mag.*. **34**, 1965, 159–76.

Uranium supply and demand

Michael Davis Ph.D., F.Inst.P., F.I.M.M.

United Kingdom Atomic Energy Authority, London, England

339.4:553.495.2

Synopsis

In the context of a changing pattern of energy usage, nuclear-generated electricity already comprising some 2% of world usage, firm plans and projections for nuclear capacity are considered. A comparison is made of the salient characteristics of nuclear with fossil-fuelled power stations, and reasons are given to substantiate a considerable growth of nuclear capacity by at least an order of magnitude in the 1970s. The present known world uranium reserves of almost 1000000 short tons U_3O_8 are outlined, and an account is given of past production, usage and likely future production. An approximate method for estimating future uranium demand is given, but more precise forecasts are described, due account being taken of uncertainties. Annual demand is likely to be around 100 000 short tons U_3O_8 in the first half of the 1980s.

Reasons are given for the present oversupply position, but, taking account of the need to have adequate forward reserves to sustain production, it is shown that it is timely to intensify the search for fresh reserves so that the discovery rate by the 1980s is enhanced threefold from about 50 000 short tons per annum secured in the past few years.

The birth of nuclear power stations for electricity generation occurred about 15 years ago more or less simultaneously in Britain, France, the U.S.A. and the U.S.S.R., but two or three years later, in 1959, an eight-year decline in the commercial market for uranium set in because of decreasing military demands. This decline also prompted the virtual abandonment of the intensive uranium prospecting of the post-war period, which lasted until the late 1960s.

The present status and future prospects of nuclear power generation are considered and related to known uranium reserves as well as to the need for further uranium discovery.

Future energy needs and the role of nuclear power

The projection to the end of the century of the world's primary energy needs, the proportion of that energy required as electricity, and the fraction of that electricity which will be generated by nuclear power stations are all forecasts fraught with uncertainties. Nevertheless, it is instructive to consider possible trends in order to form a reference framework for gauging the relative future importance of individual fuels, even if estimates which may be reasonably accurate for a quinquennium become progressively more unreliable for longer periods. Fig. 1 shows how the world's energy requirements may be met for the next 30 years. It is interesting to note that coal ceased to provide half the requirements a decade ago and that, since 1966, it has taken second place to liquid fuels. Energy consumption is, of course, dominated by North America, which takes nearly 40% of the total and now has an annual *per capita* consump-

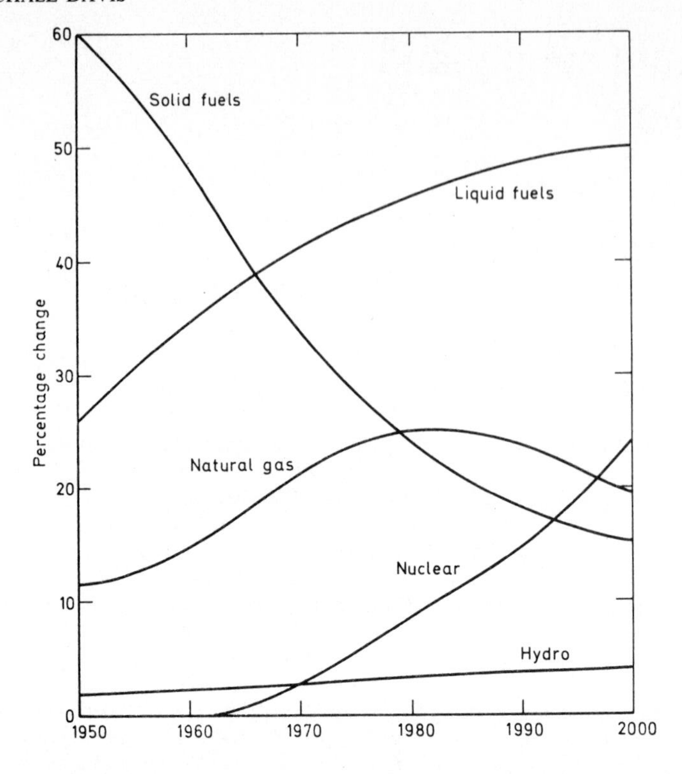

Fig. 1 Possible changes in percentage shares of primary fuels in meeting world energy requirement

tion of about 10·3 t coal equivalent (3·35 in Western Europe; world average, 1·77).

The proportion of energy consumed as electricity is increasing everywhere, and this trend is unlikely to change for some decades. In Western Europe electricity accounted for only 19% of the total primary energy consumption in 1960; the figure is now nearer 25% and may reach 30% during the 1970s. An increasing proportion of this electricity is expected to be supplied by nuclear power—perhaps as much as 20% of West European capacity will be nuclear by the end of the present decade. The overall position, estimated in 1968,[2] is indicated in Table 1. Even though there is no reason to change the forecast figures from 1975 onwards, the 24 GWE figure for 1970 is incorrect: only

Table 1 Estimated nuclear electricity generating capacity in non-Communist world. From Davis[2]

Year	1970	1975	1980	1985	1990	1995	2000
Forecast range	24	100–150	230–350	400–650	700–1200	1100–2000	1500–3000
'Best' estimate	24	115	300	580	950	1450	2000

Units are MWE × 10^3 or GWE.

16 GWE—comprising just under 100 nuclear reactors in 14 countries—were operable at the end of 1970. The others had been delayed by a few months and so were to be found among the 115 GWE of reactors then under construction or ordered. (For those not familiar with the expression of electrical generating capacity in GWE, it may help to relate it to the total electrical generating capacity of the United Kingdom—population, 55000000—which is 60 GWE, of which some 5 GWE is nuclear; one electrical gigawatt—GWE—is equal to one million electrical kilowatts—kWE.)

Nuclear electricity generating capacity is now a little in excess of 2% of the world's installed capacity. The U.S.A. dominates the world scene in forecast uranium usage for nuclear electricity generation and approximates to the rest of the non-Communist world put together. Although, by mid-1971, there were about 9 GWE of nuclear power stations operable in the U.S.A., ten times this capacity were already being built or planned. Faulkner[8] has forecast that the

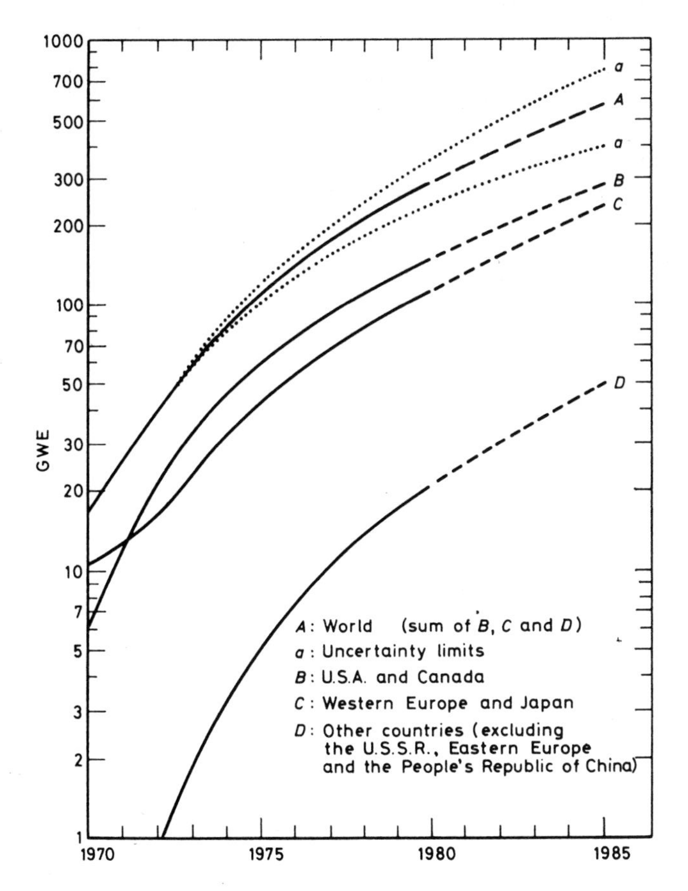

Fig. 2 Assumed growth of nuclear generating capacity used to calculate uranium demand in Fig. 4

percentage of electricity generated in the U.S.A. from uranium fuel will attain 22% by 1975, 39% by 1980 and 51% by 1985. Western Europe and Japan are the next most important areas for nuclear electricity growth over the next 15 years. Fig. 2[5] shows these projected growths.

Two questions arise from the remarks on the present status and apparently optimistic future forecast for nuclear power. (1) Why is nuclear power being preferred in so many cases to the well established fossil-fuelled electricity generation? (2) If, indeed, nuclear power is growing so fast, why is there a 'soft' market for uranium?

The rapid growth of nuclear power in industrialized countries must reflect the conviction of utilities that it is an economical means of electricity generation. Thus, during 1970, more than 14 GWE were ordered in the U.S.A. alone for operation about five years hence.* This view is supported by published plans for the extensive construction of nuclear stations. It is an affirmation by those concerned with electricity generation that nuclear power has secured an increasingly important role in their strategy.

In any particular year the number of utilities ordering a nuclear rather than a fossil-fuelled station may be said to reflect short-term factors, such as some of the recent shortages of oil and low-sulphur coal. But longer-term considerations, such as the security of supply—and future cost—of Middle East oil and the rising costs of labour-intensive coal to meet current safety and environmental requirements, are important.

It should be remembered that nuclear power stations are capital-intensive and that current high interest rates and recent high cost inflation militate against them. On the other hand, there are advantages which compensate for this capital intensiveness: (1) economy in size (reducing capital costs per unit of capacity as size is increased); (2) low fuel cost component of electricity generated; (3) very small fuel transport costs; (4) high availability; and (5) negligible pollution of the environment.

The fuel cost component in a nuclear station which employs enriched uranium is 20–25% of the total cost of generation. About one-third of this fuelling cost (i.e. 7 or 8% of electricity cost) is attributable to uranium ore concentrates. This low fuel cost provides much of the attractiveness of nuclear power: once the construction of a nuclear station is financed, the total operating costs, inclusive of fuel, are low. In other words, the utility has minimized the proportion of its total cost liable to the effects of future cost inflation. If, however, this attraction of nuclear power were diminished by fear of uranium shortage or by significant uranium price increases, fewer nuclear stations would be built and, consequently, the uranium demand would be less.

At the present time, why is there a mismatch of uranium supply and demand? The main causes are noted below.

The estimated demands for uranium made in the late 1960s did not take account of the constructional delays in completing nuclear power stations, programme slippages or delays in achieving planned load factors in operation, both in Western Europe and in the U.S.A. As a result, all losses of expected reactor operating time are reflected in effectively irrecoverable decreased uranium requirements. To a lesser extent, estimated demand was further reduced by improvements announced in fuel utilization, particularly for light water reactors. Uranium suppliers four or five years ago (and in the hope of compensation for declining sales for military programmes) were often reluctant to make contracts at a price to which purchasers were prepared to commit themselves. This resulted in some purchasers delaying or only making short-term contracts. Thus, some suppliers failed to secure a firm basis of uranium production. The suppliers evidently rather overestimated the strength of their position and perhaps failed to recall the inherent strengths of the purchaser, who could almost be sure of recourse to national stockpiles of uranium if he were caught short. This, then, turned out to be one price-depressing factor.

*A further 26 nuclear stations, of total capacity 26 GWE, were ordered in 1971 by U.S. utilities.

Since 1965 uranium production has been effectively constant—within about 10%—of 21 000 short tons U_3O_8 per annum.[4,6] During the same period the mean annual uranium requirement for power stations has only been about half this production rate. The uranium produced which has not been required for existing contracts has had to be stockpiled. It is true that some uranium suppliers have ceased production, but others have continued to produce on a reduced scale to maintain viability in the expectation of a resurgence of the market. This factor, though rather a symptom than a cause of oversupply, does, of course, tend to reinforce a buyer's market.

Uranium reserves and production

Before the likely pattern of future uranium demand is considered, together with its uncertainties, it is appropriate to stand on somewhat firmer ground and review what is known of present uranium reserves. The writer has drawn upon the publications[1,4,5] of the European Nuclear Energy Agency/International Atomic Energy Agency Working Party on Uranium Resources. It is convenient to adopt the ENEA/IAEA price categories and definitions[5] (see below).

Reserves of uranium in the mining sense are *Reasonably assured resources* in the price category (1970 U.S.$) of below $10/lb U_3O_8.

Reasonably assured resources refer to uranium which occurs in known ore deposits of such grade, quantity and configuration that it can, within the given price range, be profitably recovered with currently proven mining and processing technology. Estimates of tonnage and grade are based on specific sample data and measurements of the deposits and on knowledge of orebody habit. (As applied to uranium in the $10–$15/lb U_3O_8 category, the reliability of estimates is generally less than that for *Reserves* at $<$ $10/lb U_3O_8.)

Estimated additional resources refers to uranium surmised to occur in unexplored extensions of known deposits or in undiscovered deposits in known uranium districts, and which is expected to be discoverable and economically exploitable in the given price range. The tonnage and grade of *Estimated additional resources* are based primarily on knowledge of the characteristics of deposits within the same districts.

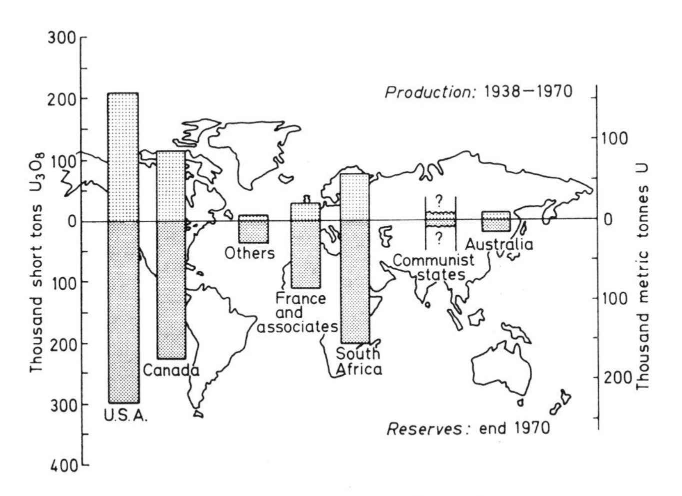

Fig. 3 Total discovered uranium

Table 2 Estimated resources of uranium (data available April, 1970)

Country	Type of resources: Price range, < \$10/lb U_3O_8: *Reasonably assured resources (reserves)*, 10^3 short tons U_3O_8	*Reasonably assured resources (reserves)*, 10^3 tonnes uranium	*Estimated additional resources*, 10^3 short tons U_3O_8	*Estimated additional resources*, 10^3 tonnes uranium	Price range, \$10–15/lb U_3O_8: *Reasonably assured resources*, 10^3 short tons U_3O_8	*Estimated additional resources*, 10^3 short tons U_3O_8	Qualifying remarks
Argentina	10	7·7	22	17	11	33	
Australia	21·7	16·7	6·7	5·1	9·2	6·6	Geological potential excellent for substantial increase in resources
Brazil	1·0	0·8	1·0	0·8			Potential for substantial increase excellent; additional resources dependent on phosphate, gold and niobium production
Canada	232	178	230	177	130	170	New data based on recent government-initiated study (January, 1970)
Central African Republic	10·4	8	10·4	8			New data in view of 1969 decision to exploit deposits
Denmark					5		As 1967 *Report:* resources are in Greenland
France	45	35	25	19	9	15·5	Increased resources following 1967–70 search; Vosges deposits now deleted
Gabon	13·5	10·4	6·5	5		6·5	Doubled resources following 1967–70 search, despite continuous mining

India						3	1	As 1967 *Report*
Italy		1·5	1·2					Prospecting 1968–69 did not change reserves; higher-priced resources linked to production of other elements
Japan		2·7	2·1			4·5		
Mexico		1·3	1			1·2		
Niger		26	20	39	29	13	13	Large increase following prospecting by French CEA; local milling 1971
Portugal	Europe	9·6	7·4	7·7	6		15	Prospecting to be renewed 1970 in Angola and Mozambique
	Angola						15	
South Africa		200	154	15	11·5	65	35	Production capability and size of reserves a function of gold production
Spain		11	8·5			10		
Sweden						350	50	Production will be limited to meeting portion of Swedish demand
U.S.A.		250	192	510	390	140	300	Does not include 90000 short tons estimated by-product U_3O_8 from phosphate and copper production in *Reasonably assured resources* up to $10. Excludes 90000 short tons of additional resources at $10 estimated in areas other than well established uranium districts
Others		3·6	2·8	11	8·5	1·5		
APPROXIMATE TOTAL		840	645			750		

Fig. 3 shows how presently known world reserves are distributed by producing countries and how this compares with cumulative past production. No information is available from the countries of Eastern Europe, the U.S.S.R. and the Peoples Republic of China. It is clear that the major reserves in the non-Communist countries occur in the U.S.A., Canada and South Africa. France and her associates, the Central African Republic, Gabon and Niger, come next. Neither the recent important discoveries in Australia nor those in South West Africa are included, because in neither case has enough information been made available for these discoveries to qualify as *Reserves*.

As of 1970, the uranium reserves and other resources estimated by the ENEA/IAEA[5] are given in Table 2. The total reserves amount to some 840000 short tons U_3O_8 and there is at least a similar tonnage in the category *Estimated additional resources* at $<$ \$10/lb U_3O_8. The U.S.A.[8] have recently announced increases in their *Reserves* of 50000 short tons U_3O_8, and of *Estimated additional resources* (at $<$ \$10/lb) of 170000 tons. This addition would raise the world uranium reserves to approximately 900000 short tons U_3O_8.* The totals do not include by-product uranium, which may be extracted from phosphate and copper production in the U.S.A. The South African *Reserves*, which are mainly a by-product of gold production, are included because they are a well established source of uranium; but it has to be remembered that the production rate of these South African figures is restricted because it is tied to gold production.

Production

As was stated above, the current annual production rate of uranium over the

Table 3 Uranium production by major producing countries, 1938–70 (short tons U_3O_8)

Year	Australia	Canada	France and associates	South Africa	U.S.A.	Other	Total
	(from 1954)	*(from* 1938)	*(from* 1949)	*(from* 1952)	*(from* 1942)		
pre 1957	531	7700	740	9895	23920		42786
1957	417	6600	480	5700	9840		23037
1958	606	13400	1050	6250	14000	2370	37676
1959	1117	15900	1165	6450	17400	2320	44352
1960	1300	12748	1379	6409	17760	1450	41046
1961	1400	9641	2141	5468	17399	223	36272
1962	1300	8430	2603	5024	17010	80	34447
1963	1200	8352	2692	4532	14218	86	31080
1964	370	7285	2113	4445	11847	144	26204
1965	370	4443	2210	2942	10442	179	20586
1966	330	3932	2223	3286	9587	162	19520
1967	330	3738	2272	3214	9125	273	18952
1968	330	3701	2235	3883	12338	295	22782
1969	330	3854	2300	3979	11873	295	22631
1970	330	4010	2250	4000	12800	320	23710
Cumulative to 1970	10261	113734	27853	75477	209559	8197	445081

*But see discussion (p. 31) by Dr. J. W. von Backström on additions to South African *Reserves*.

past five years has been about 21000 short tons U_3O_8. Table 3[4,6] gives a statement of historical production by the main producing countries. The year of maximum world production occurred in 1959, following the curtailment of military demand from 1957 onwards. The total cumulative production, inclusive of 1970, is almost 450000 short tons U_3O_8, of which perhaps 300000 t has gone to military stocks, about 80000–100000 t remains in national stockpiles destined for peaceful uses, and around 50000 t has already been used for nuclear power stations.

National stockpiles of civil uranium have been identified as indicated below.

		Short tons U_3O_8	Reference
U.S.A.	U.S.A.E.C.	50000	8
	Private	10000	8
Canada		9500	11
South Africa		9000	5
		78500	

Canada has made known[11] that a further stockpile of 3250 short tons from production in the period 1971–75 will be incurred. Decisions as to how these stocks will be used are difficult,* but the governments concerned wish to minimize the deleterious effects on the uranium industry which premature disposal of stock would entail.

Annual production capacity of some 38000 short tons, attainable by 1973, was identified by the ENEA/IAEA study,[5] but commissioning of this extra capacity and further production capacity decisions are likely to be deferred until market conditions improve.

Uranium demand

In simplest terms the annual demand is the summation of the running charges of uranium required to fuel operating reactors, plus the initial uranium inventory of new reactors. Thus, for each class of thermal nuclear reactor† operating a given regime (e.g. an 80% load factor)

Annual demand, $D = CR + IF$

where C is operating nuclear reactor capacity, GWE (1 GWE $= 10^3$ MWE $= 10^6$ kWE), I is new reactor capacity installed in a given year, GWE, R is annual running charges at given (say, 80%) load factor, short tons U_3O_8 per GWE year, and F is first fuel charge, short tons U_3O_8 per GWE.

For a typical current reactor, where $R \sim 200$ short tons U_3O_8 per GWE (at 80% load factor) and $F \sim 750$ short tons U_3O_8 per GWE, a group of, say, 10 GWE of operating reactors increased by, say, 1·3 GWE new reactor capacity in a given year would lead to a demand of

$$D = 10 \times 200 + 1{\cdot}3 \times 750$$
$$= 2000 + 1000 = 3000 \text{ short tons } U_3O_8$$

*On 13 October, 1971, a USAEC Press Release disclosed a plan for comment in which the stockpile would be sold, commencing in 1974, with the following annual quantities (expressed as thousand short tons U_3O_8): 0·8, 2·0, 4·1, 5·9, 7·5, 7·5, 7·5, 6·5, 4·5, 3·7. This plan aroused very unfavourable comment by uranium producers. An alternative proposal made on 7 March, 1972, is regarded as less objectionable. In this, the stockpile would be progressively diminished by optimizing enrichment plant operation, yielding a higher 'tails' assay. Effectively, a proportion of the stockpile used as feed would be converted into much lower-value stocks of uranium depleted to about 0·3% ^{235}U content.

†Nuclear reactors in which fission is induced by fission neutrons slowed down to 'thermal' energies by a 'moderator', e.g. graphite or water.

It will be appreciated that although this simple approach gives a rough indication of demand,[4] a more sophisticated calculation is required to make a proper estimate.[3] Due allowance has to be made for such factors as uranium usage characteristics of different reactor types, load factors, lead times for processing

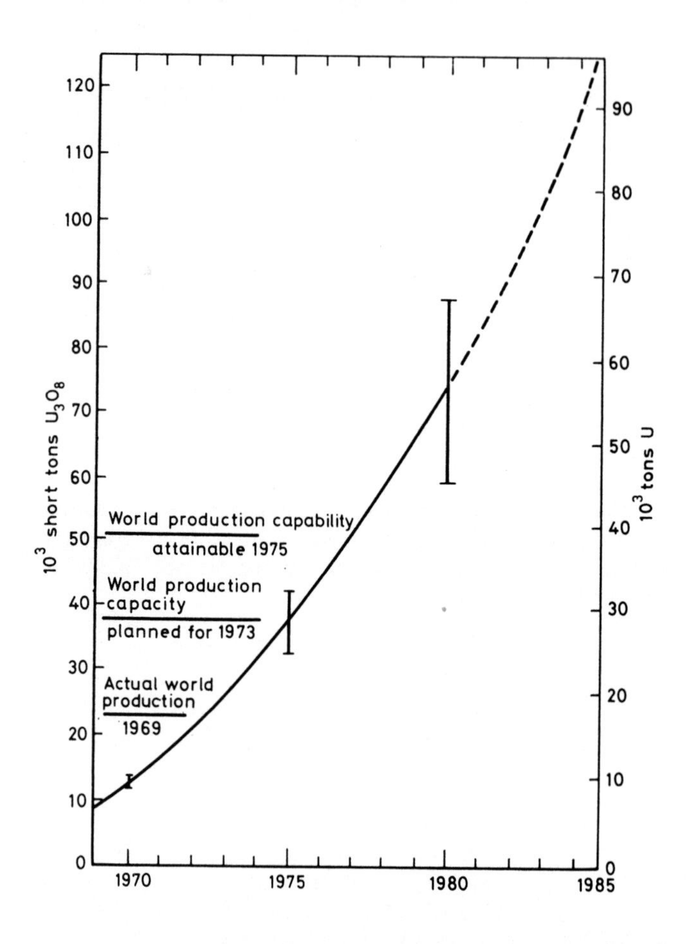

Fig. 4 Estimate of world annual uranium demand (for the years 1970, 1975 and 1980 an indication is given of the approximate range of uncertainty of the data: these ranges were estimated by taking into consideration uncertainties in the growth rate of nuclear generating capacity allotted to nuclear stations and in the choice of reactor types and variations in fuel cycle lead times and in station load factors)

uranium ore concentrates to fuel elements, new and spent fuel storage times, reprocessing and the recycling of unused uranium and recycling of plutonium extract by reprocessing. Allowance also must be made for likely uncertainties in such factors.

Fig. 4 shows the result of such an analysis of world annual uranium demand over the 15-year period covered by Fig. 2. In the progressively increasing range of uncertainty attending the estimates an attempt has been made to allow for uncertainties in the growth rate of nuclear generating capacity and the choice of reactor types, as well as for the first two factors noted above. If, however, it were desired and feasible to recycle plutonium in thermal reactors from, say, 1975, the annual demand in the ensuing five years might be diminished by up to 5%. No allowance has been made for any diminution* of uranium demand attributable to the introduction of fast neutron breeder reactors, since this is expected to be negligible in the period to 1985.

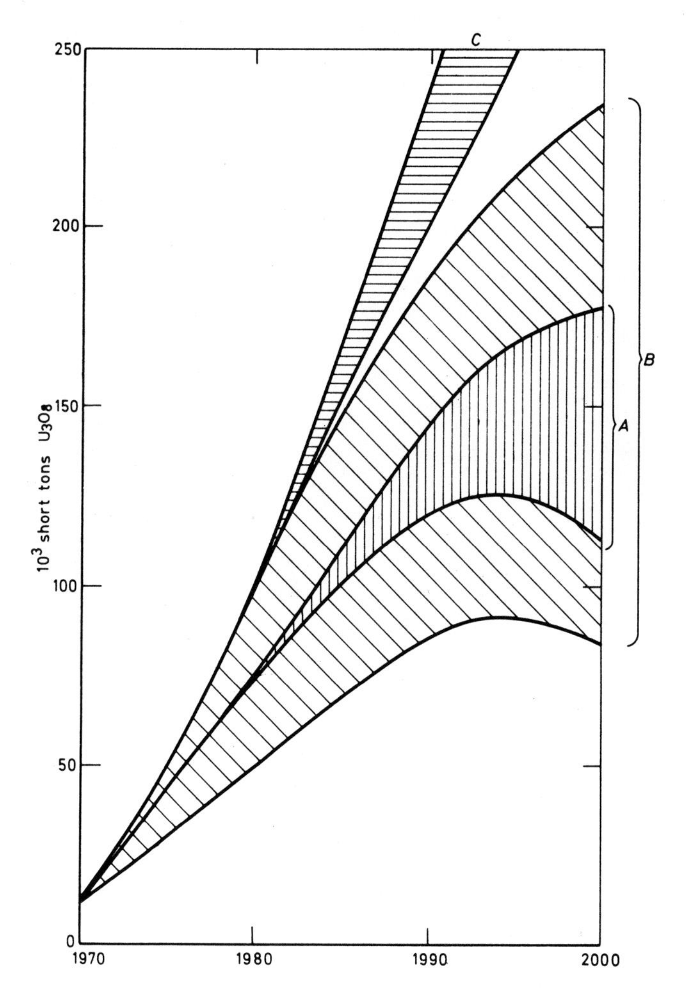

Fig. 5 Estimated annual requirements for uranium

On a longer time-scale to the end of the century, even greater problems and uncertainty of forecasting have to be faced. Based on the figures of nuclear generating capacity given in Table 1, an attempt has been made[2] to show what may reasonably be expected in annual uranium demand to the end of the

*Initially, the plutonium fuel for 'fast reactors' will come from the reprocessing of thermal reactor fuel. The core of the fast reactor is then surrounded by a blanket of uranium in which the isotope ^{238}U predominates. Spare neutrons escaping from the core are captured by this blanket to produce more plutonium. The neutron economics are such that in a fast reactor a high proportion of the ^{238}U, very little of which undergoes fission in a thermal reactor, can be converted eventually to fissile plutonium capable of fuelling more than one fast reactor. The effect of this phenomenon is clearly of prime importance in the logistics of nuclear fuel: it enables a high proportion of the 99·3% of natural uranium which consists of ^{238}U to be used.

century (Fig. 5). Range *A* corresponds to the best estimate of Table 1, assuming that some form of plutonium fuel fast breeder is introduced; range *B* shows the widening of these extremes on taking the range of installed capacity of Table 1; and range *C* shows the effect on demand of neither introducing a fast breeder reactor nor any more markedly efficient thermal reactors than are presently envisaged. Fig. 6 shows the same thing in cumulative terms and reveals that

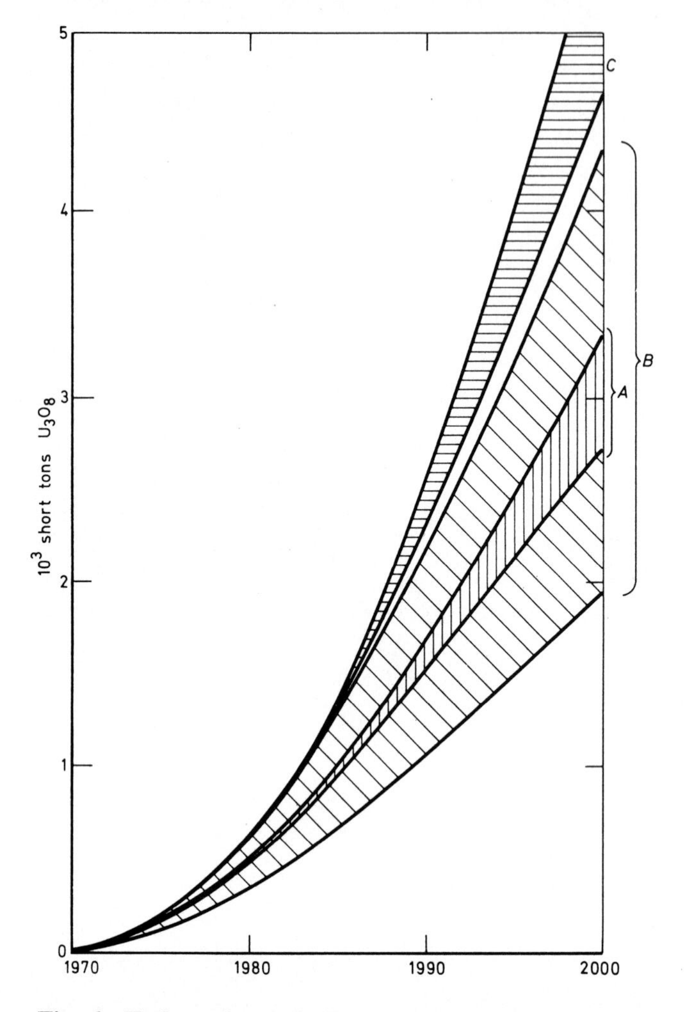

Fig. 6 Estimated cumulative requirements for uranium

between 2000000 and 4000000 short tons U_3O_8 may be required by the year 2000. This may be compared with the present uranium reserves of 900000 tons, i.e. almost 1000000 short tons U_3O_8. This demand growth from 1970 exceeds the corresponding growth rates from 1936 of copper and iron ores and even of petroleum.[12,13]

Conclusion

THE NEED FOR NEW URANIUM RESERVES

The present soft market conditions have caused several companies which participated in the upsurge of uranium prospecting activity in the late 1960s to slow down or curtail their search. In the U.S.A., where the number of feet drilled is a very convenient measure of uranium prospecting activity, a 20% drop in footage in 1970 has been reported. This has recently been noted by the Joint Congressional Committee on Atomic Energy: the Committee noted that prospecting should be sustained at a vigorous level and the desirability of the U.S. Atomic Energy Commission announcing their stockpile disposal policy without further delay. Uncertainty over the uranium stockpile disposal was considered to be a marked disincentive to those engaged in uranium prospecting.

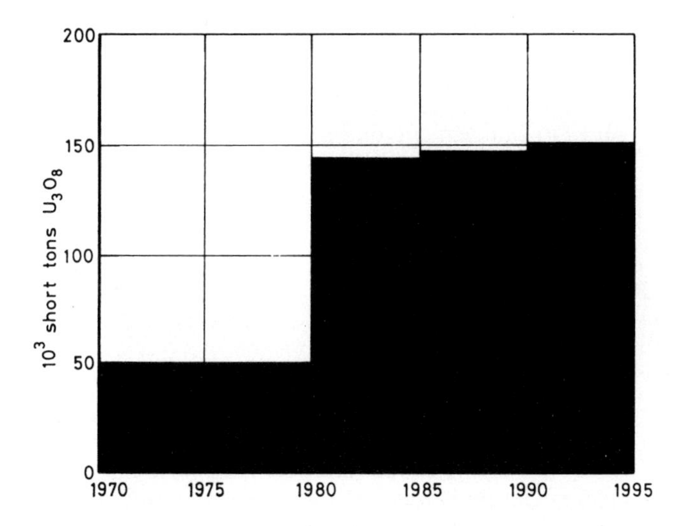

Fig. 7 Annual additions to uranium reserves to maintain eight-year forward reserves

In the context of world uranium demand (excluding, of course, the Communist countries, as before) an attempt has been made to indicate what will probably be required in terms of net annual additions to reserves in Fig. 7. The writer has adopted the widely accepted concept[7] that an eight-year forward reserve should be maintained as a satisfactory base for production. Then, taking the median of the most likely demand from Fig. 6, the mean values of net annual additions to reserves required, first, to 1980, and, subsequently, for successive five-year periods, were estimated.

It will be noted that the opening level of 50000 short tons per year approximates to the net annual increase in reserves reported between 1967 and 1970. In the three successive five-year periods after 1980, however, the annual level required is almost three times as high, i.e. approaching 150000 short tons.

How then—and over how many years—will the increase in prospecting activity necessary to prove these new reserves be achieved? Even remembering that median forecasts have been used—with their inherent uncertainties—it seems that this is no time to be cutting back on uranium prospecting activity. On the contrary, it is a time to plan how and when prospecting for uranium should be intensified. It is hoped that some of the research and development which has been carried out in the United Kingdom,[10] together with that given at this meeting, will aid in the discovery and provision of the uranium essential to the massive development of nuclear electricity generation which is in progress.

References

1. ENEA/IAEA. *Uranium resources—revised estimates* (Paris: OECD, 1967), 26 p
2. Davis M. Future uranium demands. *J. Br. nucl. Energ. Soc.*, **7**, 1968, 159–62.
3. ENEA/IAEA. *Illustrative power reactor programmes* (Paris: OECD, 1968), 36 p.
4. ENEA/IAEA. *Uranium. Production and short-term demand* (Paris: OECD, 1969), 32 p.
5. ENEA/IAEA. *Uranium. Resources, production and demand* (Paris: OECD, 1970), 56 p.
6. Williams R. M. Uranium and thorium. In *Canadian minerals yearbook 1968* (Ottawa: Information Canada, 1970), 501–13.
7. Boxer L. W. *et al.* Uranium resources, production and demand. Paper presented at 4th International Conference on Peaceful Uses of Atomic Energy, Geneva, 1971, P/678.
8. Faulkner R. L. Outlook for uranium production to meet future nuclear fuel needs in the United States. Paper presented at 4th International Conference on Peaceful Uses of Atomic Energy, Geneva, 1971, P/059.
9. Williams R. M. *et al.* Uranium and thorium in Canada: resource, production and potential. Paper presented at 4th International Conference on Peaceful Uses of Atomic Energy, Geneva, 1971, P/154.
10. Davis M. *et al.* United Kingdom research and development work in aid of future uranium resources. Paper presented at 4th International Conference on Peaceful Uses of Atomic Energy, Geneva, 1971, P/489.
11. Williams R. M. Uranium. *Can. Min. J.*, **92,** Feb. 1971, 114–20.
12. Mabile J. L'industrie des matières premières nucléaires en France—recession ou expansion. *Energie nucl.*, **11,** 1969, 541–8.
13. Nininger R. D. Uranium reserves, future demand and the extent of the exploration problem. In *Uranium exploration geology* (Vienna: IAEA, 1970), 3–19.

DISCUSSION

R. J. Chico said that the rate of uranium ore discovery in the U.S.A. had decreased since 1965, i.e. lb of U_3O_8 found per foot drilled did not continue to be as high as it was in the 1950s. In consequence, that led to a lack of corporate interest (capital invested must earn at least 10% to compete with financial yield of free enterprise). An increase in the price per pound of U_3O_8 before 1980 would be essential to cover the gap created by large exploration costs. *The approximate* costs ($) of extracting 1 lb of U_3O_8 by present-day technology were:

(*a*)	Sandstone ore deposits, U.S.A. (production, 500–1000 ton/day)	4–6 (plus vanadium)
(*b*)	Conglomerate deposits (Canada and South Africa, 500–1000 ton/day)	4–5 (plus gold)
(*c*)	Uranium mines, heap leaching	3–5
(*d*)	Phosphate deposits	40–60
(*e*)	Shales (Chatanooga Shales)	70
(*f*)	Sea-water uranium (United Kingdom)	70 (*see below—Editors*)

The author replied that the present price of uranium was around $(U.S.)6.00/lb. A few years ago it was $1.00–2.00 higher, and the many who claimed that to be insufficient would consider the present price to be even less satisfactory. As to the future, it was very difficult to predict, as the price depended on so many factors. Some producers would avail themselves of more economical methods of prospecting and of uranium ore processing;* others would benefit from higher-grade ore; correspondingly, both would earn a greater return upon investment. It was unlikely that there would be any absolute shortage of uranium in the coming decade. What was required was that, with uranium as one of the best documented metals, they should try to ensure a reasonable balance of supply and demand. They knew the sort of factors involved and should aim at achieving a balance between the relatively few users and suppliers concerned. If those two bodies of people could get together, the demand could be met at a price which would be mutually satisfactory and which would avoid market instabilities. He added that Mr. Chico's item (*f*) was about three times too high.

*See reference 10 above.

Dr. J. W. von Backström said that the President of the Atomic Energy Board of South Africa had announced* recently that 'after revaluation of our uranium reserves, the figure has now risen to 300000 tons for reasonably assured reserves of U_3O_8 recoverable at up to 10 dollars per pound. . . The increase has resulted from new discoveries in both the Republic and South West Africa, as well as from the achievement of higher extraction efficiencies coupled with the development of improved extraction processes'.

That statement therefore increased the figure of 900000 short tons U_3O_8 in the category up to $10/lb for reasonably assured reserves to 1000000 short tons for western world uranium reserves.

Dr. F. Q. Barnes asked if the author's eight-year forward figure took into account the growing demand for uranium in the eight future years, or if it were merely eight times the requirement in the year of calculation. Based on the expected growth pattern, an eight-year forward reserve would be 13–15 times the demand in the year of calculation. Furthermore, had consideration been given to the reserves required to back the production capacity in the years specified? It seemed to him that an eight-year forward figure had no practical application: the figures must relate to production capability, the types of deposits in the total world pattern of reserves dictating the level of reserve required.

The author said that his uranium requirement tonnages, based on eight times present production, were on a worldwide basis. Dr. Barnes' comments would seem to reinforce his own view and suggest that even more reserves needed to be found than he had indicated in his paper.

Dr. Barnes thought that an annual increment of 50000 tons worldwide would not be difficult to find, although 150000 tons after 1980 would be much more of a problem. His own view, however, was that 50000 tons would be required in the U.S.A. alone, and latest pronouncements from the Atomic Energy Commission seemed to agree with that.

The author said that that point had much validity. His calculations over earlier years had to include some simplified conventions, one of which was that, in the non-Communist world, uranium was freely movable from place to place. That, of course, was not true and there were various restrictions on free market movements. Those had the effect of disturbing the estimations.

Dr. E. E. N. Smith agreed that additional reserves of uranium must be found in increasing quantities up to the end of the century. Over the immediate short term, however, he suspected that it was the productive capability, rather than the reserves from which it was taken, which would be most critical. The 1970 ENEA–IAEA study† indicated that a productive capability of more than 50000 short tons U_3O_8 per year might be achieved by 1975. That figure was rather artificial in that it presupposed that all uranium deposits would be extracted at maximum rates—a condition that would not be attained until a strong demand developed for uranium, resulting in firm contracts with attractive financial incentives. The oversaturation of the present market at very low prices made it very unlikely, in his view, that production in the western world would reach 35000 short tons annually by 1975.

Whether current reserves minable at under $10/lb U_3O_8 now totalled 800000 or 1000000 tons of U_3O_8 was of little consequence unless there were some indication as to how much could be extracted annually at current price levels. It would be interesting to know how many of the tons listed by the ENEA under $10/lb were still 'reserves' at prices of $6.00–7.00/lb and how many of the remaining higher-cost tons were still 'reserves' after that $6.00–7.00 ore had been extracted.

P. G. Killeen said that he understood that thorium could be used in breeder reactors: he would like to know what effect, if any, that could have on the predictions of the demand for uranium.

The author said that he was aware that some people were keen on the use of thorium, which was reasonable in view of its relatively abundant occurrence. But he did not feel

*Roux A. J. A. Talk on South African Broadcasting Corporation programme *Top Level*, 27 June, 1971.

†See reference 5 on page 30.

that thorium was going to have any real impact in the twentieth century. The development of uranium-fuelled reactors in different parts of the world had cost billions of dollars. Movement into thorium would require further large expenditures. Rather than use thorium, he felt that the next major problem was the move to fast reactors. Thorium was more likely to be used if fusion were not attainable within a reasonable period.

Dr. Barnes asked when breeder reactors would begin to affect uranium demand figures significantly. It seemed that that date was well removed from the present and, the further it was removed, the greater would be the problem of finding sufficient uranium to meet increasing demand. If demand figures could be projected beyond 1985, they would better indicate to the mining industry the scale of the exploration requirements ahead.

The author said that that was a very important point. The agencies were, of course, often reluctant to commit themselves as to what was going to happen over the next 30 years. The market requirements mentioned were prepared from the agencies' 15-year forecast and, during that time, they expected the fast reactor to make an insignificant impact on the market—a conclusion with which he concurred. He believed, however, that during the 1990s fast reactors would have a marked effect upon demand, as was indicated in Fig. 5 (p. 27).

W. Meyer said that mineral commodities, in spite of the most optimistic forecasts, were subject to frequent 'boom and burst', and, in the past, uranium had been no exception. Once again they were faced with rising demand curves to the end of the century, but uncertainties and doubts remained. In addition, there loomed the threat of hydrogen fusion. A sudden breakthrough in controlling that reaction would send everyone prospecting for water!

The author said that his colleagues at Culham believed that a great deal more should be spent on fusion; and, round the world, there were quite big programmes in train (two or three programmes of the order of some tens of millions of dollars per year and several smaller ones). But the big breakthrough, which involved showing how to sustain the fusion reaction, had not yet come. After that had happened, there would be an immense amount of development still to be done. To develop a power-producing fusion reactor from a successful fusion reaction would require much further research into materials, magnetic fields and electrical problems; thus, for the rest of the century, it was likely that they would depend on the present general types of natural and enriched uranium and future fast reactors, which had a life of 30 years or more.

R. D. Beckinsale asked if it were possible that developments in the field of atomic weapons, such as a hydrogen bomb triggered by different methods, could result in the release on to the market of uranium stockpiled for defence purposes.

The author replied that he hoped that the balance of military power would not change too suddenly. If military material were released—as, for example, in controlled general disarmament—he believed that the material would be made available over a fairly long period. A great deal of the uranium used for military purposes had been enriched or converted into special forms, which could be used for thermal power reactors with some loss of value. In general, however, it would make more sense to use it to fuel fast reactors. But, even if one took the 300000 tons used in weapon production, that was not a great quantity when set against the 100000 tons which would be the annual requirement in the early 1980s.

Some environments of formation of uranium deposits

R. G. Dodson M.Sc., Ph.D., A.M.Aus.I.M.M., F.I.M.M.

Bureau of Mineral Resources, Geology and Geophysics, Canberra, Australia

553.061.5:553.495.2

Synopsis

Uranium minerals in Australia occur in a variety of host rocks; their wide range of occurrence is mainly due to the chemical characteristics of uranium—notably, the high solubility of the uranyl ion. Regional distribution of uranium is discussed and an account is given of conglomerate and sandstone deposits, with brief descriptions of areas in Australia which have been prospected for such deposits. Some other Australian uranium deposits are described and brief histories of their exploration are given. The ore genesis of the recently discovered deposits of Northern Australia, collectively known as the Alligator Rivers deposits, is discussed. A syngenetic origin, an origin dependent on leaching of uranium from a massive source with subsequent deposition in a suitable environment and deposition from mineralized solutions introduced by igneous activity are considered.

Introduction

Economic deposits of uranium occur in a variety of geological environments—in extrusive or intrusive rocks, in metamorphic rocks and in stratiform and non-stratiform sedimentary deposits. By far the largest known reserves of uranium occur in sediments. The variety of geological environments in which uranium mineralization is found is due to the high solubility of the uranyl ion $(UO_2)^{2+}$, its isomorphism with certain elements, such as calcium, iron, zirconium and thorium, and the stability range of uraninite.[1] Acid igneous rocks, on average, contain about ten times as much uranium as ultrabasic rocks. The average content of uranium in igneous rocks is shown below.

Type	Uranium, ppm
Ultrabasic	0·3
Basic	1
Basaltic	1
Intermediate	1·5–3
Granitic	4

Uranium provinces

The concept of regional uranium enrichment has been discussed by Davidson,[2] Saum and Link[3] and Klepper and Wyant.[4] A uranium province is defined by the present writer as a region in which anomalous concentrations of uranium occur either in a single formation or in a variety of host rocks in one general area. The uranium may have been deposited in a single epoch, or it may be the product of more than one phase of mineralization. In Australia uranium is concentrated in three regions or 'uranium provinces':[5] (1) Rum Jungle–Alligator Rivers; (2) Mary Kathleen–Westmoreland; and (3) Mount Painter–Radium Hill.

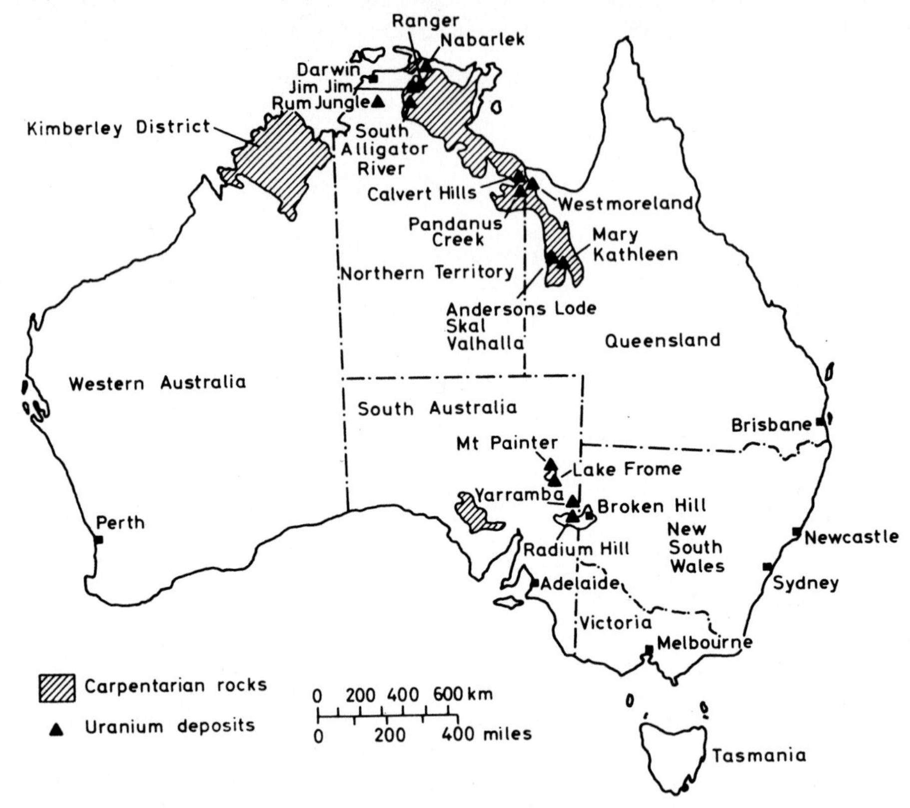

Fig. 1 Distribution of main deposits of uranium in Australia

In the Rum Jungle–Alligator Rivers province mineralization is largely contained in Lower Proterozoic (2300–1800 m.y.) sediments, in the Mary Kathleen–Westmoreland province within sediments and volcanic rocks of Carpentarian age (1800–1400 m.y.),[6] and in the Mount Painter–Radium Hill province within metamorphic rocks and granite also of Carpentarian age, as well as in early Tertiary sediments derived from the older rocks (see Fig. 1). The Rum Jungle uranium deposits in rocks of Lower Proterozoic age are about 190 km from the present outcrop edge of Carpentarian rocks, but the present land surface at Rum Jungle is believed to correspond closely with the unconformity surface from which the younger rocks have been removed by erosion. No palaeogeographic connexion between the three Australian uranium provinces is known.

Geological environments

Assuming an adequate source of uranium, concentration of ore is dependent on geological environment. The most important types of orebody throughout the world are conglomerate-type deposits, sandstone deposits and vein deposits.

In Australia no exploitable uraniferous conglomerate has been discovered, although selected areas have been investigated for this type of deposit. Uranium deposits in sandstone have been discovered in South Australia and at present are being tested. As examples of different geological environments of deposition, the uranium deposits of the Rum Jungle–Alligator Rivers province, Mary Kathleen mine and the area extending northwest from Mary Kathleen to Calvert Hills are also described.

Conglomerate-type deposits, such as those of the Elliot Lake area, Canada,[7,8] and the Witwatersrand, South Africa,[9,10] constitute by far the greatest known reserves of exploitable uranium in the western world. In the conglomerates of the Elliot Lake area and the Witwatersrand the uranium and associated gold are found in quartz-pebble beds interbedded with predominantly arkosic and other quartzose sediments derived, at least in part, from acid igneous rocks. The pebble beds typically consist of sub-rounded to rounded quartz pebbles set in a matrix of fine recrystallized quartz grains, sulphides (mainly pyrite), sericite, chlorite, secondary quartz, carbonaceous material and accessory minerals.

The origin of uranium in conglomerates has attracted considerable discussion. Two main theories have been advanced: a placer origin (Ramdohr[10]) and a hydrothermal origin (Davidson[11,12]). The scope of this paper does not permit lengthy discussion of this much argued question. Objections to the placer origin have centred around the extreme solubility of uranium oxides, the softness of uranium compounds and the virtual absence of detrital uranium minerals in present-day placer deposits. Arguments against a hydrothermal origin are the absence of mineralized channels of access to the orebody, the comparative lack of mineral zoning and the sympathetic distribution of uranium and gold—a relationship considered by some to be more difficult to explain by hydrothermal action. The writer is familiar with the mineralogy of the Witwatersrand conglomerates and favours the suggestion proposed by Derry[13] that uranium was precipitated from solutions in contact with sediments composed of a carbonaceous clay matrix and quartz pebbles during sedimentation. Of prime importance to deposition is not so much the source—the high solubility of oxidized uranium minerals would allow leaching from practically any weathered rock mass—but environmental conditions capable of precipitating uranium from mineralized solutions in contact with a pebble bed. The most likely reducing agents are hydrogen sulphide, derived from the decay of organic carbonaceous matter, and certain bacteria. The common association between uranium and carbonaceous material and the carbonaceous mineral thucholite in the Witwatersrand ore strongly suggest that organic matter played an important role in the precipitation of the uranium.

Because of certain lithological and environmental similarities between Lower Proterozoic conglomerates in northern Australia and those at Elliot Lake and Witwatersrand the conglomerates at Rum Jungle and the Kimberley district have been investigated as possible deposits of uranium. At Rum Jungle pebble conglomerates occur in the Crater Formation within the Batchelor Group. The Crater Formation south of the Rum Jungle Complex consists of a sequence of more than 600 m of arkose, quartzite and shale, dipping southeastward at between 40 and 60°. The formation contains at least three lenses of conglomerate up to 13 m thick and 1200 m long. Table 1 shows a generalized geological section of this area, adapted from Walpole and co-workers[14] and Morlock and England.[15]

At the surface, radioactivity of the conglomerates is locally as high as 30 × background (background = 0·01 mR/h). To obtain specimens of unleached conglomerate for detailed examination and analysis, two diamond drill holes were put down in 1970 by the Bureau of Mineral Resources.[15] Unweathered conglomerate core from two lenses was obtained. Superficially, the conglomerate resembles the Witwatersrand 'banket' in that both are composed of angular to sub-rounded pebbles of quartz set in a matrix of fine quartz, chlorite and sericite. In the Crater Formation conglomerate pyrite, however, is rare and carbonaceous matter has not been recorded to date. R. N. England examined the conglomerate and concluded that the source of radioactivity in the pebble conglomerates is thorium, contained in a red-brown amorphous or metamict phosphate believed to be cheralite. No uranium minerals or gold were identified.

Table 1 Summary of stratigraphy south of Rum Jungle Complex

Age	Rock unit		Thickness, m	Lithology
Lower Proterozoic	Goodparla Group	Golden Dyke Formation	3000*	Graphitic shale, chloritic shale, blue-grey siltstone, quartzite
		Coomalie Dolomite	300	Coarse-textured biohermal dolomite, magnesite, medium- to fine-grained calcilutite
	Batchelor Group	Crater Formation	600	Siltstone, sandstone, arkose, with two and, locally, three interbedded pebble conglomerates
		Celia Dolomite	300	Magnesite, dolomite, calcareous sandstone
Archaean	Red Jungle Complex			Granite, gneiss, banded ironstone

*The uranium deposits at Rum Jungle are contained in graphitic and chloritic shale of the Golden Dyke Formation.

Three mining companies have briefly investigated pebble conglomerates in the King Leopold Sandstone of the Carpentarian Kimberley Group, and to a lesser extent pebble beds in the Lansdowne Arkose and O'Donnell Formation, both of the Speewah Group, which underlies the Kimberley Group, in the southeastern part of the Kimberley district, northern Western Australia. Cuttings and core from the drill-holes known to have intersected conglomerate bands from the three formations investigated have been found to be devoid of uranium minerals, their anomalous radioactivity being due to the presence of heavy mineral concentrations of thorium-bearing minerals, such as thorogummite.

Sandstone deposits are typified by the numerous complex uranium deposits of the Colorado Plateau area, the Wyoming Basin area and the Texas Gulf coast in the U.S.A. Other notable examples are the deposits of Catamarca and Mendoza Provinces in Argentina, the deposits of the Republic of Niger, and of the Ferghana Basin in the U.S.S.R. The Colorado Plateau contains by far the greatest known reserves of uranium in the U.S.A., and in an area of about 380000 km^2 more than 2000 individual deposits are known. The geology of the Colorado Plateau deposits has been described in detail.[16–19]

In the sandstone deposits a close relationship exists between the concentration of ore and sedimentary features, such as ancient stream channels and deltaic deposits. The most common type of host rock is coarse sandstone with moderate permeability, usually overlain by an impervious unit. The host rocks also include conglomerate and, less commonly, siltstone. Factors favourable to ore deposition are thickening of the sandstone host rock, a high ratio of sandstone to mudstone in the sequence and the presence of carbonaceous matter in the host rock.

The recent discoveries in South Australia of sandstone-type deposits of Tertiary age at Lake Frome, about 500 km north of Adelaide, and at Yarramba, about 75 km northwest of Broken Hill, New South Wales (see Fig. 1), are examples of successful selective exploration following an environmental study. The Lake Frome uranium occurs in flat-lying sediments of Tertiary age. The mineralization occurs in sandstone, siltstone and gravels about 100 m below the surface. The uranium is present as finely divided pitchblende, mainly in carbonaceous siltstone and to a lesser extent in siltstone and consolidated gravels. The sediments at Lake Frome are derived partly from granite and metamorphic rocks near Mount Painter which contain uraniferous breccia deposits, and partly from slightly metamorphosed and unmetamorphosed rocks of Adelaidian (1400–600 m.y.) age.[6] Exploration at Lake Frome, reported by Petromin N.L. in an interim report released on 20 April, 1971, has indicated the presence of two deposits separated horizontally by about 730 m of untested ground. It is possible that the two deposits may, in fact, be part of a single more extensive deposit. The reserves given in April, 1971, are:

Prospect Beverley 5600 short tons U_3O_8 (average ore grade, 0·24% U_3O_8)
Prospect 37A 4400 short tons U_3O_8 (average ore grade, 0·15% U_3O_8)

Similar deposits occur at Yarramba in Tertiary sediments derived from the Radium Hill area, the site of the former uranium mine.

Vein deposits are the most common form of uranium ore. They vary considerably in size, shape and mineralogy, and because they are normally unrelated to the country rock in which they occur, uranium-rich vein deposits are found in a variety of host rocks, usually of Precambrian age. The veins are often rich in sulphides. Perhaps the best known vein deposits in the western world are those in the Goldfields area of Saskatchewan, Canada,[20,21] and the Shinkolobwe deposits of the Katanga Province in the Congo Republic.[22] Less extensive vein deposits are found in most parts of the world. No vein deposits closely comparable with those of the Goldfields area, Canada, or of the Shinkolobwe area, Congo Republic, occur in Australia.

Uranium deposits of Rum Jungle–Alligator Rivers province, Australia

The most important uranium deposits occur in a roughly semi-circular area in north central Australia. The region includes the uranium mines at Rum Jungle, South Alligator River and the newly discovered deposits of Nabarlek, Ranger and Jim Jim. The deposits which have been mined occur in folded and sheared low-grade metamorphosed sediments of Lower Proterozoic age. Uranium minerals also occur in dolomite and in sandstone and volcanic rocks of Carpentarian age. The main deposits of the region are noted below.

RUM JUNGLE

Uranium has been mined from three deposits of moderate size and two minor deposits at Rum Jungle, about 64 km south of Darwin, Northern Territory. The deposits[23,24] occur in carbonaceous black shale or chloritic shale of the Golden Dyke Formation, near to the contact with dolomite–magnesite of the Coomalie Dolomite (see Table 1). The deposits are concentrated in a broad syncline, each located close to, or in, shears, faults or folds. A summary of the geology of the deposits is given in Table 2.

About 8 km northwest of the Rum Jungle treatment plant a low-grade uranium prospect is located in a shallow syncline. The ore is contained mainly in magnesite, dolomite and breccia of the Coomalie Dolomite and in the overlying Golden Dyke Formation shale. In the carbonate rocks uranium is present

Table 2 Summary of main features of Rum Jungle mines

Name of mine	Approximate dimensions of orebody, m	Approximate production of ore, tons	Remarks
White's	100 × 10	396000 (0·31% U_3O_8)	Steeply dipping, layered uranium–copper deposit in black shale with lead, cobalt and nickel minerals. Secondary torbernite, autunite, saleeite, gummite, phosphuranylite and johannite. Pitchblende occurs about 9 m below surface
White's Extended	Small	100 (0·2% U_3O_8)	Located 500 m east of White's in chloritic slate. Mainly secondary gummite, autunite and saleeite
Dyson's	40 × 25	154000 (0·39% U_3O_8)	Steeply dipping ovoid orebody in black pyritic shale. Mainly saleeite, with rare autunite and sklodowskite
Rum Jungle Creek South	270 × 70	653000 (0·48% U_3O_8)	Elongated irregular-shaped orebody in chloritic shale. Sooty pitchblende
Mt Burton	Small	6000 (0·21% U_3O_8)	Uranium and copper minerals in black shale. In the oxidized zone torbernite, malachite, chalcocite and, rarely, native copper. Primary pitchblende, pyrite and chalcopyrite

as thucholite in fine veinlets; in the shale it occurs as both thucholite and fine grains of pitchblende in veinlets along cleavage planes and minor fractures. Diamond drilling has proved a low-grade orebody of about 800000 tons (0·16% U_3O_8).

Uranium was discovered at Rum Jungle in 1949 by a prospector, J. H. White, who recognized secondary uranium minerals illustrated in a Bureau of Mineral Resources prospectors' handbook. Following the initial discovery, the Rum Jungle area was subjected to intensive investigation by the Bureau of Mineral Resources and Territory Enterprises Pty, Ltd. The investigations consisted of detailed mapping, geochemical surveys and radiometric and other geophysical surveys. Interpretation of data, together with information derived from drilling, has revealed that at Rum Jungle (1) uranium deposits can be related to structure: they are either close to or fill faults or shears (the Rum Jungle Creek South lode is located in a tight fold); (2) the uranium tends to be concentrated in black shale or chloritic shale of the Golden Dyke Formation, near the contact with Coomalie Dolomite; (3) the uranium occurs close to the surface, seldom deeper than 100 m; and (4) at White's and Mt Burton uranium occurs with copper minerals: most radiometric anomalies are coincident with copper–lead–cobalt–nickel surface anomalies.[25]

The most successful method of exploration in the Rum Jungle area has been radiometric survey, followed by test drilling. Experience proved the most favourable lithological target to be the Golden Dyke Formation shale, close to the contact with the Coomalie Dolomite (see Table 1). Experimentation showed that initial testing by auger drilling is advisable, as numerous radiometric anomalies, particularly those located in areas covered by lateritic soils, are surface features which rapidly decline in intensity below the surface. Auger drill-holes were put down on surveyed grids and were also used for the collection of geochemical samples, radiometric probing and, experimentally, for radon detection surveys.

SOUTH ALLIGATOR RIVER

Uranium deposits in the South Alligator River area occur in a geological environment comparable with that at Rum Jungle. The mines are smaller and individual lodes are of higher grade: for example, El Sherana, the largest of the mines, produced 59800 tons of ore (0·73% U_3O_8). Uranium has been mined at thirteen small mines.[26,27] The mines extend over a distance of about 21 km along an almost vertical major fault zone striking at about 300°. Most of the deposits occur in steeply dipping carbonaceous shale and ferruginous siltstone of the Lower Proterozoic Koolpin Formation close to the unconformity with the overlying flat-lying Kombolgie Formation of Carpentarian age.

Table 3 Summary of stratigraphy, South Alligator River Valley

Age	Rock unit		Thickness, m	Lithology
Carpentarian	Katherine Group	Kombolgie Formation	1700	Conglomerate, greywacke, sandstone, with interbedded volcanic rocks
		Unconformity		
Lower Proterozoic	South Alligator Group	Fisher Creek Siltstone	5000	Siltstone, micaceous siltstone, greywacke–siltstone
		Gerowie Chert	1000	Chert, siliceous siltstone (facies variant of underlying Koolpin Formation)
		Koolpin Formation	1700	Algal dolomite, pyritic carbonaceous siltstone
		Unconformity		
Archaean		Stag Creek Volcanics		Altered basalt

Minor uranium deposits also occur just above the unconformity separating Lower Proterozoic and Carpentarian rocks—for example, in the Kombolgie Formation (Palette mine) and in the Edith River Volcanics (Coronation Hill). At Scinto 6 uranium occurs in much altered dolerite below the unconformity. Pitchblende is the predominant uranium mineral in most of the deposits, Saddle Ridge being the only deposit composed almost exclusively of secondary uranium minerals. The ore at El Sherana, Palette and Coronation Hill has a significant gold content.

In the South Alligator Valley secondary minerals were first recognized in the field at Coronation Hill in 1953. Subsequent discoveries were made by geological mapping and low-level airborne radiometric surveys.

ALLIGATOR RIVERS AREA

The newly discovered uranium deposits of Nabarlek, Ranger and Jim Jim to the NNE of South Alligator River occur in a geological environment similar to that of the South Alligator River deposits and, to some extent, to Rum Jungle. The limited information available about the Nabarlek occurrence indicates that it is a vein deposit, probably occupying a slip shear. The main deposit is a vein, part of which is massive pitchblende, which extends over a strike length of about 270 m and is concordant with the schistosity of the host rock (chlorite–muscovite schist). The vein apparently cuts out before a gently inclined dolerite dyke at a depth of about 65 m. Massive pitchblende occurs within 1·5 m of the surface, and is coated with gummite, autunite, saleeite and rare torbernite, which are also disseminated in the schist and hornfelsed schist near the dolerite contact. In addition to disseminated ore, narrow veinlets of secondary minerals follow joint planes and cleavage planes. There does not appear to be any connexion between the dolerite beneath the deposit, other dolerites in the Carpentarian rocks near Nabarlek, and the uranium mineralization.

Ranger deposit, situated about 60 km southwest of Nabarlek, consists mainly of fine veinlets or stringers of uranium minerals in quartz–chlorite schists (Fisher Creek Siltstone?). At a surface exposure the host rock dips steeply to the east, but drilling results indicate that below the surface the band flattens out to become nearly horizontal. The quartz–chlorite schist typically consists of fine laminae of quartz and chlorite, and, locally, the chlorite bands contain widely dispersed, rare, fine flakes of graphite. The uraniferous stringers (secondary minerals at the surface, sooty pitchblende below the zone of oxidation) are seldom more than a few millimetres thick and are concentrated in brecciated parts of the host rock. The veinlets are mainly crosscutting, often thickening around brecciated lumps of rock. In addition to the pitchblende, the veins contain rare pyrite and galena, with traces of copper and vanadium. The gold content is variable, ranging from a trace to 5 dwt/ton. The host rock conformably overlies a band of siliceous cherty breccia (Gerowie Chert?). The presence of fragments of silicified siltstone retaining their original bedding indicates that the rock was formed by the silicification of a disrupted argillite. A band of sericite–chlorite schist (Koolpin Formation?) underlies the chert, apparently conformably. The sericite–chlorite schist overlies a band of pegmatitic–granitic rock composed essentially of quartz and feldspar with abundant prisms of tourmaline, the pegmatitic rock containing veins of pitchblende, seldom more than 2 cm thick. In the Ranger area highly chloritic rocks, believed to be altered dolerites, are patchily mineralized, but the less altered dolerites, believed to be of Carpentarian age, do not contain uranium minerals.

The writer's information about Jim Jim uranium deposit, located about 20 km SSW of Ranger, is limited. As far as is known, like Ranger, the deposit consists of uraniferous veinlets in a mica schist. A band of highly folded quartzite apparently occurs within the mica schist.

In July, 1971, Pancontinental Mining, Ltd., announced the discovery of a uranium prospect about 40 km southwest of Nabarlek and about 20 km north of Ranger.[32] The announcement described uranium mineral deposits in both Lower Proterozoic metasediments (grab samples ranging from 0·1 to 3·4% U_3O_8) and in conglomerate of the Carpentarian Kombolgie Formation (grab samples 0·03 to 2·5% U_3O_8).

The uranium deposits of Nabarlek, Ranger and Jim Jim were discovered by airborne gamma-ray spectrometer surveys. The radiometric anomalies were first located on the ground by detailed radiometric gridding, the defined anomalies being tested by drilling, trenching and costeaning. Testing of the large number of radiometric anomalies outlined by airborne surveys over areas which include the three deposits and other adjoining areas will not be completed for some years. An interesting feature of the exploration in the region is that detailed ground investigations have established a number of anomalies not recorded by airborne survey. These anomalies were located by hand-held scintillometer, usually, but not always, on steep slopes where overburden had concealed the radioactivity from airborne gamma-ray spectrometry. Most of the area west of the newly discovered prospects is at least partly covered with superficial deposits, the area to the east being overlain by sediments of the Kombolgie Formation.

Although a vast amount of work is required before an accurate assessment can be made of the field, the following observations are drawn from available information. (1) The region is an important uranium field, which will take many years to test fully. (2) The bulk of the uranium deposits discovered so far occurs in metasediments of Lower Proterozoic age near the surface, recently stripped of a cover of Carpentarian Kombolgie Formation rocks. The deposits are grouped close to the present-day erosional edge of Kombolgie Formation rocks. Uranium minerals also occur in the basal beds of the Kombolgie Formation. (3) Primary uranium has been intersected in granitic rock. Future exploration of the Alligator Rivers area will be greatly assisted if the genesis of uranium ore is established. The source of uranium and manner of deposition is probably similar to the deposits at South Alligator River and Rum Jungle. In the light of admittedly insufficient available data, three possible origins for the uranium are considered: (*a*) syngenetic deposition in a favourable host rock with later concentrations of ore; (*b*) concentration from a massive source rock; and (*c*) hydrothermal–mesothermal deposition from an underlying igneous host rock.

SYNGENETIC ORIGIN

Reference has been made to the preferential concentration of uranium ore in carbonaceous shale, of the Golden Dyke Formation at Rum Jungle, and the Koolpin Formation at South Alligator River. The high background content of uranium in carbonaceous shale, ~125 ppm,[28] would ensure a more than adequate source of uranium. Concentration of the uranium would require leaching and redeposition, possibly induced during the late consolidation stage of sedimentation, mineralized solutions being squeezed into fractures. Although this hypothesis is attractive, its applicability to the deposits of Nabarlek, Ranger and Jim Jim might be questioned on the grounds that carbonaceous shale is not, in the present state of knowledge, known to occur near any of these deposits. The high solubility of uranium and the fractured nature of the country would, however, allow transportation of uranium over considerable distances, particularly under the stress of tectonic pressure. Relocation of syngenetic uranium may explain a characteristic feature of the deposits—the concentration of ore in the Lower Proterozoic rocks near the unconformity with Carpentarian rocks. If the uranium pre-dates the Kombolgie Formation sedimentation, progressive lowering of the Lower Proterozoic surface by erosion would tend to concentrate ore near the surface by successive solution and accretion. (Age determinations carried out[29] on pitchblende from South Alligator River and Nabarlek have indicated age ranges of between 550 and 700 m.y. and 710 and 815 m.y., respectively, but there is no proof that the pitchblende tested had not been redeposited.)

CONCENTRATION FROM A MASSIVE SOURCE ROCK

The relationship between the distribution of Australian uranium deposits and rocks of Carpentarian age has been noted. Furthermore, in the Alligator Rivers area, the close spatial relationship between known deposits and the Kombolgie Formation has been pointed out. Concentration of ore from a massive overlying source rock, such as the Kombolgie Formation, consisting of greywacke, sandstone and conglomerate, would simply require leaching of soluble uranium minerals, prolonged interstratal migration and deposition at the unconformity separating the underlying sediments. The Kombolgie Formation is invaded by, and interbedded with, a variety of volcanic rocks of the Katherine River Group, which may also be considered as possible source rocks of uranium. The status of the volcanic rocks as source rocks is enhanced by the fact that they are contemporaneous with, and might be correlatable with, uraniferous volcanic rocks in the Westmoreland–Mary Kathleen uranium province to the southeast of the Alligator Rivers area (see Fig. 1). This hypothesis is attractive because of certain known features of the uranium deposits: uranium is typically concentrated in the upper part of the Lower Proterozoic rocks and locally in the basal beds of the Kombolgie Formation rocks. The deposits at South Alligator and Rum Jungle tend to narrow below the surface of the older rocks. Reducing conditions are likely to have prevailed at the unconformity, after deposition of up to 1700 m of overlying sediments.

Some objections can, however, be levelled at this theory. Taylor[26] noted that the ages of pitchblende at South Alligator River indicate that uranium ore was deposited between 550 and 700 m.y. ago. He considered redistribution of uranium by meteoric waters during the interval unlikely. The gold content of Ranger and certain of the South Alligator River deposits would also present some difficulty to a concept requiring deposition from meteoric water leaching an overlying source rock.

HYDROTHERMAL–MESOTHERMAL DEPOSITION

At Ranger deposit primary uranium has been intersected in granitic rock below the host rock. Further work in this area may prove the presence of channelways from depth rather than from an overlying source. Taylor[26] described 'hydrothermal' wallrock alteration around ore in the South Alligator River in unoxidized rock. The relationship between structures and ore deposition at Rum Jungle and South Alligator Valley may be related to the introduction of mineralized solutions from the igneous source at depth along shears and faults. The relationship between ore deposition and the Kombolgie Formation may be accounted for by the assumption that ascending mineralized solutions moved freely up to the cap of nearly horizontal Kombolgie Formation sediments, which acted as a restricting control, channelling movement of the solutions along the plane of the unconformity. A relationship between igneous activity and ore deposition may explain the frequent coincidence of ore and anomalous ground magnetism.

Future exploration in the Alligator Rivers area will require shallow drilling to investigate areas blanketed by superficial deposits. The shallowness of the water-table in this area, probably less than 10 m, will be an aid to radiometric and hydrogeochemical investigations.

Other types of uranium deposit

MARY KATHLEEN MINE

Mary Kathleen mine[30] is situated about 52 km east of Mount Isa. Mining operations by open-cut methods ceased in 1962 when the mine had reached a depth of about 61 m and had yielded a total of about 3714 tons of U_3O_8.

The deposit is contained in the predominantly calc-silicate upper part of the Corella Formation. Host rock to the ore is a breccia skarn composed mainly of quartzitic and feldspathic fragments in a fine-grained matrix composed mainly of garnet, scapolite and feldspar. The orebody is located in the axis of a syncline and is bounded by two roughly north–south shears. The mineralization is considered to be the result of deposition of late-phase metasomatic emanations from the Mt Burstall granite—an environment unique in Australia and rare throughout the world. The primary mineral is uraninite, and it is generally accompanied by the rare-earth minerals allanite and stillwellite.

WESTMORELAND AND MARY KATHLEEN AREA DEPOSITS

In the area between Mary Kathleen and the Calvert Hills, a belt straddling the Queensland and Northern Territory boundary, numerous, mostly small, uranium deposits have been discovered.[31] At Westmoreland pitchblende is present in marginal shears of a trachyandesite dyke and in sandstone and conglomerate adjacent to the dyke. (Secondary minerals have been transported as far as 1700 m from their original source.) At *Pandanus Creek* pitchblende occurs in acid volcanics. The deposits northwest of the Mary Kathleen mine, typified by *Anderson's lode*, *Skal* and *Valhalla*, are mainly contained in sheared sediments within the Eastern Creek Volcanics.

Although a number of the deposits in this region are of little or no economic significance, their widespread distribution provides proof of an overall uranium enrichment in the Carpentarian volcanics. Exploration for deposits of uranium in the region should be governed, first, by investigation of porous interbedded sediments or pyroclastic layers in the volcanics capable of containing concentrations of relocated uranium ore, and, secondly, by investigation of the larger tectonic features, such as breccias, shears and folds in the volcanics, in which fill or replacement lodes could be located.

Acknowledgment

The writer is indebted to Mr. J. Elliston, Executive Geologist, Geopeko Limited, and Dr. E. Rod, Chief Geologist, Queensland Mines, Ltd., for providing information about the geology of the uranium deposits of Ranger and Nabarlek, respectively. This paper is published by permission of the Director, Bureau of Mineral Resources, Geology and Geophysics, Canberra, A.C.T.

References

1. Heinrich E. W. *Mineralogy and geology of radioactive raw materials* (New York: McGraw-Hill, 1958), 614 p.
2. Davidson C. F. The distribution of radioactivity. *Min. Mag., Lond.*, **85**, 1951, 329–40.
3. Saum N. M. and Link J. M. Exploration for uranium. *Colo. Sch. Mines Miner. Ind. Bull.*, **12**, no. 4 1969, 1–23.
4. Klepper M. R. and Wyant D. G. Geology of uranium. *Bull. geol. Soc. Am.*, **66**, 1955, 1585–6. (Abstract)
5. Dunn P. R. and Dodson R. G. Uranium and thorium in Australia. (In preparation).
6. Dunn P. R., Plumb K. A. and Roberts H. G. A proposal for time-stratigraphic subdivision of the Australian Precambrian. *J. geol. Soc. Aust.*, **13**, 1966, 593–608.
7. Joubin F. R. Comments regarding the Blind River (Algoma) uranium ores and their origin. *Econ. Geol.*, **55**, 1960, 1751–6.
8. Roscoe S. M. and Steacy H. R. On the geology and radioactive deposits of Blind River Region. In *Proc. 2nd U.N. Conf. peaceful Uses atomic Energy, Geneva 1958* (Geneva: U.N., 1958), **2**, 475–83.
9. Hiemstra S. A. The mineralogy and petrology of the uraniferous conglomerate of the Dominion Reefs Mine, Klerksdorp area. *Trans. geol. Soc. S. Afr.*, **71**, 1968, 1–67.
10. Ramdhor P. New observations on the ores of the Witwatersrand in South Africa and their genetic significance. *Trans. geol. Soc. S. Afr.*, **61**, 1958, Annex., 173 p.
11. Davidson C. F. On the occurrence of uranium in ancient conglomerates. *Econ. Geol.*, **52**, 1957, 668–93.
12. Davidson C. F. Uranium in ancient conglomerates: a review. *Econ. Geol.*, **59**, 1964, 168–77.
13. Derry D. R. Evidence of the origin of the Blind River uranium deposits. *Econ. Geol.*, **55**, 1960, 906–27.
14. Walpole B. P. *et al.* Geology of the Katherine–Darwin region, Northern Territory. *Bull. Bur. Miner. Resour. Aust.*, **82**, 1968, 2 vols, 304 p.
15. Morlock J. S. and England R. N. Results of drilling of Crater Formation, Rum Jungle. *Rec. Bur. Miner. Resour. Aust.* 1971/65, unpublished.

16. Fischer R. P. Uranium–vanadium–copper deposits of the Colorado Plateau region. In *Proc. Int. Conf. peaceful Uses atomic Energy, Geneva, 1955* (New York: U.N., 1956), 605–14.
17. Fischer R. P. Similarities, differences, and some genetic problems of the Wyoming and Colorado Plateau types of uranium deposits in sandstone. *Econ. Geol.*, **65**, 1970, 778–84.
18. Kelley V. C. Regional structure and uranium distribution on the Colorado Plateau. Paper presented at Nuclear Engineering Sciences Congress, Cleveland, Ohio, 1955, preprint 254.
19. Kelley V. C. Influence of regional structure upon the origin and distribution of uranium in the Colorado Plateau. In *Proc. Int. Conf. peaceful Uses atomic Energy, Geneva, 1955* (New York: U.N., 1956), **6**, 299–306.
20. Buffam B. S. W. Uranium deposits, Beaverlodge area, Saskatchewan, Canada. *Bull. geol. Soc. Am.*, **62**, 1951, 1427. (Abstract)
21. Robinson S. C. Mineralogy of uranium deposits, Saskatchewan. *Bull. geol. Surv. Can.* **31**, 1955, 128 p.
22. Derriks J. J. and Vaes J. F. The Shinkolobwe uranium deposits: current status of our geological and metallogenic knowledge. In *Proc. Int. Conf. peaceful Uses atomic Energy, Geneva, 1955* (New York: U.N., 1956), **6**, 94–128.
23. Spratt R. N. Uranium ore deposits of Rum Jungle. In *Geology of Australian ore deposits, 2nd edn* McAndrew J. ed. (Melbourne: Congress and Australasian IMM, 1965), 201–6. (*Publications 8th Commonw. Min. Metall. Congr., Australia N.Z., 1965*, vol. 1)
24. Berkman D. A. The geology of the Rum Jungle uranium deposits. In *Symposium: uranium in Australia* (Melbourne: Australasian IMM, 1968), 12–31.
25. Dodson R. G. and Shatwell D. O. Geochemical and radiometric survey Rum Jungle, Northern Territory, 1964. *Rec. Bur. Miner. Resour. Aust.* 1965/254, unpublished.
26. Taylor J. Origin and controls of uranium mineralization in the South Alligator Valley. In *Symposium: uranium in Australia* (Melbourne: Australasian IMM, 1968), 32–44.
27. Prichard C. E. Uranium ore deposits of the South Alligator River. In *Geology of Australian ore deposits, 2nd edn* McAndrew J. ed. (Melbourne: Congress and Australasian IMM, 1965), 207–9. (*Publications 8th Commonw. Min. Metall. Congr., Australia N.Z., 1965*, vol. 1)
28. Robertson D. S. and Douglas R. F. Sedimentary uranium deposits. *CIM Trans.*, **73**, 1970, 109–18.
29. Cooper J. A. On the age of uranium mineralization at Nabarlek, N.T. *Rec. Bur. Miner. Resour. Aust.*, unpublished.
30. Hughes F. E. and Munro D. L. Uranium ore deposits at Mary Kathleen. In *Geology of Australian ore deposits, 2nd edn* McAndrew J. ed. (Melbourne: Congress and Australasian IMM, 1965), 256–63. (*Publications 8th Commonw. Min. Metall. Congr., Australia N.Z., 1965*, vol. 1)
31. Brooks J. H. The uranium deposits of north-western Queensland. *Publ. geol. Surv. Qld* 297, 1960, 50 p.
32. *The Australian Financial Review*, 29 July, 1971.

DISCUSSION

Dr. E. E. N. Smith asked if any age determinations had been made on the Nabarlek or Ranger pitchblendes. In reply, the author said that he was somewhat dubious about the one result from Nabarlek, because of the solubility of uranium and the possibility of redistribution. The range obtained was 710–815 m.y. (550–700 m.y. at South Alligator River). Some of the ore could have been reworked, possibly during a later period of igneous activity.

I. R. Wilson enquired, first, as to the precipitating influence, if any, of the carbonaceous matter present in the sediments: it had been shown by many workers that gold might be precipitated from solutions by carbon-bearing materials. Secondly, was there any relationship between the uranium and gold contents in carbonaceous sediments?

The author replied that there was a close relationship between organic matter and uranium. There had been a heated controversy over the alleged precipitation of gold. At Ranger and at Nabarlek there was no carbonaceous matter present, but that was not so at Rum Jungle and South Alligator River. At Rum Jungle there was no gold; at Ranger there was gold but not much hydrocarbon; in the South Alligator Valley there were rich gold values but no constant interrelationship with the uranium.

R. J. Chico enquired as to the extent to which the ore textures studied in samples from the Alligator River district pointed to the need to continue with the Davidson–Ramdohr ore genesis disputations, especially those concerning the Witwatersrand, Blind River and Brazilian uranium–gold conglomeratic deposits.

The author said that there was no comparison between the Alligator River deposits, which were vein-type deposits, and the conglomeratic deposits in Canada and South Africa. The rest of the uranium deposits were in a black shale environment, as at Rum Jungle.

In reply to a question by W. T. Meyer regarding the possibility that Australian uranium deposits could be due to surface enrichment through capillary action, the author said that the Northern Territory area was extremely humid: 60 in of rain fell in three months, it rained for another two months and then remained dry, but not arid, for the rest of the year. There had been suggestions that the deposits were concentrated near the surface by peneplanation, the eroded deposit being concentrated at depth, but in proximity to the original deposit.

G. R. Ryan said that geologists working on the Ranger deposit favoured a hydrothermal origin for the following reasons: veins of pitchblende and hematite, with accompanying chloritization, had been encountered below the orebody at depths of about 220 m; the deposit contained gold, pyrite, galena and chalcopyrite; chalcopyrite, in particular, was not normally found in supergene deposits; and the entire rock mass appeared to have been very extensively chloritized. Systematic studies by Geopeko Limited geologists in the overlying sandstones had failed to reveal any signs of remnant uranium in that formation; and no mineralization was known in the immediately underlying rocks elsewhere in the area. What sulphides were present at Rum Jungle?

The author said that the sulphides at Rum Jungle included abundant pyrite at all properties; also, in two of the orebodies (White's and Mt Burton) there was copper mineralization as well as an unusually high concentration of lead. There was also some nickel. Of course, many base-metal deposits in the area had no uranium mineralization at all.

Dr. F. Q. Barnes noted with some satisfaction that the author was reluctant to refer to the new Australian uranium deposits as veins. Of course, there were true vein deposits in the five-element types of uranium deposits which must be considered of hydrothermal origin: however, was it not too fortuitous to have so many of the former types associated with overlying sandstones without there being some genetic significance, particularly when so much uranium was known to originate from sandstones? He would welcome a description of the petrology of the Kombolgie Sandstone, with particular reference to the feldspars and the presence, or absence, of acidic pyroclastics and volcanics in the area of the new finds.

The author replied that there were geologists in Australia who believed that the source of the uranium was the Kombolgie Sandstone. There was a dearth of information on the petrography of the sandstone.

Dr. H. D. Fuchs, with regard to the Lake Frome deposit, sought further details on the lithology of the uranium ore-bearing beds. At the uranium deposit in the Republic of Niger and elsewhere in Africa the mineralization occurred in similar lithological environments. The author replied that there was very little information available at present. Drill-holes showed that the deposit was in silty carbonaceous rocks, but also in siltstone bands and in reconstituted gravel.

Prospecting criteria for sandstone-type uranium deposits

E. W. Grutt Jr.

U.S. Atomic Energy Commission, Grand Junction, Colorado, U.S.A.

550.8 : 553.495.2 : 552.513(73)

Synopsis

A number of criteria which are commonly used as guides in the exploration for uranium are tabulated. The regional geologic settings and local features regarded as important to the problems of uranium in sandstone are checked against nine U.S. uranium areas. The significance of the various factors is evaluated and the types of deposits are described.

The uranium geology of the nine areas is related to possible modes of origin, and the guides most useful in exploration programmes are briefly stated. A number of topics which warrant future investigation are listed.

Webster defines the word 'criteria' as 'standards on which a decision or judgement may be based'. In this paper criteria are pointed toward exploration, which is a systematic and sequential collection, development and interpretation of geologic data for the purpose of discovering mineral deposits. The term 'sandstone-type deposits' is a general term and needs explanation because it does not fit into a genetic classification. It includes epigenetic uranium impregnations in sandstone where uranium coats sand grains, fills interstitial spaces or replaces earlier minerals. The rocks are friable or cemented, but not metamorphosed.

The subject matter in this paper is based on observations in the U.S. uranium areas; however, the criteria can be applied broadly, because the sandstone deposits have characteristics in common worldwide.

The topic of sandstone-type uranium deposits has been popular among geologists for more than 20 years, and a wealth of literature exists. The origin of these deposits is still not rigorously demonstrated, despite tens of thousands of field investigations, considerable geochemical research and much propounding of theory. On the other hand, the literature continues to grow and better knowledge of the habitat of deposits is being gained from observations made as the ores are being mined.

This paper is organized largely around Tables 2 and 3, which summarize selected criteria for nine areas on regional and local scales, respectively. The criteria, theoretical considerations and types of deposits are discussed briefly. This is followed by summaries of the selected uranium areas, with the idea in mind that criteria must be relatable to the environment of deposits in order to be meaningful.

The areas were selected on the basis of different host rocks, resources and geographic dispersion in order to provide a representative cross-section of the U.S. uranium deposits in sandstones. All but one of the examples are in one or another of the three major U.S. uranium regions, i.e. in the Colorado Plateau, Wyoming Basins and Texas Gulf Coastal Plain. Although differences among areas are apparent, overall geologic similarities are striking, and, collectively, the areas typify the kinds of settings that offer the best exploration targets.

Table 1 lists the ages and names of host rocks and gives the reserves of the selected areas (Fig. 1).

Table 1

Area	Age of host rock	Principal host formations	Production plus $8.00 reserves at 1 January, 1971, tons U_3O_8
Lisbon Valley, Utah	Triassic	Chinle	32440
Monument Valley–White Canyon, Arizona–Utah	Triassic	Chinle	9860
Uravan Mineral Belt, Colorado	Jurassic	Morrison	38670
Ambrosia Lake, New Mexico	Jurassic	Morrison	118620
Black Hills, Wyoming–Dakota	Cretaceous	Lakota Fall River	2890
Gas Hills, Wyoming	Eocene	Wind River	48910
Powder River Basin, Wyoming	Paleocene Eocene	Fort Union Wasatch	32760
Texas Gulf Coast	Eocene Miocene	Whitsett Oakville	13750
Maybell–Baggs, Colorado–Wyoming	Miocene	Browns Park	2330
Total			300230

The areas listed account for approximately 65% of the 1 January, 1971, U.S. reserves plus production. Of these areas, only Monument Valley–White Canyon and Maybell–Baggs are now inactive due to depletion of economic reserves.

It is appropriate to mention here that the AEC is still evaluating the results of the U.S. uranium industry exploration effort of the 1966–70 period, during which record amounts of drilling were done. In the five-year period 94000000 ft of drilling added more than 100000 tons of U_3O_8 to the $8.00 reserves, in spite of 60000 tons' production in the period. This very successful effort did not result from breakthrough of technology but was due to systematic application of generally accepted geologic criteria, methods and theory, and especially to saturation-type exploration in known favourable areas.

Review of selected criteria

No attempt was made to formulate a complete list of all possible criteria because such a list would be unmanageable. Only those judged to be most relevant or applicable to important areas were considered for inclusion.

REGIONAL GEOLOGIC SETTINGS (TABLE 2)

Age of host rock

In the western interior of the U.S. the period from the Permian through the Tertiary was one during which geologic conditions were especially suited for deposition of hosts favourable for uranium, because the period was one of general emergence and orogeny. Widespread and thick continental sandstones

Fig. 1 Uranium resource regions, western United States

containing detrital vegetal materials were deposited in all periods, but are especially prevalent in the Triassic, Jurassic and Tertiary sections. Numerous acidic-rock volcanic centres were active; the ash falls blanketed entire regions and local areas received heavy accumulations.

Sedimentary rocks of Jurassic, Tertiary and Triassic ages, in that order, account for about 95% of the U.S. production and reserves to date. Moreover, they are credited with 90% of the 1970 production and 98% of the 1970 reserve.

Depositional environment of host rock

The host rocks for significant sandstone-type deposits seem to be limited to fluvial, marginal marine or aeolian sandstones. All have the requisite permeability, but only the fluvial and marine deltaic or lagoonal sandstones are likely to contain vegetal organics and carbonaceous shale interbeds. Aeolian sandstone, however, can be host for uranium if reductant from an extraneous source is introduced.

Provenance of host rock

Review of the various uranium areas shows a range in provenance. In most of the examples erosion of granitic and metamorphic rocks and sandstones and shales has created the facies and textures favourable for uranium deposition.

Table 2 Regional geologic setting

Environmental factors	*Age of host rock*					*Depositional environment of host rock*				*Provenance of host rock*				*Special strati-graphic factors*			*Type of host rock*			*Physio-graphic province*			*Tectonic element*	
	Permian	Triassic	Jurassic	Cretaceous	Tertiary	Terrestrial–aeolian	Fluvial–coalesced alluvial fan	Fluvial–stream channel and flood plain	Marginal marine: deltaic–lagoonal barrier bar	Granite and metamorphic rocks	Sandstones and shales	Limestones	Acid volcanic centres–tuffs	Unconformity superjacent to host	Unconformity subjacent to host	Tuffaceous sediments in section above host sandstone	Feldspathic or arkosic sandstone	Quartzose sandstone	Tuffaceous or bentonitic sediments present within host sandstone	Colorado Plateau	Wyoming Basins	Texas Gulf Coastal Plain	Intracratonic basins in Cordilleran Foreland Belt	Geosynclinal
Uranium areas Lisbon Valley, Utah	O	X						X		X		O	O		X	X	X		X	X			X	
Monument Valley–White Canyon, Arizona–Utah		X						X		X			O	X	X	X	X		O	X			X	
Uravan Mineral Belt, Colorado			X				X	X		O	X		O			X	O	X		X			X	
Ambrosia Lake, New Mexico			X				X	O		X	X					X	X			X			X	
Black Hills, Wyoming–South Dakota				X				X	O		X	O		X	X	X		X					X	
Gas Hills, Wyoming					X		X	O		X	O		O	X	X	X	X		O		X		X	
Powder River Basin, Wyoming					X		O	X		X	O			X	X	X	X		O		X		X	
Texas Gulf Coast					X			X	X	O	X	O	X	X	X	X	O	X	X			X		X
Maybell–Baggs, Colorado–Wyoming					X	O		X		O	X		O		X	X		X	X		X		X	

X, Dominant or of primary importance; O, subordinate or of secondary importance.
Blank space, factor is of doubtful significance, absent or does not apply.

The presence of acid-volcanic centres is also a factor common to several areas. Theoretically, the granitic and tuffaceous rocks could be important as sources of uranium.

Special stratigraphic factors

The prevalence of unconformities in some important areas suggests that they may have a role in the formation of deposits. Unconformities enhance regional transmissivity and facilitate groundwater movement over large areas. Truncated beds in the subcrop may be pathways along which ascending groundwaters, hydrocarbons or gases reach aquifer systems. The bevelling of beds by erosion permits deep weathering of rocks, recharge of groundwater in aquifers, and commingling of groundwaters originating in different beds.

Tuffaceous sediments in the stratigraphic section above the host are a feature in all of the nine districts. Such rocks are commonly sources of carbonate groundwaters having higher than normal uranium content.

Type of host rock

The composition of the host is dependent on the provenance. Statistically, permeable feldspathic (typically, 15% feldspar) and arkosic (typically, 40% feldspar) sandstones create the most favourable hosts, but quartzose (mainly quartz) sands can also be favourable, and tuffaceous sandstones (typically more than 10% tuffaceous material) contain important deposits. This range in rock types indicates that perhaps the type of host rock alone is not as important as combinations of several geologic factors.

Physiographic province

Four of the nine selected areas are in the Colorado Plateau, three in the Wyoming Basins, one in the Texas Coastal Plain, and one, the Black Hills, is on the eastern margin of the Wyoming Basins. Together, these three provinces account for about 95% of the total U.S. production plus reserves. These three physiographic provinces also seem to be provinces in the metallogenic sense, because of the prevalence of uranium.

Tectonic element

The Wyoming Basins and Colorado Plateau are part of the Cordilleran foreland tectonic unit, which is characterized by anticlinal mountains separated by large asymmetrical basins. Eight of the nine areas are in foreland basins and one, the Texas Gulf Coast, is in the Gulf Coast geosyncline. The structural characteristics of the three physiographic provinces (see above) are inherited from these underlying tectonic elements.

LOCAL FAVOURABLE CRITERIA (TABLE 3)

Types of deposit

There are three main types, depending on the relationships of the deposits to bedding or structure: peneconcordant or so-called 'blanket' or 'trend' deposits;

Table 3 Local criteria regarded as favourable

	Types of deposit			Features of host rocks within uranium areas									Elements associated with U							Reducing agents				Other factors					
Types of deposit and criteria of favourability	Peneconcordant: blanket-like	Transgressive: rolls	Rectangular along faults: 'stack'	Fine-grained sandstone	Medium to coarse, poorly sorted sand	Grey, green or tan sandstones with/inter-bedded grey and green mudstones	Unoxidized sandstone—contains pyrite	Outcrops of host sandstone commonly are limonite- and/or hematite-stained	Normally tan to reddish host sandstone is bleached	Sand–shale ratio 1/1 to 4/1	Beds dip less than 5°	Small to medium-scale faults are an ore control	Vanadium	Molybdenum	Selenium	Arsenic	Phosphorus	Manganese	Copper	Vegetal carbonaceous material	Structureless humic compounds	'Dead oil'	H_2S-bearing gas or water	R/A anomalies >5 x background	R/A in upper part of lignitic beds	Sampling of host rock outcrops shows anomalies >5 ppm U_3O_8	Anomalous uranium values peripheral to calcareous concretions	Groundwater anomalies >10 ppb U_3O_8	Oxidized uranium minerals in outcrop
Uranium areas Lisbon Valley, Utah	X				X	X	X	X	O	X			X	O						X	O			X		?		?	O
Monument Valley—White Canyon, Arizona–Utah	X				X	X	X	X		X	X		X						X	X				X		?		?	X
Uravan Mineral Belt, Colorado	X	X		X		X	X	X	X	X	X		X		O					X	O			X		?		?	X
Ambrosia Lake, New Mexico	X		X		X	X	X	X		X	X	X	O	X	X			O		O	X			X		X		X	X
Black Hills, Wyoming–South Dakota	X	O		O	X	X	X	X		X	X		X	O						X			O	X			X	X	X
Gas Hills, Wyoming	X	X			X	X	X	X		X	X	O		X	X	X	X	O		X	O	O	O	X	X	X		X	X
Powder River Basin, Wyoming	X	X			X	X	X	X	O	X	X		O		O			O		X				X	X	X	X	X	X
Texas Gulf Coast	X	X		X		X	X	X		X	X	X		X	O	O				X		O	X	X	X	X		X	X
Maybell–Baggs, Colorado–Wyoming	X	O			X	X	X	X		X	X	O		X	X	O	O					O	O	X		X	O	X	X

X, Dominant or of primary importance; O, subordinate or of secondary importance;
Blank space, criterion is of doubtful significance, absent or does not apply; ?, not determined.

transgressive or 'rolls'; and rectangular along faults or 'stacks'. All of the unoxidized deposits contain coffinite and/or uraninite as the primary minerals.

Blanket or trend deposits are flat-lying or gently dipping bodies of ore essentially concordant with the sandstone bedding. In plan, they may be amoeba-shaped and about equidimensional (blankets) or they may be characteristically elongate in one direction (trend). The habit varies from district to district, but such deposits can be as much as 50 ft thick, ranges from 3 to 15 ft being more common, and as much as several thousand feet long and hundreds to a thousand feet or more wide. Tenor ranges between 0·15 and 0·40% U_3O_8 and sharp grade cutoffs are the rule at margins of orebodies. Classic blanket deposits occur in Lisbon Valley and trends at Ambrosia Lake, where single deposits contain millions of tons of ore.

Roll deposits cut across sandstone bedding and, ideally, are *C*-shaped in vertical section. The deposits form at the contact between large tongues of altered sandstone and unaltered sandstones. The limits of the altered sands are considered to be the most forward penetration of the oxidizing groundwater that also transported uranium to the site of deposition. The term 'solution front' is commonly used to describe the complex three-dimensional shape of the periphery of the altered sandstone tongue. Roll deposits contain as much as several hundred thousand tons of ore, range from a few to 50 ft thick, from a few to hundreds of feet wide, and may be a mile or more in length along sinuous solution fronts. Tenor is between 0·10 and 0·50% U_3O_8; grade cutoff is sharp, within inches, on the concave side of the roll, but is gradational on the convex side. Roll deposits are important in the Uravan Mineral Belt, and in the Wyoming Basins and Texas Gulf Coast almost all the uranium is in rolls.

These deposits deserve special mention because a large amount of work has been done on them in the last ten years, and many papers describing special features of roll deposits have been written in the last five or six years. Most of the effort has been focused on the geochemistry and on recognizing altered sandstone by colour, mineralogy, associated elements and pyrite morphology. As there are many subtle differences in rolls and in characteristics of altered ground from one area to another, a synthesis of the criteria for rolls was deemed beyond the scope of this paper.

Stack deposits, commonly equidimensional or rectangular-shaped in vertical section, occur in and along fault and fracture zones in the Ambrosia Lake area. In plan, they are linear and confined to the permeable fault zones, and to the adjacent favourable host rock. These are secondary deposits and clearly result from oxidative destruction of the original trend ore and subsequent redeposition of the uranium in and along nearby fault zones. Deposits can contain more than 1000000 tons of ore; be more than 100 ft thick and as much as 300 ft wide; and follow fault zones for several thousand feet. The tenor ranges between 0·10 and 0·25% U_3O_8 and the ore may grade imperceptibly into the parent trend deposits.

Features of host rocks

Uranium is found in fine to coarse sandstones, but medium to coarse, poorly sorted, sandstones seem to be the most favourable. Degree of sorting may be very important because it is a factor controlling permeability.

Grey, green or tan sandstones with interbedded grey and green mudstones are common to all deposits and manifest the reducing environment necessary for their deposition and preservation. The presence of pyrite also manifests reducing conditions. The outcrops of host sandstones within uranium areas commonly are stained by limonite and/or hematite. This local feature, which has been noted in all the areas, results from the oxidation of pyrite contained in the host sandstone. In a few areas heavy gossans have formed on the outcrops up-dip from deposits.

Normally tan to reddish sandstones are bleached. This is a criterion noted in several areas, and is important locally. In the Uravan Belt it is regarded as manifesting the reducing conditions that accompany deposition of uranium.

Sand–shales ratios range from 1:1 to 4:1. Ratios within this range seem to prevail in all areas. Alternating sandstones and shales are regarded as important in establishing patterns of groundwater migration and influencing the rate of groundwater movement in the individual sandstone aquifers within the host. The carbonaceous shales also may be effective in establishing and maintaining reducing conditions which favour precipitation and retention of uranium.

Beds dip less than 5°: this yardstick applies in all areas except Lisbon Valley, and evidence shows that the steeper dip there is due to post-mineral uplift in the Tertiary. The dip of the host beds in all areas at the time of ore formation probably ranged from less than 1° to 5°. Gentle dips assure slow migration of groundwaters, at a rate low enough to prevent flushing of reductants, but still allow ample uraniferous groundwater throughput for formation of deposits. Low dips also assure a large area of outcrop for groundwater recharge of aquifers.

Small- to medium-scale faults were ore controls in some areas by creating permeable pathways for uraniferous solutions, as in Ambrosia Lake; or by creating graben structures that influence migration of uraniferous groundwater, as in the Texas Gulf Coast; or by creating pathways along which reducing gases, waters or hydrocarbons from underlying formations can enter the host formation, as in the Texas Gulf, in Maybell–Baggs and in the Gas Hills.

Elements associated with uranium

Many elements are associated with uranium in deposits, but only vanadium, molybdenum, selenium, arsenic, phosphorus, manganese and copper appear to be common enough in the nine selected areas to warrant mention as possible guides. Even these are so closely associated with uranium that it is probable that uranium or its radioactivity would be detected first anyway. Even so, vanadium, molybdenum, selenium and copper may be useful indicators: vanadium, because the yellow secondary compounds with uranium are relatively stable in outcrops; molybdenum, because the hydrous oxide may form blue crusts; selenium, because it is stable in oxidized soils, where the easily recognizable selenium indicator plant, *Astragalus*, may thrive; copper, because the green secondary minerals are easily recognized in outcrop.

Reducing agents

These compounds effect reduction of the uranium, which is transported by groundwater in the soluble sexavalent state, to the insoluble quadrivalent minerals uraninite or coffinite. Vegetal carbonaceous material is a common constituent of fluvial sandstones. In an intermediate stage of degradation this organic material can precipitate uranium directly from solution. Examples of wood replaced by uraninite are common. Vegetal carbon can also be the nutrient for *Desulfovibrio* bacteria, which produce hydrogen sulphide, a powerful reducing agent.

Structureless humic compounds are found in a number of areas, but are the primary ore control only at Ambrosia Lake. This epigenetic dark organic was probably carried by groundwater as soluble humates or fulvates, and at Ambrosia Lake it was precipitated in great quantities. These organic compounds fix uranium by chelation within the molecular structure, according to Schmidt-Collerus,[1] and may also support bacterial action.

Dead oil in a semi-oxidized state seems capable of either effecting the direct precipitation of uranium from solution by reduction or furnishing nutrient for bacteria. Such organic material is probably a factor favouring uranium deposition in certain deposits in Gas Hills, Texas Gulf Coast and Maybell–Baggs.

Hydrogen sulphide bearing gas or water originating in leaking sour oil or gas structures in underlying rocks is postulated to be an important reductant in several areas. Hydrogen sulphide bearing natural gases are present in the Texas Gulf Coast, Gas Hills and Baggs areas,[2] and hydrogen sulphide in water is present in wells in the Black Hills.[3]

Other factors

Radioactivity anomalies greater than five times background were found in all nine areas, and this probably applies to all known uranium areas in the U.S.

Radioactivity in the upper parts of lignitic beds, with or without uranium present, is a possible guide to deposits in sandstone in the same general area. Numerous coal outcrops in the Powder River Basin area are radioactive in their upper surface, and outcrops of some lignitic shales in the Gas Hills are radioactive.

Sampling of host rock outcrops shows that anomalies greater than 5 ppm U_3O_8 are common in Ambrosia Lake, Gas Hills, Powder River Basin, Texas Gulf Coast and Maybell–Baggs. The sampling in these areas was done on many square miles of outcrop and it was demonstrated that sampling directly for uranium may be more productive of results than geochemical prospecting for other indicator elements.

In the Black Hills, according to Gott,[4] areas marginal to large bodies of carbonate-cemented sandstone are favourable exploration targets. A possible geochemical explanation recently described by Hagmaier[5] postulates that the soluble uranyl–carbonate complex, which is stable in bicarbonate groundwater facies, becomes unstable when calcite is precipitated by commingling of carbonate and sulphate groundwater facies. The uranyl ion thus released is then easily precipitated under reducing conditions. Additional possible evidence for such a reaction is provided by thin layers of uranium ore in the form of shells surrounding calcareous concretions in a part of the southern Powder River Basin, Wyoming, and in some deposits in the Baggs, Wyoming, area.

Also, in the southern Powder River Basin mapping by Meschter[6] indicated that the long axes of the ellipsoidal-shaped calcareous concretions reflected the direction of movement of mineralizing groundwaters. Exploration has demonstrated that the mapped pattern of concretions defines the shape of a large solution front, and, furthermore, that the concretions are within a tongue of altered red sandstone that has uranium roll deposits on its periphery.

Groundwater anomalies greater than 10 ppb* U_3O_8 are found in host rocks in all but three of the nine areas. The method is being used for prospecting, and at least one important discovery is attributed to the method. The technique is best applied to the search for deposits undergoing oxidation.

Oxidized uranium minerals occur in outcrops of the host rock in all nine areas, and the statement could be expanded to include all important U.S. uranium areas.

Theoretical considerations

The problem of origin of the uranium in these deposits, although minor to the

*Parts per thousand million.

main topic of this paper, is probably always a conscious or unconscious element in exploration thinking. Working hypotheses include provision for genesis of ores, and the selection of prospecting guides may be influenced by a preferred theory of origin. The main theories of origin are noted below.

Magmatic hydrothermal was the preferred theory until discoveries of uranium in sandstones in the Wyoming Basins, far removed from sites of plutonic activity, stretched the idea to the limits of credibility. Today, only a few geologists in the U.S. advocate this origin for sandstone-type deposits.

Weathering and leaching of granitic rocks in situ is a concept that relates the uranium in sandstones to the uranium in granitic rocks in the areas of provenance. It postulates that deep oxidation and groundwater leaching of jointed granites solubilizes much of the loosely bound uranium, which migrates into the aquifer systems.

Leaching of uranium from the arkosic sandstone that is also the ore host was one of the prominent ideas of the 1950s and it is still a popular theory.

Leaching of uranium from tuffaceous sediments was proposed in the early 1950s. By the late 1950s the theory had gained wide acceptance among field geologists in Wyoming and elsewhere. The concept holds that uranium is leached by groundwater from tuffaceous rocks either within or above the host sandstone. The theory was strengthened by observations of Garrels[7] and work by Denson, Zeller and Stephens[8] that reported the abnormal uranium content of waters issuing from tuffaceous rocks as well as the presence of abnormal amounts of selenium, arsenic, vanadium, phosphorus and molybdenum in the waters. This theory is currently preferred by most geologists exploring for uranium in the U.S.

The *multiple migration–accretion concept* of Gruner[9] deserves mention. It postulates that several stages of oxidation–solution–migration–accretion are required for the formation of ore-grade deposits. The idea seems to be particularly applicable to the sandstone-type deposits, regardless of the origin of uranium.

No one theory may be the final answer, and a combination of sources of uranium may be the most realistic postulate.

Descriptions of areas

The following brief descriptions of each of the nine areas are included to illustrate the criteria in the light of their interrelationships to the geology and events that seem to be important to the formation of sandstone-type deposits.

LISBON VALLEY, UTAH

The Lisbon Valley area (Fig. 2) is within the Paradox salt basin, a structural basin in eastern Utah and western Colorado that was formed in Pennsylvanian time. The most important deposits in this major uranium area are in the Moss Back Member of the Chinle Formation, which is the basal Chinle member in this area. The Moss Back in the vicinity of the deposits was deposited on a surface of unconformity that truncated the Cutler Formation of Permian age on the Lisbon Valley anticline, the site of this important uranium area.

The Moss Back, ranging from 13 to 80 ft in thickness, is fluvial, and was deposited by anastomosing streams in flood-plain environments. It is a fine- to coarse-grained, poorly sorted, arkosic, calcareous cemented sandstone containing tuffaceous material and abundant carbonaceous trash. The provenance was mainly one of granitic rocks and volcanic activity. The major deposits are in the

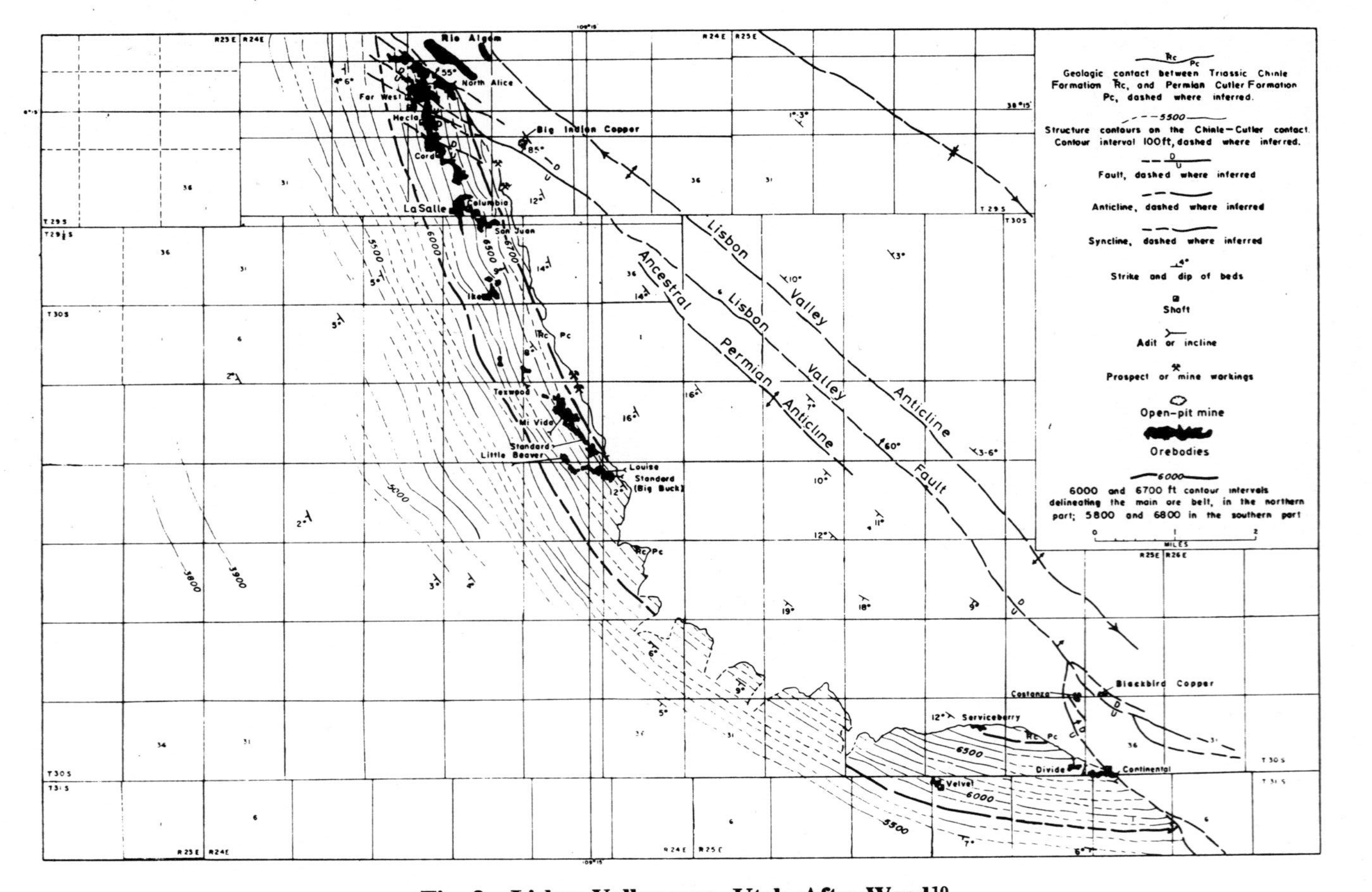

Fig. 2 Lisbon Valley area, Utah. After Wood[10]

Moss Back, but the underlying Cutler is a minor ore host in the same general area. The outstanding feature of the area is the localization of blanket-like deposits in the Moss Back in an arcuate belt about 15 miles long superjacent to truncated Cutler sandstone on the southwest flank of the Lisbon Valley anticline. The Cutler sandstone beneath the deposits in the Moss Back is bleached from a normal red to grey or white. Individual deposits, which are clusters of orebodies and pods, are irregular in shape, and range from 1 to 40 ft in thickness, and from a few hundred to 3000 ft long; the main ore trends are parallel to the Cutler subcrop.

The average grade of ore produced has been about 0·40% U_3O_8—the highest for any sandstone-type area in the U.S.; molybdenum is present in some of the deposits in the central and southern parts of the area, and vanadium content,

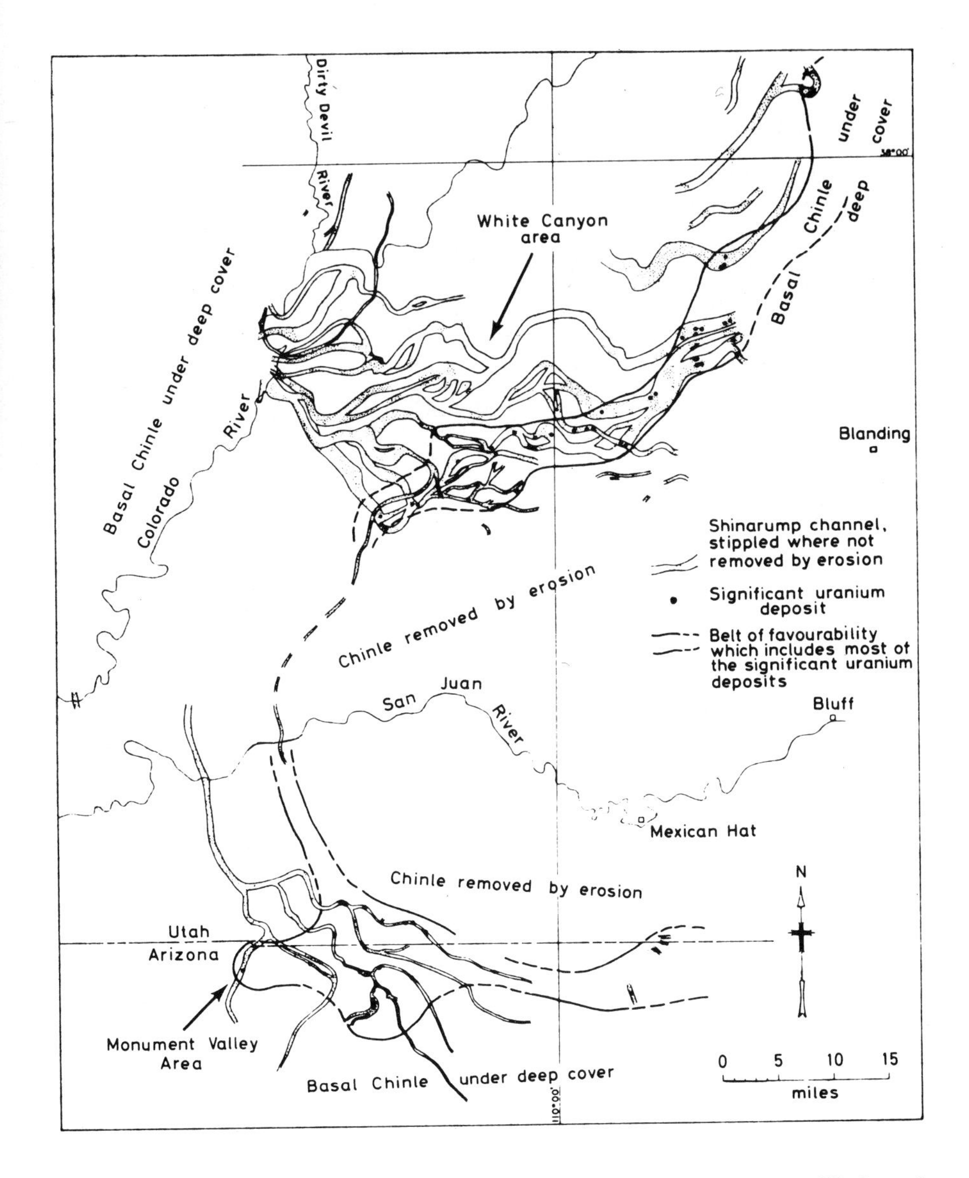

Fig. 3 Palaeo-channels in Monument Valley–White Canyon area, Utah and Arizona. After Malan[11]

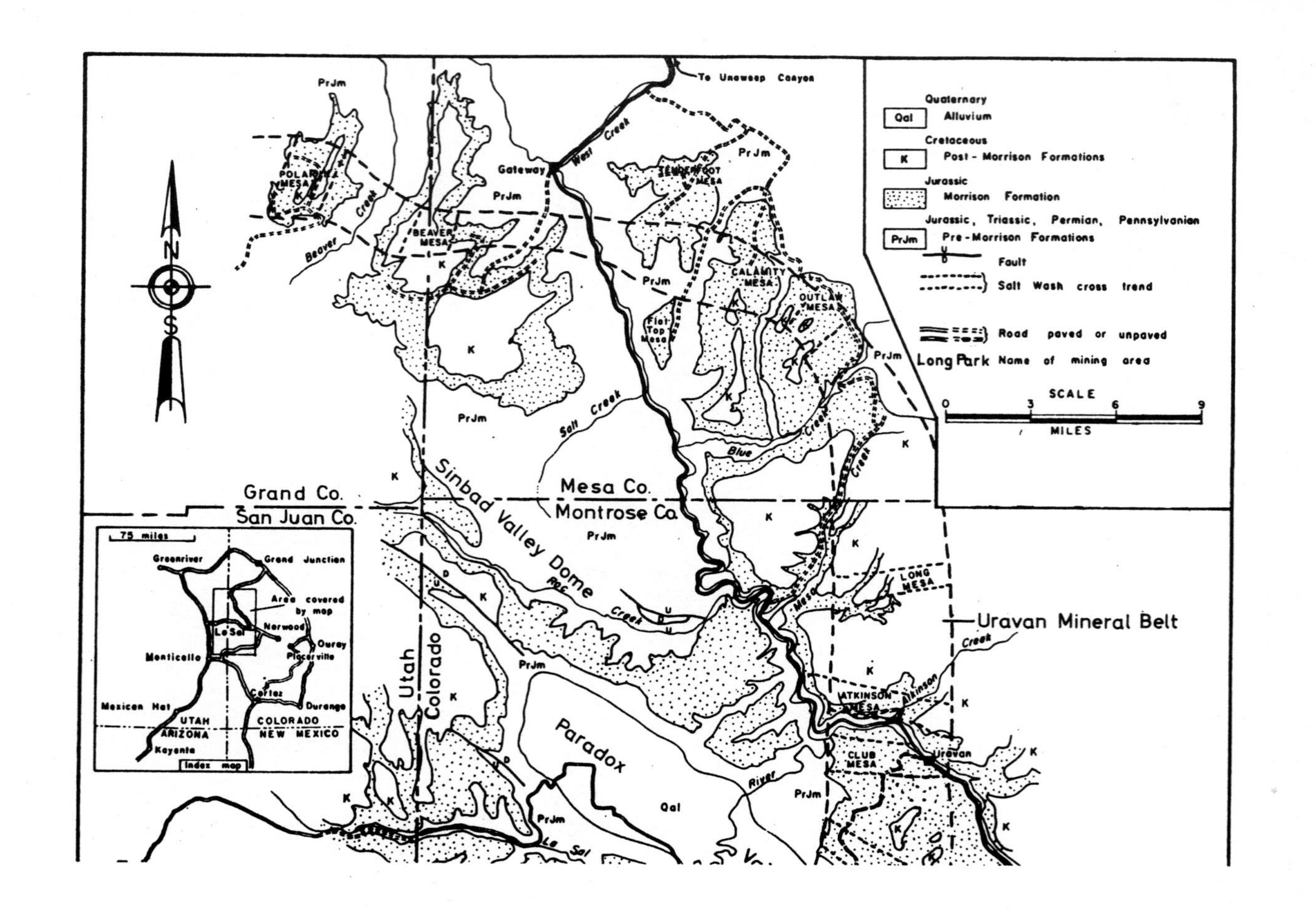
Quaternary
Qal Alluvium
Cretaceous
K Post-Morrison Formations
Jurassic
Morrison Formation
Jurassic, Triassic, Permian, Pennsylvanian
PrJm Pre-Morrison Formations
Fault
Salt Wash cross trend
Road paved or unpaved
Long Park Name of mining area
SCALE
0 3 6 9
MILES
Uravan Mineral Belt
To Unaweep Canyon
Gateway
West Creek
Mesa Co.
Montrose Co.
Grand Co.
San Juan Co.
Sinbad Valley Dome
Paradox
River
Utah
Colorado
Uravan
LONG MESA
ATKINSON MESA
CLUB MESA
OUTLAW MESA
CALAMITY MESA
Flat Top Mesa
BEAVER MESA
POLAR MESA
Salt Creek
Blue Creek
Mesa Creek
Atkinson Creek
Beaver Creek
75 miles
Greenriver
Grand Junction
Area covered by map
Norwood
La Sal
Ouray
Placerville
Monticello
Cortez
Durango
Mexican Hat
UTAH
ARIZONA
COLORADO
NEW MEXICO
Kayenta
Index map

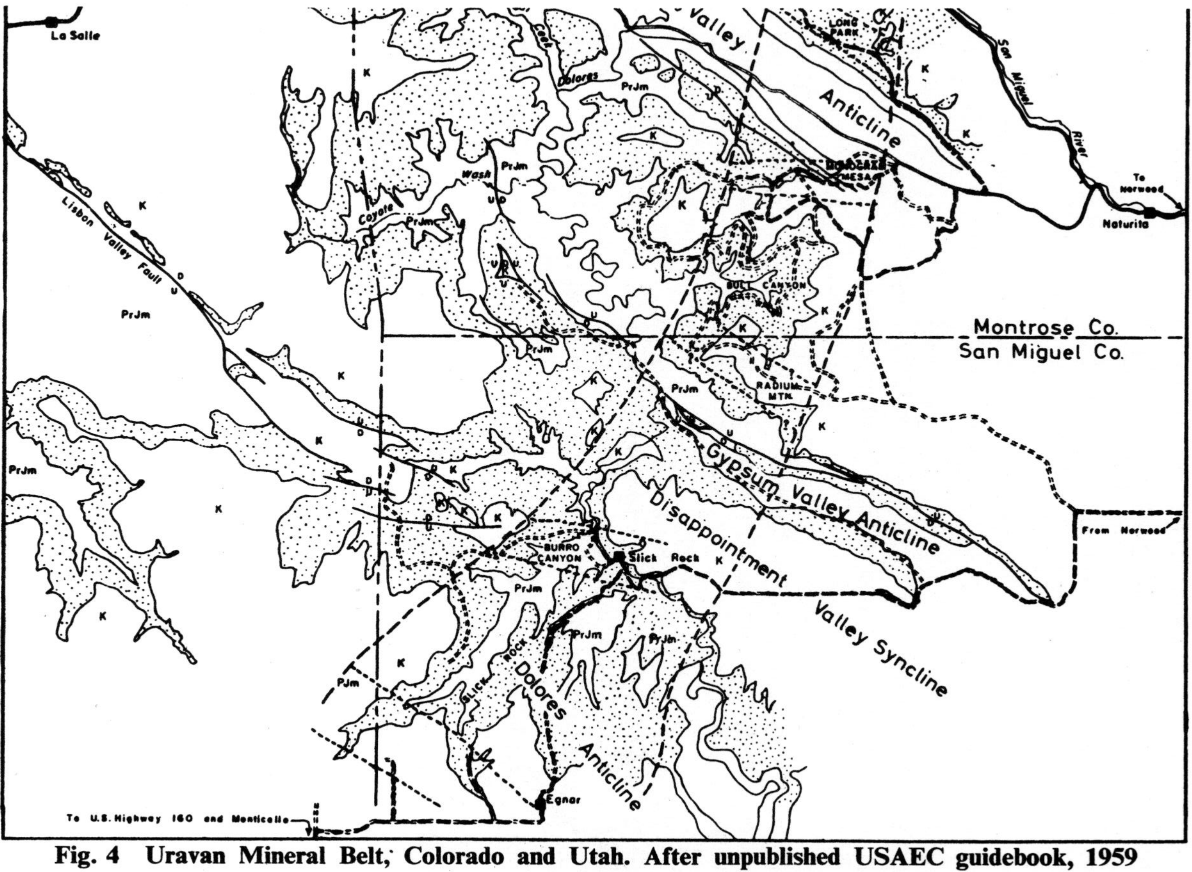

Fig. 4 Uravan Mineral Belt, Colorado and Utah. After unpublished USAEC guidebook, 1959

which ranges from one-third to three times the uranium values, is highest in the southern part.

There is no common agreement on the origin of these deposits, but Wood[10] postulated that the uranium, originally in the tuffaceous material in the Chinle, was solubilized during diagenesis, transported by groundwaters migrating in the permeable Moss Back and reduced to form deposits in places where the Moss Back overlies permeable sandstone in the Cutler subcrop. The necessary reducing environment could have been created by the carbonaceous material in the Moss Back or perhaps by hydrogen sulphide bearing gases or waters, originating deep in the salt basin, travelling upward along permeable Cutler sands and gaining entry into the Moss Back across the surface of unconformity.

Exploration has been pointed toward the search for similar deposits on the northeast limb of the Lisbon Valley anticline, where a recently discovered deposit is being prepared for mining. Other salt anticlines in the Paradox Basin regarded to have similar geologic histories are also targets.

MONUMENT VALLEY–WHITE CANYON, ARIZONA AND UTAH

All of the important deposits are in the arkosic Shinarump Member, the basal unit of the Chinle Formation of Triassic age. The provenance was granitic and, in part, sedimentary terrain in which volcanism was prevalent. The Shinarump, ranging from 10 to 250 ft in thickness, is fluvial and was deposited by anastomosing streams on flood plains above a widespread surface of unconformity (Fig. 3). The Shinarump is unconformably overlain by the Monitor Butte Member, comprised of bentonitic mudstones, conglomerate and sandstone.

This area is characterized by deposits in channels, incised by streams into the underlying red siltstones of the Moenkopi Formation. The deposits consist of closely spaced pods of ore that are associated with vegetal carbonaceous debris. A notable feature is the common association of copper sulphide with uranium in White Canyon. Copper is present in minor amounts in Monument Valley, but all ores are vanadiferous, ranging in grade from 0·24 to 1·4% V_2O_5 and averaging about 0·30% U_3O_8.

Malan[11] postulated that the uranium was leached from the granitic and tuffaceous detritus during the deposition and subsequent fluvial reworking of the Chinle sediments. Control of uranium deposition was brought about by the reducing effects of carbonaceous materials in the sandstones that fill channels.

Exploration is based on reconstructing the channel systems from all available evidence and drilling fences of holes at right angles to the projections of individual channels. Most of the shallower, easily explored, target areas have been drilled.

URAVAN MINERAL BELT, COLORADO

The Uravan Mineral Belt area also is within the Paradox salt basin. Nearly all of the deposits are in the Salt Wash, the basal member of the Morrison Formation of Jurassic age. The Salt Wash ranges from 300 to 400 ft in thickness, is fluvial and was deposited in stream channels and on flood plains forming a very large alluvial fan. It consists of interbedded red, brown and grey feldspathic to quartzose sandstones and reddish and grey mudstones derived from a provenance of mainly sedimentary rocks. The Salt Wash is overlain by the Brushy Basin Member, which is 300–700 ft thick and comprised of bentonitic clay of volcanic origin, sand and silt.

The outstanding features of the area are the large number of small to medium deposits, about 1300 in all, and the mineral belt concept originated by Fischer and Hilpert[12] (Fig. 4). The belt is a westward-opening crescent about 70 miles long, and 2–8 miles wide, in which numerous deposits are clustered in a number of cross trends, each about 2 miles wide, that are nearly normal to the main belt. The cross trends are thought to be meander belts of the larger streams. Three southeast-trending salt anticlines, which were positive elements during Morrison time, diverted the direction of stream flow, and thick favourable sandstones that are sites for important deposits were deposited along the flanks.

Most of the uranium is in light grey to tan fine-grained, well sorted bleached sandstone in the upper sandstone bed of the Salt Wash Member. Abundant carbonaceous material is disseminated throughout the ore sandstones; deposits are irregularly shaped and consist of numerous pod-like orebodies. They range from a few to, rarely, nearly 1 000 000 tons of ore and may be as much as 30 ft thick, but the average is 4 ft. Roll-type orebodies that are *C*-shaped in cross-section are common and contain the highest-grade ore. Rolls may be as high as 30 ft and are bottomed and topped by impervious mudstones. The thicker parts of sandstone beds are the most favourable.

Average uranium ore grades are between 0·20 and 0·30% U_3O_8; vanadium content of the ores is variable, but averages approximately five parts vanadium to one part uranium.

The genesis of these deposits is unknown, but many workers, including Keller,[13] consider it reasonable that the uranium was leached from volcanic ash in the overlying Brushy Basin Member and moved by groundwater into the subjacent Salt Wash sandstones. It may be coincidence, but the upper sandstone bed of the Salt Wash is the most prolific for uranium.

Exploration drilling programmes are directed at the determination of the locations of thick sandstone channels and bleached and organic rich sandstone. Projections of channel cross trends within the Belt are preferred targets.

AMBROSIA LAKE, NEW MEXICO

The Ambrosia Lake area deserves special mention as it is the single largest uranium area in the U.S. (Fig. 5). It occupies the central part of the Grants Mineral Belt in northwestern New Mexico. The belt, which was credited as of 1 January, 1971, with production and reserves of nearly 200 000 tons U_3O_8, trends some 80 miles southeasterly from Gallup to Laguna, New Mexico, parallel to the southern margin of the San Juan Basin.

The largest deposits are in the arkosic Westwater Canyon Member of the upper Morrison Formation. The Westwater Canyon ranges from 30 to 270 ft in thickness, is fluvial and was deposited in the form of a large coalesced alluvial fan. The provenance was sedimentary and granitic rocks. The Brushy Basin Member of the Morrison Formation, which is mainly a montmorillonitic clay facies derived from devitrification of volcanic ash, is superjacent to the Westwater Canyon.

The outstanding geologic features are the size of individual trend deposits (Fig. 6), as much as several million tons of ore in deposits more than one mile long, half a mile wide and averaging about 10 ft thick, and the ubiquitous association of uranium with epigenetic structureless humic and fulvic compounds. These organics impregnate large volumes of sandstone and constitute the primary control for uranium occurrence, because the organics and impregnations of uranium are everywhere coextensive, according to Granger.[15] These deposits are peneconcordant and have a marked tendency to cluster in southeasterly trends, parallel to the main direction of sedimentation. Large stack deposits are controlled by steeply dipping faults of Tertiary age that strike across the main ore trends. These secondary deposits are rectangular or equidimensional in cross-section.

The ores grade between 0·10 and 0·40% U_3O_8. Vanadium is a constituent of the ores, especially in the western part of the district; molybdenum is common immediately peripheral to ore in trend deposits, but is absent in stacks; selenium is widespread, being associated with both stack and trend ores.

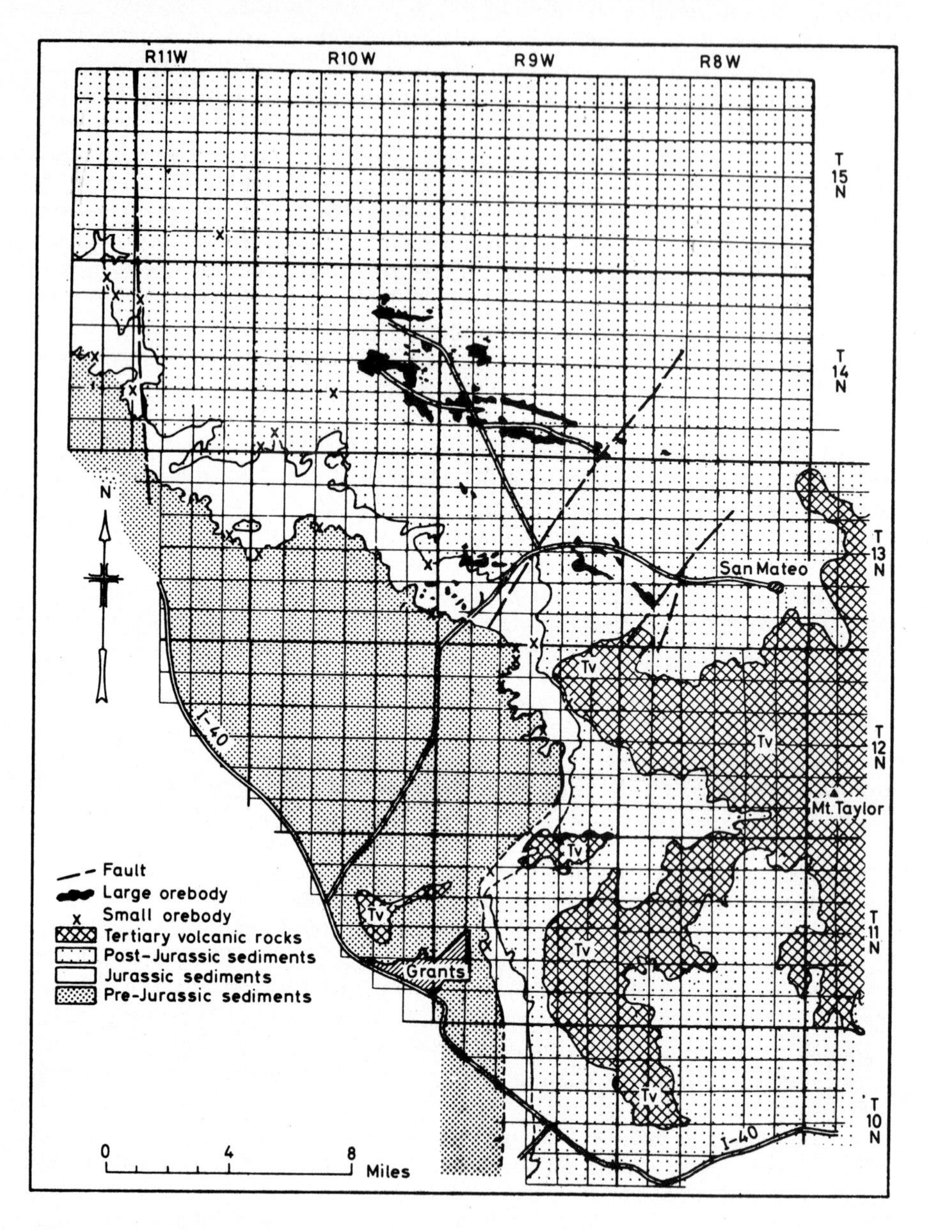

Fig. 5 Ambrosia Lake, New Mexico. After H. K. Holen, USAEC (unpublished)

The Westwater Canyon is universally altered in the Ambrosia Lake area, even several miles from known ore, as reported by Austin.[16] Alteration has produced kaolinite, montmorillonite, chlorite, anatase and quartz overgrowths. Heavy detrital minerals in the sandstone have been destroyed and only shells of sanidine remain.

Many workers consider it plausible that indigenous uranium was leached by groundwater from the huge prism of Westwater Canyon or from the overlying Brushy Basin during or soon following Morrison time. The long period of pre-Dakota erosion is suspected as the time of introduction of the organics and the uranium into the Westwater Canyon.

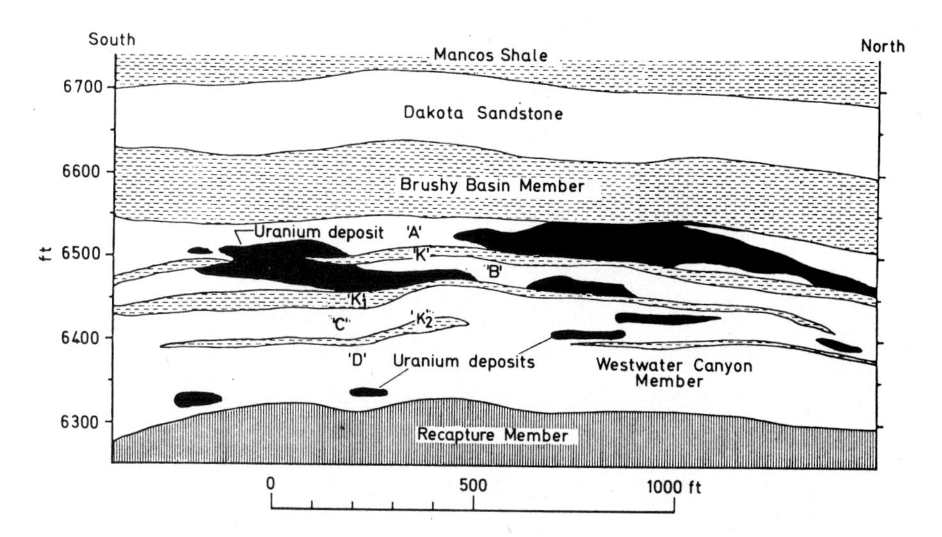

Fig. 6 Cross-section through Section 30 mine, Ambrosia Lake, New Mexico. After Clary and co-workers[14]

Exploration programmes are based largely on drilling for the stratiform deposits along projections of established trends, for stack deposits along mineralized faults and for possible new trends within the favourable sandstone facies of the Westwater Canyon. The coarsest sandstones, where intercalated with grey mudstones, are good targets, and thick sands enhance favourability.

BLACK HILLS, WYOMING AND SOUTH DAKOTA

The northern, northwestern and southwestern gently dipping flanks of the Black Hills domal uplift are the main areas of uranium deposits (Fig. 7). Host rocks are the Lakota and Fall River Formations, which are the lower and upper formations, respectively, of the Inyan Kara Group of early Cretaceous age; both formations were derived from erosion of older sedimentary rocks.

The Lakota ranges in thickness from 100 to 300 ft and is comprised mainly of medium- to coarse-grained quartzose sandstones, conglomerates, siltstones and coals. It is fluvial, being deposited in stream and flood-plain environment. The unconformity that marks the change from fluvial to marginal marine deposition separates the Lakota from the overlying Fall River. The Fall River, ranging from 120 to 150 ft in thickness, is a fine-grained ferruginous sandstone, with shales and siltstones. It is fluviomarine, being deposited along shorelines of transgressive seas.

Subsequent to deposition of the Fall River, more than 7000 ft of Cretaceous and Tertiary sedimentary rocks was deposited. The Black Hills Uplift was formed in the early Tertiary and by Oligocene erosion had truncated the sediments draped over the dome. During the Oligocene period tuffaceous beds of the areally extensive White River Formation were deposited across the breached structure.

Deposits are most prevalent in the carbonaceous, cross-bedded, thick Lakota

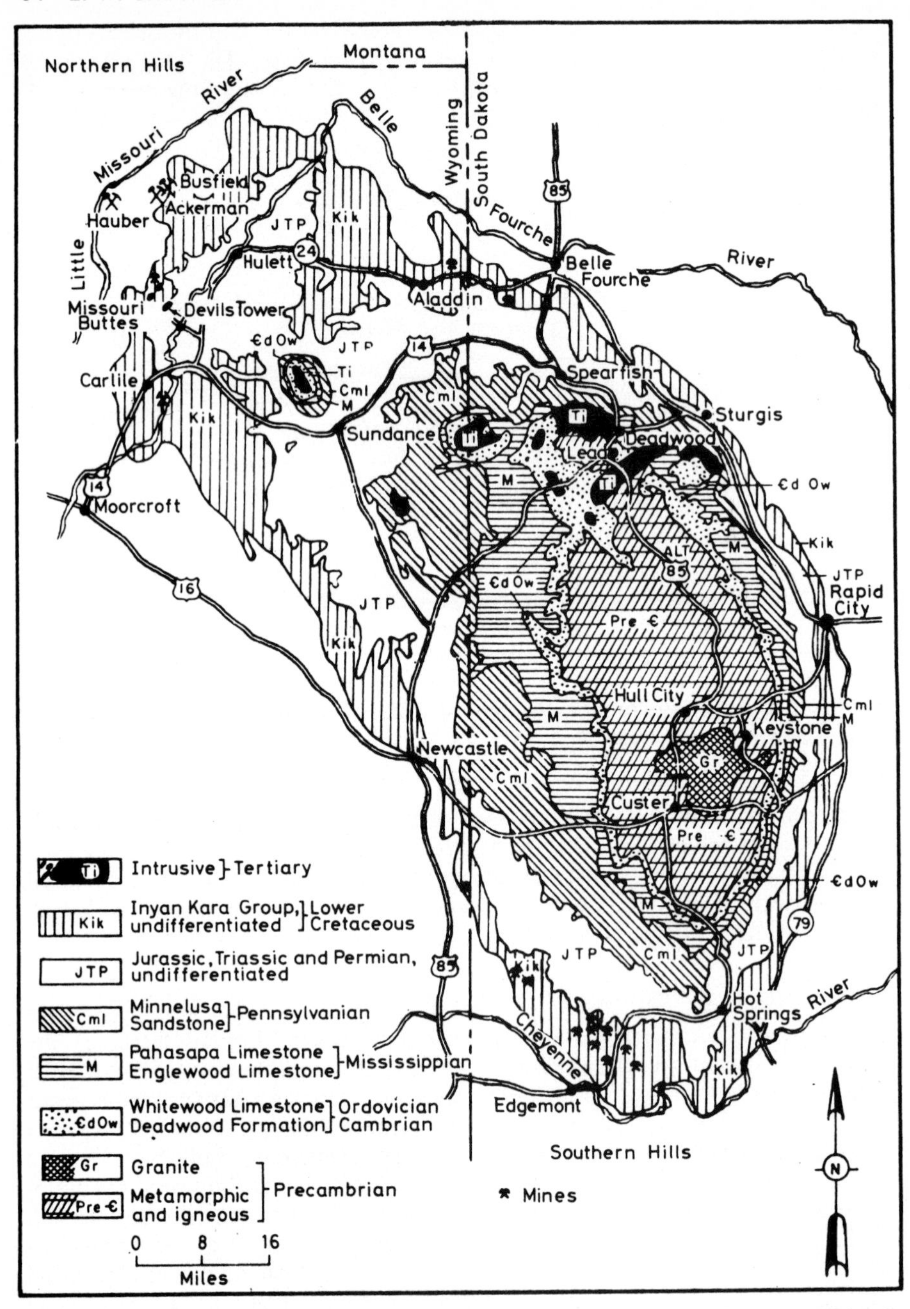

Fig. 7 Black Hills, South Dakota and Wyoming. After unpublished USAEC guidebook, 1959

channel sandstones, but significant deposits also occur in fine-grained, thin-bedded, fluviomarine Fall River sandstones that are interbedded with carbonaceous shales. Some deposits contain carbonaceous material, but others do not.

The blanket deposits are comprised of clusters of irregular-shaped pods. The tenor of ore ranges from 0·15 to 0·30% U_3O_8, the vanadium to uranium ratio is about 1·5 to 1, and small amounts of molybdenum are present.

A pink hematite staining in the vicinity of, and contiguous with, uranium ore is a common feature. This alteration feature is commonly related to emplacement of *C*-shaped rolls. The pink sandstone is most prevalent on the up-dip side of the roll deposits and the contact with ore is sharp. The occurrence of uranium, vanadium and pyrite marginal to calcareous concretions or large masses of calcareous cemented sandstone is a feature of some deposits. Gott and Schnabel[17] indicated that other favoured sites for deposits are in structural terraces adjacent to the monoclinal axis in the western flank of the uplift.

The genesis of the deposits is unknown, but two main theories have been proposed. Renfro[18] proposed that the uranium, indigenous to the Lakota and Fall River sandstones, was mobilized by oxidizing groundwaters and transported down-dip, where it was precipitated along the oxidation–reduction boundary. On the other hand, Hart[19] proposed that uranium was leached by groundwater from the tuffaceous beds of the White River Formation that were unconformably deposited across the truncated Black Hills Dome. Migrating groundwater carried the uranium into the permeable host rocks, where it travelled down-dip into reducing environments.

Exploration is mainly by drilling to locate contacts of pink and unaltered sandstones. Other targets are structural terraces and thick sandstones in the large Lakota channels.

GAS HILLS, WYOMING

The host rock is a thick, but local, prism of arkosic, coarse-grained, fluvial sandstones within the Wind River Formation of Eocene age, which was deposited on an irregular surface of truncated Palaeozoic and Mesozoic sedimentary rocks. The Wind River was derived locally from a provenance of dominantly

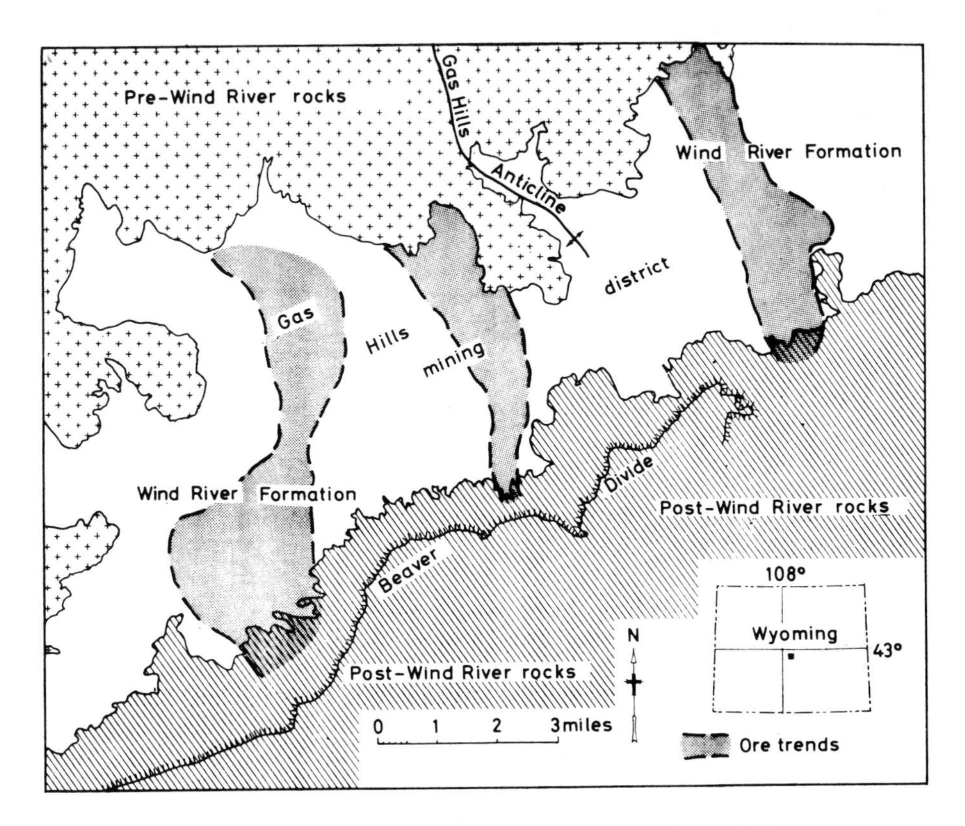

Fig. 8 Gas Hills, Wyoming. After King and Austin[21]

granite and metamorphic rocks. Deposition was in a coalesced alluvial fan and stream channel environment. Vegetal carbonaceous material is present in the sandstone, but is not everywhere mineralized.

The dominant feature is the concentration of uranium in three northerly trending belts defined by clusters of roll deposits (Fig. 8). The width of the belts ranges from one-half to more than 2 miles, and they are separated from one another by 1–4 miles of barren or poorly mineralized sandstone. Work by Soister[20] showed that all of the deposits are in the middle or Puddle Springs Member of the Wind River, which ranges from 400 to 800 ft in thickness. Sandstones are limonite-stained, tan to buff in outcrop and light to dark grey and pyritic in the subsurface. Interbedded with the sandstones are lenses of conglomerate, siltstone and mudstone and locally carbonaceous shale and coal.

The roll ore, which ranges in grade from 0·10 to 0·5% U_3O_8, is commonly between 10 and 20 ft thick. The deposits are floored and topped by impervious carbonaceous mudstones and occur at the interface of bleached altered sandstone with dark grey sandstone (Fig. 9). Elements associated with uranium are

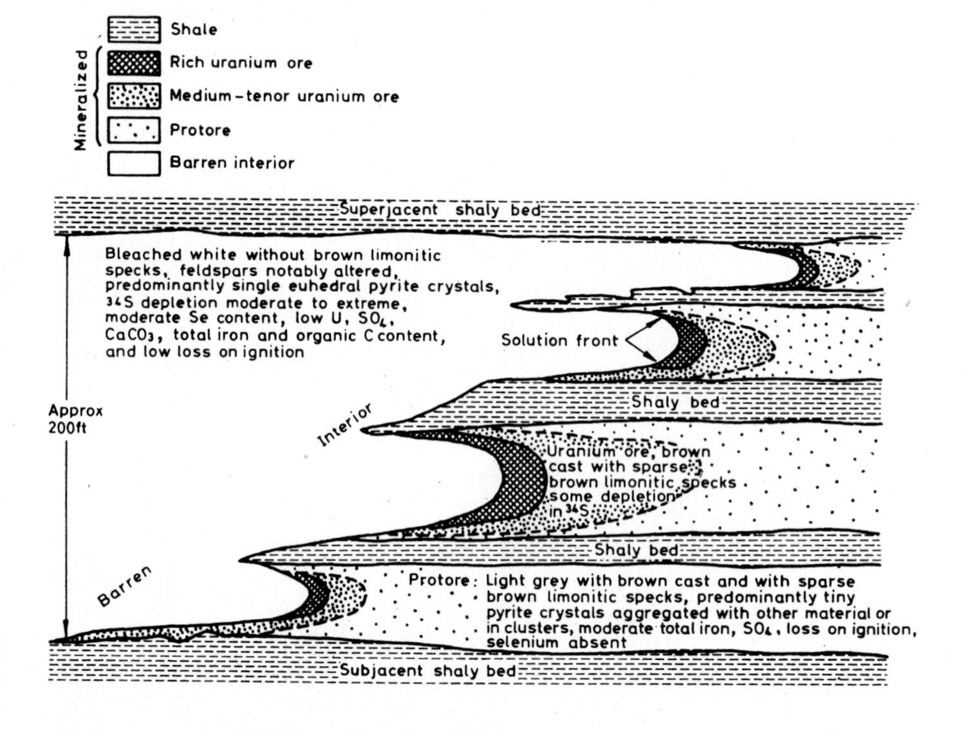

Fig. 9 Idealized vertical section across a solution front, Gas Hills, Wyoming. After King and Austin[21]

molybdenum, selenium, arsenic and phosphorus. Vegetal carbonaceous fragments are common in the host, but are not everywhere mineralized. King and Austin[21] reported that structureless carbonaceous material has been incorporated into the roll structures and may be a factor in the reduction of uranium. Hydrogen sulphide bearing groundwaters entering the Wind River from truncated Cretaceous sandstones may also be an ore control. Miocene and later faulting is present, and in the eastern ore trend a graben may have influenced the location of deposits by creating a favourable setting for uranium deposition.

Generally, two main schools of thought are advanced for the formation of these deposits: leaching from arkose and leaching from tuff.

The more favoured theory is that uranium was leached from the overlying thick tuffaceous beds of the Oligocene White River Formation and was carried downward by groundwater into the permeable strata and reducing environment of the Wind River. Zeller[22] noted that two large and deep White River channels have scoured down into the Wind River Formation, thereby facilitating direct entry of groundwater from the White River into the Wind River. By coincidence or otherwise, the largest channel is exposed in the Beaver Divide escarpment almost on line with the southern projection of the central ore trend, which is the largest and richest of the three.

Exploration programmes are directed at determination of the contacts between the bleached altered and the unaltered grey sandstone and the configurations of the sinuous solution fronts along which the roll deposits occur.

POWDER RIVER BASIN, WYOMING

The host rocks are coarse-grained, poorly sorted, fluvial, arkosic sandstones in the upper part of the Fort Union Formation of Palaeocene age and in the Wasatch Formation of Eocene age (Fig. 10). The Fort Union, which crops out on the basin margin, is about 3000 ft thick, but only local coarse-grained sandstones in the uppermost part are important for uranium; the Wasatch, which fills the central basin area, is comprised of more than 1000 ft of mudstones interbedded with thick beds of sandstone containing vegetal carbonaceous material. The provenance of the uranium-bearing sandstones was granitic, metamorphic and sedimentary rocks in nearby uplifts. Deposition was in an alluvial fan and stream and flood-plain environment. Dips of host beds in the uranium areas range between 1 to $2\frac{1}{2}°$ to the north. The tuffaceous White River Formation, of Oligocene age, isolated remnants of which are about 200–250 ft thick, unconformably overlies the Wasatch and all older rocks. Love[24] showed that, prior to removal by erosion, the tuffaceous White River and Miocene and Pliocene formations blanketed the entire basin. The present relief of more than 1000 ft was due to excavation during the late Tertiary and Pleistocene.

There are two separate clusters of roll deposits—one in the southern basin and one in the central basin. The deposits, which range in grade from 0·1 to 0·3% U_3O_8, are at the periphery or edges of large oxidized tongues of red hematitic sandstones that can be guides to ore down-dip from outcrops. Davis[25] has observed that near the deposits hematite commonly changes to limonite, and zones of white bleached sandstone are commonly present adjacent to ore. The contact between ore in the roll and oxidized sandstone is distinct, but on the down-dip convex edge of the roll uranium content grades into barren unaltered grey sandstone. Orebodies range from 2 to more than 20 ft in thickness and are hundreds of feet wide and thousands of feet long. Systems of roll deposits are present in a number of different sandstone beds in a stratigraphic range of 800 ft. The optimum conditions prevail in sandstone beds 80–100 ft in thickness and impervious mudstones above and below the ore sand are essential for formation of rolls.

Elements closely associated with reduced ores are vanadium, about one part to six parts uranium, and selenium. Small pods of oxidized ore commonly contain manganese oxides.

The origin is regarded as due to either leaching of tuffs or arkose, or perhaps both.

Exploration programmes are directed at locating the contact between the red tongues of altered and the grey unaltered sandstones. Uranium analysis of groundwater from drill-holes has been successful in locating favourable areas.

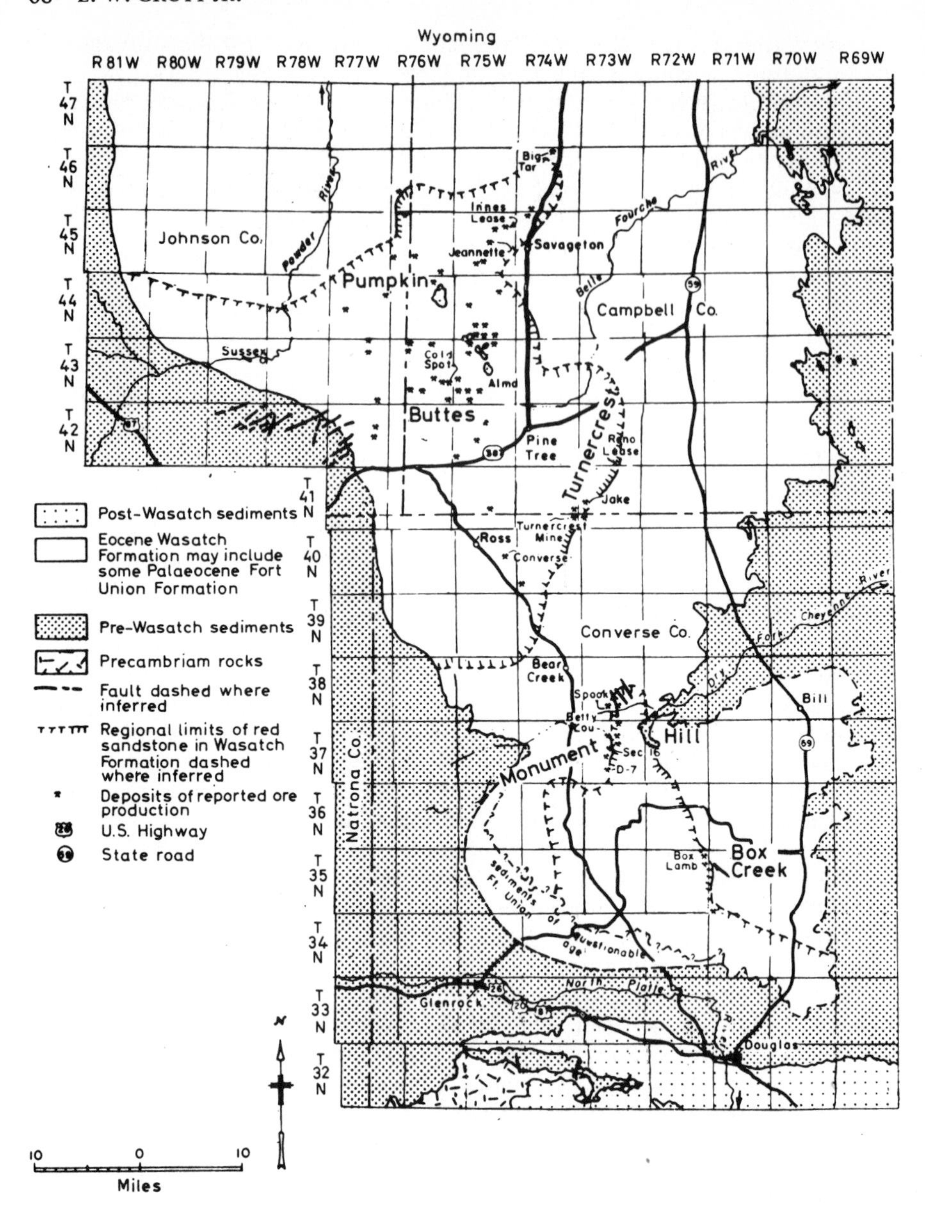

Fig. 10 Powder River Basin, Wyoming. After Love and co-workers;[23] uranium geology by D. L. Curry and D. Gobel, USAEC (unpublished)

TEXAS GULF COAST

The known uranium occurrences define a 300-mile long belt of favourable sediments that parallels the Texas Gulf Coast (Fig. 11). Deposits occur in the Whitsett Formation of Eocene age, Catahoula Tuff and Oakville Sandstone of Miocene age and Goliad Sand of Pliocene age, but the deposits in permeable sandstones of the Whitsett and Oakville have been the most important. The provenance, to the west, was one of sedimentary rocks, tuff, volcanics and minor granitic and metamorphic rocks. Norton[27] has reviewed conditions of deposition of the uranium host rocks.

The Whitsett was deposited in a marginal marine environment as deltaic, lagoonal and barrier bar sediments. Fine-grained quartzose sandstones ranging from 20 to 70 ft in thickness are interbedded with tuffaceous sandstones, bentonites and lignite. Vegetal carbonaceous materials are present in the sandstones.

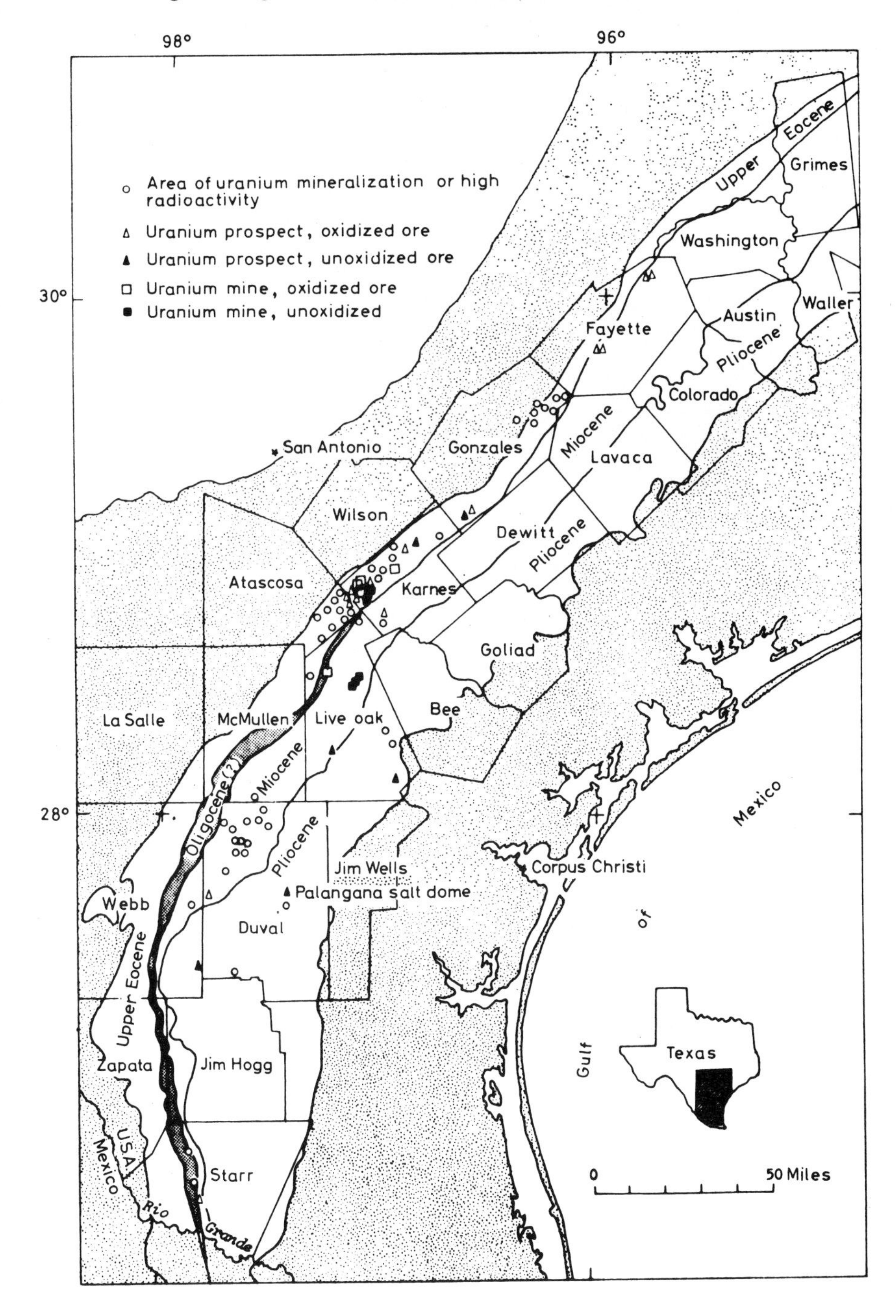

Fig. 11 Texas Gulf Coast area. After Eargle and co-workers[26]

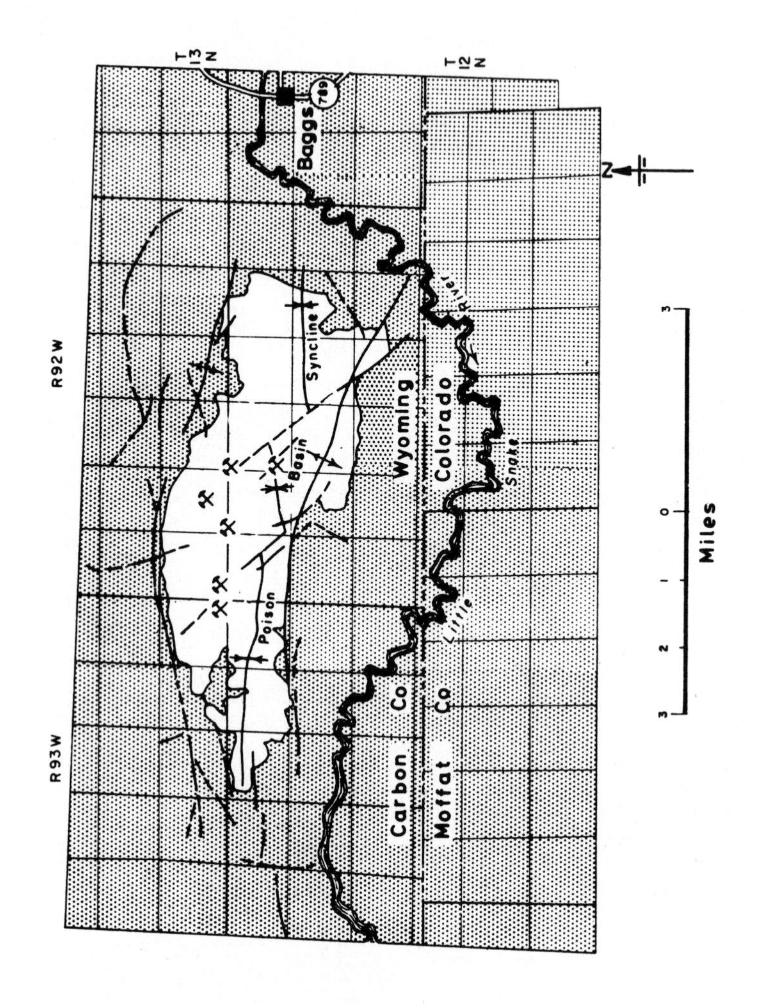
R92W
R93W
T 13 N
T 12 N
Baggs
789
Syncline
Basin
Poison
Wyoming
Colorado
Carbon Co
Moffat Co
Little Snake River
N
3 2 1 0 3
Miles

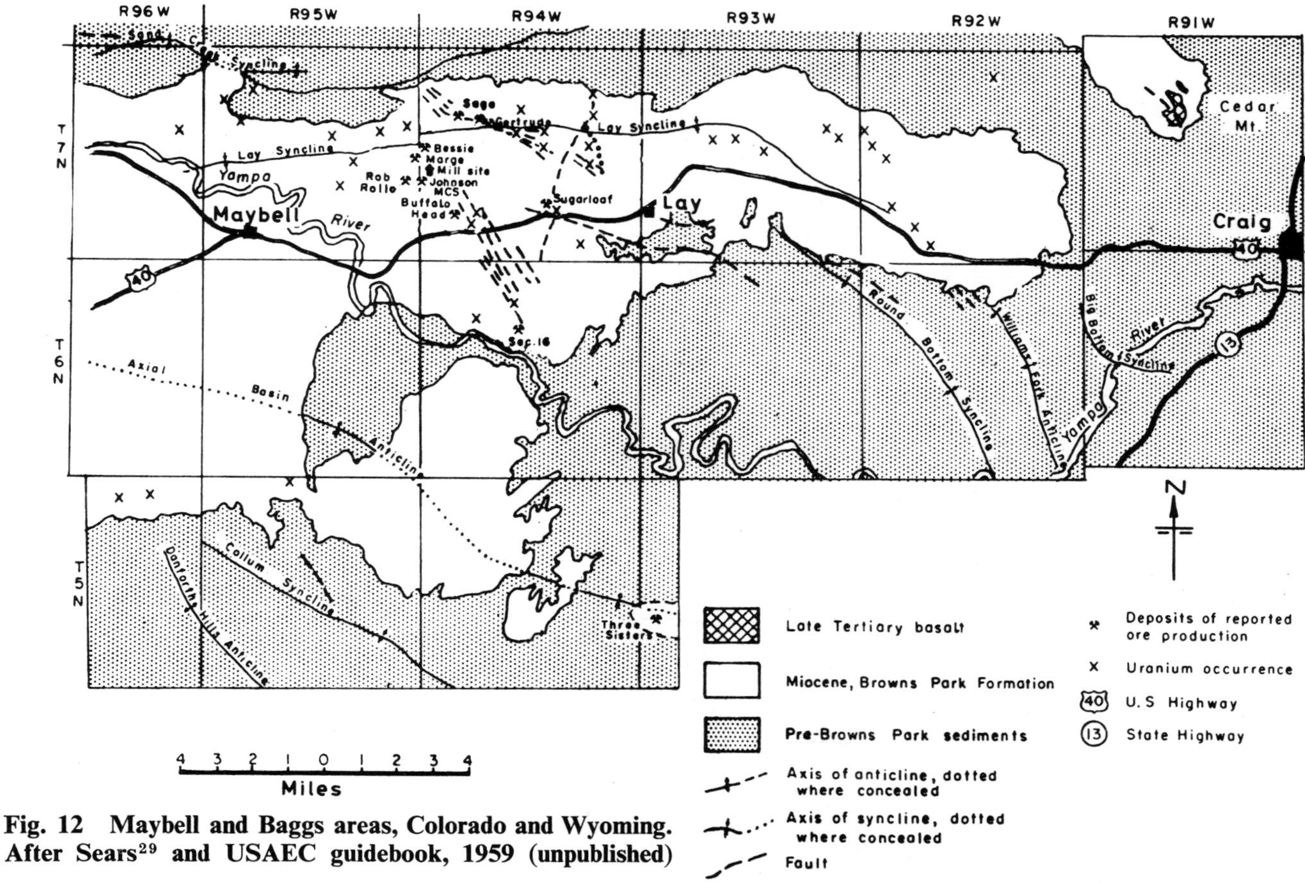

Fig. 12 Maybell and Baggs areas, Colorado and Wyoming. After Sears[29] and USAEC guidebook, 1959 (unpublished)

The Oakville Sand is an alluvial fan and stream and flood-plain deposit. The sandstones are fine-grained and feldspathic and contain little vegetal carbonaceous material.

Most of the formations dip toward the Gulf at one degree or less. The Whitsett is unconformably overlain by the tuffaceous Catahoula in the southern area, and the Goliad laps over all the upper Tertiary units, including the Oakville and Catahoula.

The uranium is in roll deposits similar in many respects to those in the Wyoming Basins. Faulting is an ore control for the deposits situated in grabens and other deposits aligned with faults. Deposits are clustered in two separate areas of the central coast; the Whitsett is host in the Karnes County area and the Oakville and Goliad are hosts in the Duval County area (Fig. 11).

The tongues of altered sandstone inside the ore rolls are white to light grey or brown as compared with the normal grey pyritic sand and dark grey to black of the ore deposits. Molybdenum is common adjacent or peripheral to ore and in beds of lignitic shale. Selenium is present in the ore sand, especially on the oxidized interior of rolls. Individual deposits may contain as much as several million pounds U_3O_8, range between 10 and 25 ft thick, have widths of several hundred feet and lengths of a mile or more.

The most accepted theory of origin is that uranium was leached from the plentiful tuffaceous rocks. Hydrogen sulphide originating in deep oil and gas fields in underlying Cretaceous formations and entering the uranium host rocks along faults is the preferred reductant, according to Eargle and Weeks.[28] In the Whitsett Formation vegetal carbonaceous material is sufficiently abundant to provide additional reducing environment.

Lines of faulting in favourable formations and grabens are good exploration targets, because locally thick sands in grabens are preferred sites of deposition. The contacts between altered and unaltered sandstones are used to guide exploration for roll-type deposits, and the thickest parts of the most continuous sandstones are the most favourable.

MAYBELL–BAGGS, COLORADO AND WYOMING

The Maybell, Colorado, area is in the southern part of the Sand Wash Basin and the Baggs, Wyoming, area is in the southern part of the Washakie Basin. Although about 40 airline miles apart, the geology of each area is essentially the same (Fig. 12).

The uranium host rocks are sandstones of the Browns Park Formation of Miocene age, which unconformably overlie truncated beds of Cretaceous and Eocene ages. The Lay Basin at Maybell and the Poison Basin at Baggs, which are structural basins filling topographic lows in the pre-Browns Park palaeosurface, are sites for the deposits. The areas of Browns Park Formation at each location are erosional remnants, and, due to erosional thinning, maximum thickness probably does not exceed 1200 ft. The Browns Park is a fine- to medium-grained quartzose sandstone containing tuff and bentonite derived from a provenance of mainly sedimentary and metamorphic rocks and volcanic activity. The sediments are fluvial, dominantly flood-plain and in part aeolian.

Outcrops contain secondary uranium minerals and are intensely limonite-stained, but unoxidized sandstone is dark grey to blue due to abundant finely divided pyrite. The sandstones are nearly devoid of vegetal carbon, but some 'dead' oil is present along faults, according to Grutt and Whalen[30].

Ore is in amoeba-shaped peneconcordant deposits, which range in size from a few tons of ore to more than 250000 tons. Thickness of ore is from a few feet to 30 ft and tenor ranges from 0·05 to 0·25% U_3O_8. A characteristic of the

Maybell district is the large amount of sub-commercial ore present. The ores are pyritic and also contain molybdenum and selenium. Faults are present in both areas and have influenced location of deposits, especially in Poison Basin. Although uncommon, a few of the deposits in Poison Basin are rolls with prominent molybdenum concentrations peripheral to the uranium. Elliptical calcareous concretions, surrounded by thin uraniferous and pyritic shells, seem to be related to emplacement of some deposits in Poison Basin.

Most geologists who have worked in this area think that the uranium was indigenous to the tuffs in the Browns Park. It has been postulated that uranium was transported by groundwater migrating through the permeable beds; the reducing environment was created by hydrogen sulphide bearing gas from underlying formations, which entered the Browns Park along the surface of unconformity and the numerous faults.

Prospecting is guided by radioactivity and limonite-stained outcrops. The axes of local basins and fault lines are worthy of investigation. As there is no distinctive alteration adjacent to or surrounding the stratiform deposits, grid drilling is a common resort.

Summary

A number of selected criteria have been checked for applicability against nine U.S. uranium areas. The probable significance of the criteria has been reviewed and related to the principal geologic aspects of the nine areas.

Examination of the categories checkmarked in Table 2 shows, in the light of present knowledge, that for optimum favourability host rocks should be (1) Triassic, Jurassic or Tertiary continental sandstones; (2) fluvial sandstones; (3) derived from provenance of granitic and metamorphic rocks or clastic rocks; (4) near an unconformity; (5) stratigraphically lower than tuffaceous measures; (6) feldspathic, arkosic or quartzose sandstones; (7) in the Colorado Plateau, Wyoming Basins (or Texas Gulf Coast); and (8) in intracratonic basins (or certain(?) geosynclines).

Similar study of the favourable local criteria indicates the optimum conditions for host rocks to be (1) medium to coarse sandstones; (2) grey, green or tan sandstones with interbedded grey and/or green mudstones; (3) finely divided pyrite in unoxidized sandstones; (4) limonite- and/or hematite-stained outcrops; (5) ratios of sandstone to shale of 1:1 to 4:1; (6) sandstones dipping less than 5°; (7) sandstones containing anomalous vanadium, molybdenum and selenium; (8) sandstones containing vegetal carbonaceous material; (9) radioactivity anomalies greater than 5 times background; (10) U_3O_8 geochemical anomalies greater than 5 ppm; (11) U_3O_8 in groundwater greater than 10 ppb; and (12) oxidized uranium minerals in outcrops.

The above factors may not have been assembled in this way before, but, in one place or another, most have been discussed previously in the literature.

An important point to be made is that the criteria should be used within the framework of geologic history of the areas to be most meaningful. The sequence of events is of prime importance to the formation of these deposits, and the time of ore deposition is a key factor. Unfortunately, the mobility of uranium in sandstones results in age determinations that are of questionable value in fixing the time of primary ore emplacement.

In conclusion, there are a number of topics, important to the uranium problem, that warrant additional study. A few of these are listed below.

(1) Tuffs seem to play a part in the formation of the deposits, possibly by being a source of uranium and/or a source of alkaline carbonate waters. Study of the way uranium is held in volcanic ash and the behaviour of uranium during devitrification is needed.

(2) Structureless organics of vegetal origin are a controlling factor in major deposits. Research has identified the organics as humic and fulvic compounds and established their effectiveness in precipitating uranium. More study is needed to determine the nature of the bonding and whether organic acids have a role in the transportation of uranium in groundwater.
(3) The role of groundwater as a medium in the formation of deposits is now accepted and the role of carbonate groundwater in transportation of uranium is indicated. The possibility that uranium is preferentially deposited at the interfaces between carbonate and sulphate groundwater facies should receive more study.
(4) Perhaps as a corollary of item (3), the location of calcareous concretions or calcareous cemented sandstones in relation to the deposits is worthy of much more study and field investigation as a possible exploration guide.
(5) Research into the palaeohydrology of uranium areas in terms of the probable direction and dynamics of groundwater flow at the time of ore formation is warranted as a possible guide to exploration.

The search for new uranium areas is a formidable challenge. All possible knowledge and skill is needed to maximize the chances of success.

Acknowledgment

Liberal use was made of the vast body of literature pertaining to the sandstone-type uranium deposits. Cited references are noted in the text and topical and general references are included in a separate listing. The author is indebted to the AEC geologic field staff, who provided data and reviewed criteria, and to R. T. Russell and W. L. Chenoweth, who furnished valuable assistance with text and illustrations.

References

1. Schmidt-Collerus J. J. Investigations of the relationship between organic matter and uranium deposits. Part II. Experimental investigations. Denver Research Institute publication 2513, University of Denver, Denver, Colorado, 1969.
2. Grutt E. W. Jr. Environment of some Wyoming uranium deposits. In *Advances in nuclear engineering* (Oxford, etc.: Pergamon, 1957), vol. 2, 313–23.
3. Bowles C. G. Present day groundwater, a possible guide to uranium exploration in the southern Black Hills of South Dakota and Wyoming. *Open-file Rep. U.S. geol. Surv.*, 1967.
4. Gott G. B. Inferred relationship of some uranium deposits and calcium carbonate cement in southern Black Hills, South Dakota. *Bull. U.S. geol. Surv.* 1046-A, 1956, 1–8.
5. Hagmaier J. L. The relations of uranium occurrences to groundwater flow systems. *Earth Sci. Bull., Wyoming geol. Ass.*, June 1971, 19–24.
6. Meschter D. Y. A study of concretions as applied to the geology of uranium deposits. *Bull. geol. Soc. Am.*, **69**, 1958, 1736–7. (Abstract)
7. Garrels R. M. Geochemistry of "sandstone-type" uranium deposits. In *Advances in nuclear engineering* (Oxford, etc.: Pergamon, 1957), 288–93.
8. Denson N. M., Zeller H. D. and Stephens J. G. Water sampling as a guide in the search for uranium deposits and its use in evaluating widespread volcanic units as potential source beds for uranium. *Prof. Pap. U.S. geol. Surv.* 300, 1956, 673–80.
9. Gruner J. W. Concentration of uranium in sediments by multiple migration–accretion. *Econ. Geol.*, **51**, 1956, 495–520.
10. Wood H. B. Geology and exploitation of uranium deposits in the Lisbon Valley area, Utah. In *Ore deposits of the United States, 1933–1967* Ridge J. D. ed. (New York: AIME, 1968), 770–89.

11. Malan R. C. The uranium mining industry and geology of the Monument Valley and White Canyon Districts, Arizona and Utah. In *Ore deposits of the United States, 1933–1967* Ridge J. D. ed. (New York: AIME, 1968), 790–804.
12. Fischer R. P. and Hilpert L. S. Geology of the Uravan mineral belt. *Bull. U.S. geol. Surv.* 988-A, 1952, 13 p.
13. Keller W. D. Clay minerals in the Morrison Formation of the Colorado Plateau. *Bull. U.S. geol. Surv.* 1150, 1962, 90 p.
14. Clary T. A., Morbley C. M. and Moulton G. F. Jr. Geologic setting of an anomalous ore deposit in the Section 30 mine, Ambrosia Lake area. *Mem. New Mex. St. Bur. Mines* 15, 1963, 72–9.
15. Granger H. C. Localization and control of uranium deposits in the southern San Juan basin mineral belt, New Mexico—an hypothesis. *Prof. Pap. U.S. geol. Surv.* 600-B, 1968, 60–70.
16. Austin S. R. Alteration of Morrison sandstone. Geology and technology of the Grants uranium region. *Mem. New Mex. St. Bur. Mines* 15, 1963, 38–44.
17. Gott G. B. and Schnabel R. W. Geology of the Edgemont NE quadrangle, Fall River and Custer Counties, South Dakota. *Bull. U.S. geol. Surv.* 1063-E, 1963, 62 p.
18. Renfro A. R. Uranium deposits in the Lower Cretaceous of the Black Hills. *Wyoming Univ. Contr. Geol.*, **8,** Summer 1969, 87–92.
19. Hart O. M. Uranium in the Black Hills. In *Ore deposits of the United States, 1933–1967* Ridge J. D. ed. (New York: AIME, 1968), 832–7.
20. Soister P. E. Stratigraphy of the Wind River Formation in south-central Wind River Basin, Wyoming. *Prof. Pap. U.S. geol. Surv.* 594-A, 1968, 50 p.
21. King J. W. and Austin S. R. Some characteristics of roll-type uranium deposits at Gas Hills, Wyoming. *Min. Engng, N.Y.*, **18,** May 1966, 73–80.
22. Zeller H. D. The Gas Hills uranium district and some probable controls for ore deposition. *Wyom. geol. Ass. Guidebook, 12th ann. Field Conf.*, 1957, 156–60.
23. Love J. D., Weitz J. L. and Hose R. K. Geologic map of Wyoming. Scale 1:500000. (Washington D.C.: U.S. Geological Survey, 1955).
24. Love J. D. Preliminary report on uranium deposits in the Pumpkin Buttes area, Powder River Basin, Wyoming. *Circ. U.S. geol. Surv.* 176, 1952, 37 p.
25. Davis J. F. Uranium deposits of the Powder River Basin. *Wyom. Univ. Contr. Geol.*, **8,** Summer 1969, 131–41.
26. Eargle D. H., Hinds G. W. and Weeks A. D. Uranium geology and mines, South Texas. *Univ. Texas., Bur. econ. Geol., Guidebook* no. 12, 1971, 8 p.
27. Norton D. L. Uranium geology of the Gulf Coastal area. *Bull. Corpus Christi geol. Soc.*, **10,** no. 5, 1970, 19–26.
28. Eargle D. H. and Weeks A. D. Possible relation between hydrogen sulfide-bearing hydrocarbons in fault-line oil fields and uranium deposits in the southeast Texas Coastal Plain. *Prof. Pap. U.S. geol. Surv.* 424-D, 1961, 7–9.
29. Sears J. D. Geology and oil and gas prospects of part of Moffat County, Colorado, and southern Sweetwater County, Wyoming. *Bull. U.S. geol. Surv.* 751-G, 1924, 269–319.
30. Grutt E. W. Jr. and Whalen J. F. Uranium in northern Colorado and southern Wyoming. Guide book to the geology of north-west Colorado. In *Intermountain Ass. Petrol. Geol., Guidebook, 6th ann. Field Conf.*, 1955, 126–9.

Topical and general references*

Adler H. H. The conceptual uranium ore roll and its significance in uranium exploration. *Econ. Geol.*, **59**, 1964, 46–53.
Armstrong F. C. Geologic factors controlling uranium resources in the Gas Hills District, Wyoming. *Wyom. geol. Ass. Guidebook, 22nd ann. Field Conf.*, 1970, 31–44.
Austin S. R. Some patterns of sulfur isotope distribution in uranium deposits. *Earth Sci. Bull., Wyoming geol. Ass.*, **3,** no. 2 1970, 5–22.

*This list contains some of the more pertinent topical and general papers.

Craig L. C. *et al.* Stratigraphy of the Morrison and related formations, Colorado Plateau region: a preliminary report. *Bull. U.S. geol. Surv.* 1009-E, 1955, 43 p.
Finch W. I. Geology of epigenetic uranium deposits in sandstone in the United States. *Prof. Pap. U.S. geol. Surv.* 538, 1967, 121 p.
Fischer R. P. Similarities, differences, and some genetic problems of the Wyoming and Colorado Plateau types of uranium deposits in sandstone. *Econ. Geol.*, **65**, 1970, 778–84.
Flawn P. T. Uranium in Texas—1967. *Geol. Circ. Texas Univ. Bur. econ. Geol.* 67–1, 1967, 16 p.
Garrels R. M. and Larsen E. S. Geochemistry and mineralogy of the Colorado Plateau uranium ores. *Prof. Pap. U.S. geol. Surv.* 320, 1959, 236 p.
Harshman E. N. Genetic implications of some elements associated with uranium deposits, Shirley basin, Wyoming. *Prof. Pap. U.S. geol. Surv.* 550-C, 1966, 167–73.
Uranium exploration geology (Vienna: IAEA, 1970), 384 p.
Kelley V. C. Regional tectonics of the Colorado Plateau and relationship to the origin and distribution of uranium. *Publs Geol. Univ. New Mex.* no. 5, 1955, 120 p.
Klohn M. L. and Pickens W. R. Geology of the Felder uranium deposits, Live Oak County, Texas. Paper presented at 1970 AIME annual meeting, Denver, Colorado, preprint 70-I-38.
Langen R. E. and Kidwell A. L. Geology and geochemistry of the Highland uranium deposit, Converse County, Wyoming. Paper presented at 1971 AIME annual meeting, New York, preprint 71-I-37.
New Mexico Bureau of Mines and Mineral Resources. Geology and technology of the Grants uranium region. *Mem. New Mex. St. Bur. Mines* 15, 1963, 277 p.
Osterwald F. W. and Dean B. G. Relation of uranium deposits to tectonic pattern of the central Cordilleran foreland. *Bull. U.S. geol. Surv.* 1087-I, 1961, 54 p.
Rackley R. I., Shockey P. N. and Dahill M. P. Concepts and methods of uranium exploration. *Earth Sci. Bull., Wyoming geol. Ass.*, **1**, no. 3 1968, 23–4.
Ridge J. D. ed. *Ore deposits of the United States, 1933–1967* (New York: AIME, 1968), 2 vols, 1880 p.
Robinson C. S., Mapel W. J. and Bergendahl M. H. Stratigraphy and structure of the northern and western flanks of the Black Hills uplift, Wyoming, Montana and South Dakota. *Prof. Pap. U.S. geol. Surv.* 404, 1964, 134 p.
Santos E. S. Stratigraphy of the Morrison Formation and structure of the Ambrosia Lake district, New Mexico. *Bull. U.S. geol. Surv.* 1272-E, 1970, 30 p.
Schultz L. G. Clay minerals in Triassic rocks of the Colorado Plateau. *Bull. U.S. geol. Surv.* 1147-C, 1963, 71 p.
Shawe D. R., Simmons G. C. and Archbold N. L. Stratigraphy of Slick Rock district and vicinity, San Miguel and Dolores Counties, Colorado. *Prof. Pap. U.S. geol. Surv.* 576-A, 1968, 1–108.
Shawe D. R. Possible exploration targets for uranium deposits, south end of the Uravan mineral belt, Colorado–Utah. *Prof. Pap. U.S. geol. Surv.* 650-B, 1969, 73–6.
Stewart J. H. *et al.* Stratigraphy of Triassic and associated formations in part of the Colorado Plateau region. *Bull. U.S. geol. Surv.* 1046-Q, 1959, 89 p.
Wyoming uranium issue. *Wyoming Univ. Contr. Geol.*, **8**, no. 2, pt 1 1969, 149 p.
Van Houten F. B. Tertiary geology of the Beaver Rim area, Fremont and Natrona Counties, Wyoming. *Bull. U.S. geol. Surv.* 1164, 1964, 99 p.
Waters A. C. and Granger H. C. Volcanic debris in uraniferous sandstones, and its possible bearing on the origin and precipitation of uranium. *Circ. U.S. geol. Surv.* 224, 1953, 26 p.
Weeks A. D. and Eargle D. H. Relation of diagenetic alteration and soil-forming processes to the uranium deposits of the southeast Texas Coastal Plain. In *Clays and clay minerals, volume* 10 (New York, etc.: Macmillan/Pergamon Press, 1963), 23–41.

DISCUSSION

R. J. Chico said that Fig. 1 supplemented the author's excellent illustration (Fig. 9). It showed not only the geometrical relationship between altered and bleached ore and fresh zoning in a typical roll-front case history but also the distribution of U_3O_8

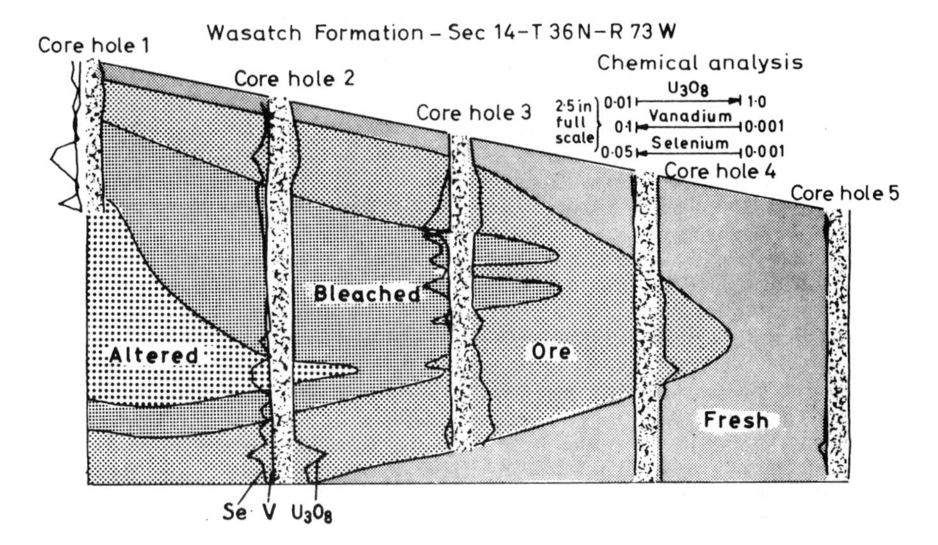

Fig. 1 Uranium roll front, Powder River Basin, Wyoming: datum, approximately 550 ft; vertical scale, 1 in = 1 ft; well interval, 100 ft. Data reproduced by courtesy of Teton Exploration Drilling Co. and Chemical and Geological Laboratories

and associated elements accurately determined in core sections. The field exploration effort of American geologists in Wyoming continually focused attention on investigations leading to uranium roll-front discoveries. Valuable references dealing with that field approach were those of Rackley and co-workers* and Rubin.†

N. R. Newson asked the author to give the approximate mineral compositions, ranges and modes of the rocks associated with the uranium deposits. Mr. Grutt said that the feldspathic sandstone would contain 10–15% feldspar; arkosic sandstone, 25–40% feldspar; quartzose sandstone, 75% quartz; a tuffaceous sandstone was so called if it contained more than 10–15% vitric material. Modal analyses were seldom produced for those coarse rocks. The parent materials of the tuffs were acidic in composition.

D. L. Dowie asked if the uranium deposits were related to the present surface. Was the same population of deposits being discovered by deeper exploration, and were there any areas which had reached exploration saturation?

The author replied that there was, of course, a tendency to get deeper and deeper in all exploration areas. In the Wyoming Basin targets were shallow open-pit deposits, and that also applied to the Texas Gulf Coast. In the Grants Mineral Belt the drilling was far deeper (3000–4000 ft) and had had a great deal of success. Certain areas (e.g. Monument Valley–White Canyon and Maybell–Baggs) were pretty much depleted at the current economic price of $6–$8/lb. The Grants area had not been completely explored and might take up to 20 years before it was completely explored.

Dr. A. G. Darnley asked if there were any evidence that the granitic and metamorphic source areas which supplied the sediments enclosing the uranium deposits possessed an above-average radioelement content. The author replied that most of

*Rackley R. I., Shockey P. N. and Dahill M. P. Concepts and methods of uranium exploration. *Earth Sci. Bull., Wyoming geol. Ass.*, **1**, no. 3 1968, 23–4.

†Rubin B. Uranium roll front zonation in the southern Powder River basin, Wyoming. *Earth Sci. Bull.*, **3**, no. 4 1970, 5–12.

the surrounding granites in central Wyoming had about 8–15 ppm, which was higher than most of the granites in the west. They could not sample the sources of most of the other deposits because they were not certain where those rocks were. A Precambrian framework study had been initiated, but results were still being accumulated.

Dr. Darnley said that his question arose from the fact that they had observed that the Canadian Shield contained belts, up to several hundred miles in length and tens of miles in width, with above-average radioelement concentrations. Those belts contained the known mineral deposits and the bulk of the radioactive mineral occurrences.

W. Meyer said that numerous sandstone uranium deposits were closely associated in space, if not time, with various evaporites. Any marine evaporite cycle brought to completion must deposit considerable amounts of metal: what happened to those? Could it be that the metals in sandstone deposits had been derived from sea water via an evaporite cycle?

The author said that Bowles* had postulated such an origin for the Black Hills deposits in that the water would have derived the uranium from the anhydrite that underlay the area.

Dr. F. Q. Barnes said that the provenance for the Western States sandstone deposits was of some significance, but he suggested that it was overemphasized because many of the Front Range rocks were not granites. In fact, much was sediment of at least one generation; and the terrain, in general, probably did not have abnormal clarkes for uranium and thorium. In the Texas fields the sediments containing the deposits were probably of second or third or even later generations, even though they were originally of Precambrian granitic origin.

He further would suggest that the mobility of the various elements involved in the deposits, particularly of uranium itself, had been underemphasized in the geochemical environment, particularly as it related to Eh and pH conditions at the critical stage which could be reached under normal surface and near-surface conditions.

The author said that that point was well taken. The White River Formation, for example, normally contained 4–6 ppm of uranium and, in certain areas, small concentrations of uranium had been found due to evaporation of groundwater at the surface. That problem had been dwelt on a little in the paper, where the importance of having a source, a carbonate-rich alkaline groundwater for transport and, finally, a precipitating zone at the boundary between the carbonate and sulphate water facies, had been emphasized.

*Bowles C. G. Present-day groundwater, a possible guide to uranium exploration in the southern Black Hills of South Dakota and Wyoming. *U.S. geol. Surv. open file rep.*, 1967.

Uranium exploration costs

F. Q. Barnes M.A., Ph.D., P.Eng.

David S. Robertson and Associates, Ltd., Blind River, Ontario, Canada

338.58 : 550.8 : 553.495.2

Synopsis

Considerable emphasis, in uranium exploration, is currently placed on the new developments in geophysics and geochemistry. Expenditures on uranium exploration, however, are primarily incurred in the collection of geological information and in the testing of geological ideas. The importance of understanding the geochemical environment relating to the occurrences of economic uranium deposits cannot be overemphasized in the application and interpretation of the new geophysical and geochemical techniques. Gamma-ray spectrometry, which measures the chemical favourability of an area *vis-à-vis* uranium concentration, or chemical methods which directly measure the concentration of uranium in the groundwater regime, are all important in the selection of prospecting areas. But, regardless of the favourable geochemical evidence found over extensive areas, the latest round of worldwide uranium exploration has again indicated that economic uranium deposits are restricted to well-known geological situations.

The cost of finding uranium relates directly to the understanding of uranium occurrences, to improvements in the field application of geophysics and geochemistry, and to the efficiency of the supporting services. In all, these aspects of cost are improving. The availability of accessible new areas for exploration, however, is diminishing and the expenditures in time and service per pound of uranium found are steadily mounting. In North America the finding cost has increased from a weighted average of about 30 cents per pound to about 80 cents per pound in the latest period of exploration (1965–70).

Finding cost, as distinct from concentrate cost, relates indirectly to uranium mining and milling technology, to concentrate prices and to legislation. Improvements in extraction technology, and increases in concentrate prices with the forthcoming demand for uranium, will permit the utilization of lower-grade deposits, many of which are already known and are larger and more numerous than the presently sought high-grade deposits. Government legislation, in the net importing countries, will favour the producer, whereas there is a tendency in the exporting nations to obtain maximum benefit from resources. In a strong domestic market for raw materials true exploration costs can be largely hidden by favourable government legislation.

The costs of finding uranium in the past were extremely low in comparison with those of the present time because the radioactive properties of the element permitted its ready detection in surface deposits in the industrial areas of the world. Future finding costs, in the industrial areas, will be governed, partly, by the geological occurrence of uranium—a subject which we are only beginning to understand—and by our success in applying new knowledge to the discovery of more ore.

The technical aspects of uranium extraction have been a major factor in determining the cost of uranium concentrates, and improvements in extractive technology could change the whole economic picture of uranium. With sharply reduced milling costs, or substantial price increases for uranium, large-tonnage, low-grade deposits presently known could immediately become part of world reserves. The future finding cost, then, will also be partly governed by extraction technology, and costs for the first generation of low-grade deposits could be minimal. Given present conditions, however, there can be little doubt that exploration costs will continue to rise.

In developing exploration costs it has been the practice to apply total expenditures to the pounds of recoverable uranium located during an exploration period, or to uranium found in a particular area of exploration effort. Needless to say, virgin territory in which uranium ore occurs will, initially, yield cheap exploration pounds if it has easily found surficial showings, whereas, in the same time period, areas formerly explored may yield pounds at a substantially higher cost. Exploration costs for uranium from the recent finds in Australia as compared with recent exploration costs in North America provide a case in point. The ease of finding uranium, however, is often not the main concern of management. The company with an existing mill, representing a substantial investment, can improve its competitive position in the industry provided that it continues to have ancillary reserves. The finding of new pounds nearby, even at high unit cost, can present a more attractive financial return than cheap exploration pounds in a new or remote area.

It is true, then, that the exploration finding cost is not necessarily the most important factor in the overall picture of price and profit, which are the main considerations of management in formulating exploration policy. To evolve a policy for uranium exploration based on purely geographical considerations or on the basis of present mill locations is beyond the scope of this paper because it would have to take into account the technical and business requirements peculiar to each producer rather than only exploration costs *per se*.

Definition of exploration costs

The price of uranium concentrate, once supply and demand are in general balance, will be the sum of exploration, development, mining and milling costs, taxes, royalty payments, overhead and administration, capital and interest charges, and a profit commensurate with the risks involved, all for the marginal producer. It is mining practice to charge mining and milling, administration and royalty costs to current production, but capital charges for plant and development work are charged on a time or ore-reserve basis, together with the appropriate portion of the accumulated interest. In the case of exploration, the expenditures are commonly written off against current production profits, provided that one is fortunate enough to be a profitable producer! For the non-producer, at least in North America, expenditures on exploration are accumulating non-interest-bearing expenses, which, if the explorer is successful, can be written off against production at a future date. Capital investments normally earn interest according to their risk, but for exploration expenditures such an approach is not currently the method of cost treatment. Exploration costs are certainly returned to the producer—otherwise he could not continue in business in the long run. The cost of the product assures him a profit on his total outlay and the price of the product finds its own level in the competitive market.

How does the producer recover exploration cost? In part, at least, he receives it from the purchaser, but in some countries, if not all, the producer receives bonuses by means of allowances, notably depletion, which encourage him to expand his operations with new finds through further exploration. This is a worthwhile incentive for the industrialized nations, as well as for others, which pays off in higher employment through mining and the development and multiplier effect of greater business activity. In Canada, a net exporter of mineral products, the effective tax, after the initial three-year tax-free period afforded new mines, is about 20% lower than for other industries. Changes in mine taxation policies throughout the world, and presently under legislative review in Canada, could eventually exert an upward pressure on prices, attributable, for the most part, to a reassessment of exploration costs. It seems unlikely that the cost of exploration, treated as an investment, will ever be applied directly to the selling price of mineral products, because the highly industrialized countries will, from the consideration of national supply, continue to underwrite exploration through favourable taxation and subsidies, thereby requiring the exporting nations to be competitive.

If, however, we apply the uranium finding cost evenly to each recoverable pound of U_3O_8 in a deposit, collecting on each pound as it is sold, with interest accruing from the time of discovery at a rate of 8%, the average initial finding cost escalates 3·85 times in 25 years. The 25-year period would cover the time from discovery, in the 1950s, of many United States ore deposits until they are depleted in the 1970s.

Under the current national systems of subsidies to the mineral industry calculation of interest-bearing exploration costs is impractical, and this discussion is therefore restricted to simple 'finding costs'. Finding costs are thereby a measure of exploration efficiency rather than figures which bear directly on the evolution of exploration philosophy. The Canadian figures on exploration costs cited here are correct to the extent that one can calculate the level of exploration expenditures on a nation-wide basis. With the detailed information available from company and government records, the Canadian figures are considered to be reasonably correct. In the case of the United States, the inclusion of development pounds with new exploration pounds in the available statistics and difficulties involved in calculating total exploration charges on a per foot drilled basis, as distinct from drilling costs alone, permit a high factor of error in the calculation of exploration costs.

There is a certain flexibility in assigning expenditures to exploration or to development of a mineral body. Generally, where exploration drilling and/or surface trenching and sampling have indicated, to a reasonable level of probability, a reserve of sufficient size and grade to support a profitable mining operation, costing to exploration ceases. Notwithstanding the value of underground work in reducing the risk of a mining enterprise, underground work is undertaken in a manner which lends itself to the recovery of the indicated ore reserve. Although, in a minority of cases, the underground work may indicate that the reserve is not economically recoverable and its costs are therefore applicable to exploration, the exploration costs herein considered will cover expenditures only to the point of indicating an economic mineral concentration.

Operating methods and cost in uranium search

All but one of the United States uranium producers operate deposits in permeable Triassic or younger sandstones, which have a more or less uniform manner of occurrence and mode of origin. Uranium search areas in the United States are reasonably accessible and have had an appreciable amount of prior work. Techniques, therefore, follow a common procedure. Canadian deposits, on the other hand, have distinctly different modes of origin and manner of occurrence. Search techniques are modified according to the amount of previous work, the character of the country, the nature of the geology and the accessibility of the region. Costs and search patterns are therefore treated separately for Canada and the United States. Most exploration programmes for uranium deposits in other countries can follow, in general, one or other of the procedures described.

Canada

The search techniques in Canada are largely dependent on the amount and type of information available on the search areas. In general, systematic airborne radiometric surveys, or, in the case of extensive overburden, geochemical surveys, are used for the reconnaissance of lightly explored areas, followed by geological investigations of anomalies and, finally, should the findings warrant, ground geophysical and geochemical surveys, geological mapping, trenching and diamond drilling. Because of the size of the country and the many favourable areas yet to receive serious exploration effort, however, work of a detailed nature, such as that applied to much of the European land mass, remains to be done, except in the immediate areas of known deposits.

Geological work is commonly thought of as a map-making function. Credits for geology applied against mining claim assessment work require the submission of a geological map and report to the appropriate government department. Geological knowledge applied before and during an exploration programme, however, is the most effective approach in mineral search. The application of geological principles and pragmatic knowledge of ore deposit formation is essential, luck notwithstanding, to successful exploration. It is fair to say that the successful explorers in the latest period of uranium search have been primarily those with experienced uranium exploration staffs. Obviously, if practical geological knowledge or correct theoretical considerations result in the selection of the more favourable search areas, and the appropriate search techniques are effectively and efficiently applied, the chances of success are multiplied. It is proper to think of bad exploration direction as a double loss. Poorly utilized funds are not only wasted but the time has been lost when the same funds could have been better channelled.

The role of the geologist in uranium exploration is to select favourable areas for search, to follow the field survey programmes, whether they be geological, geophysical or geochemical, to be able immediately to follow up discoveries or interesting anomalies or to determine whether the programmes, as constituted, are effective or can be modified or eliminated.

Uranium exploration can be considered in three stages—reconnaissance, follow-up examinations and detailed work on prospects.

RECONNAISSANCE

Geophysics and geochemistry are used in uranium search from reconnaissance through various stages of detailed work. The largest expenditures on these techniques have been at the reconnaissance stage, particularly in regions in which there is little geological knowledge. The gamma-ray spectrometer, the geophysical tool most commonly employed, measures portions of the spectrum characteristic of potassium, uranium and thorium as well as total radioactivity. The intensity of the radiation through each selected channel is recorded as a continuous reading or as a digital printout on a chart or tape, along with a ground profile beneath the aircraft, usually measured by a radar altimeter. Height and speed of aircraft, surface area of crystal or crystal combinations and digital time constant are important in determining the sensitivity and discrimination of the survey. Approximately 400 in^3 (6500 cm^3) of crystal connected with a four-channel gamma-ray digital spectrometer in an aircraft capable of flying 150–200 ft (45–60 m) above the ground, at speeds of 100 miles/h (160 km/h) or less, is considered an acceptable reconnaissance exploration standard. Many surveys flown are too high above the ground or the volume of crystal is too low for the detection of most economic occurrences of uranium, and the knowledge that previous airborne surveys have been made in an area should not necessarily be a deterrent to further airborne work.

In some airborne systems the Th : U ratio can be plotted directly on the chart simultaneously with the individual readings. A positioning camera is synchronized with the chart. Much of the recorded data is primarily of geological interest, but this is of value to the explorer in areas of scant geological knowledge. For this reason, and as a general exploration policy for all metals, a magnetometer and electromagnetic equipment are commonly carried in the aircraft to give additional geological knowledge with some possible economic significance. The amount of information gathered in these surveys, for some areas, is greater than can possibly be assimilated on a single exploration programme, and the charts represent valuable information for future retrieval.

Table 1 Equipment costs and airborne survey lease and contract rates

Fixed wing: 200 ft (60 m) above ground; 90 miles/h (145 km/h)	
(*a*) Gamma-ray spectrometer, 'window' type, four-channel, 400 in³ (6500 cm³) crystal, digital output. Price dependent on location and size of contract	$7–10/line mile
(*b*) With magnetometer carried in aircraft	$10–12/line mile
(*c*) With magnetometer and electromagnetic system	$18–22/line mile
(*d*) Dry lease with all three systems	$17 000/month
Helicopter: height above ground, 200 ft (60 m); lower if not towing an EM transmitter–receiver; 90 miles/h (145 km/h); camera film fiducials and radar altimeter to survey recorder	
(*a*) As above	$9–12/line mile
(*b*) As above	$12–14/line mile
(*c*) As above	$20–25/line mile
(*d*) As above	$20 000/month
Equipment costs	
Portable Geiger counters and scintillometers	
Lease price/month, 12–18% of purchase price	$100–2000
Portable gamma-ray spectrometers suitable for helicopter follow-up, detector unit separate, with 7 in³ (115 cm³) crystals, from	$2000
with 50 in³ (820 cm³) crystals, from	$3800
Lease price/month, 12% and less of purchase price	$3950
Airborne spectrometers: installation extra	
Crystals about 450 in³ (7300 cm³); six-channel recorder with the following spectrometers	
Analogue spectrometer, from	$28 000
With analogue–digital spectrometer, from	$27 000
Lease price/month, 12% and less of purchase price	

It is the author's opinion, based on examination of routine private and government airborne surveys, that the ground follow-up of anomalies should be reserved to geologically favourable areas, if the survey itself was not so restricted, in order to maintain exploration efficiency. Even so, the number of anomalies is likely to be too great for effective individual investigation unless discrimination is possible through U:Th ratio comparisons. The discrimination offered by the four-channel gamma-ray spectrometer with either visual comparison of the U and Th response on the chart, or a direct chart print of the ratio, is invaluable in reconnaissance airborne surveying.

Airborne surveys in Canada are commonly conducted by geophysical contractors, rather than by individual exploration companies, because the equipment and aircraft need to be in fairly continuous operation for cost efficiency. Most exploration companies have insufficient work to keep an aircraft flying on an economical basis. Also, geophysical equipment soon becomes obsolete with the application of new techniques and refinements, and most exploration companies find it beneficial to contract their geophysical work, or rent or lease their equipment. Contract and lease rates for airborne work and rental rates for equipment are given in Table 1. Lease rates on a per mile flown basis with aircraft fitted with geophysical equipment are generally lower than contract rates, provided that the client has sufficient work and is favoured by reasonable flying conditions.

Geochemical reconnaissance for uranium has been widely applied in Canada. For example, radon gas determinations in waters have been used effectively in areas where extensive overburden limits the use of radiometric surveying, particularly where vein-type occurrences are to be found. Reconnaissance costs for the method vary between $10 and $20 per square mile, depending on ground conditions, which influence the effective use of helicopters. The density of water sampling is about one per square mile. Although coverage by the airborne radiometric technique costs $30–$40 on a square mile basis, and it may be demonstrated that water sampling could have indicated the known producing areas, there has been no significant move to the radon method of exploration in Canada. The widespread use of geophysics may be attributed to uncertainties in interpreting radon values. Moreover, airborne radiometry is faster and is more specific in outlining the source of the radioactivity, thereby reducing costs at the follow-up stage. Other geophysical equipment may be carried in the survey aircraft at very little additional cost wherever there is exploration advantage in so doing.

FOLLOW-UP INVESTIGATIONS

Whereas the requirement of the reconnaissance stage is to outline areas or locate points of high radioactivity—a function that can be performed by a technician—the follow-up stage requires a geologist to analyse the anomalies found. Through low-level helicopter or ground follow-up, anomalies caused by radiation mass effect can be eliminated and the nature of the uranium or thorium mineralization determined by inspection or sampling. The scintillation or Geiger-tube equipment used can be light and portable. The scintillometer crystal can be relatively small, $1\frac{1}{2}$ in × 2 in or 4 in × 4 in (7–50 in^3) (115–820 cm^3), for ground or low-level helicopter work; the smaller size is common for ground work and, depending on the degree of rock exposure and the operating height of the helicopter, can be suitable for follow-up work. Operating heights of 50 ft or less in Arctic and lightly forested areas are common. The 4-in crystal is, of course, capable of much more resolution and is preferable in helicopter checking. The instrumentation described above for reconnaissance is likewise suitable, particularly if reconnaissance and follow-up work run continuously. The use of a 'squealer' device on follow-up equipment has been found useful. Purchase and lease costs for small spectrometer units suitable for helicopter follow-up work are given in Table 1, and helicopter rental charges for a three-place machine, with a useful load of 1100 lb (500 kg) and a 250-mile (400-km) range, are shown in Fig. 5. Helicopter leasing for areas distant from a reliable supply centre generally require the hirer to supply the fuel and accommodation and board for the pilot. The effective use of a helicopter depends, to a major degree, on the density of the vegetation in the search area. Desert areas, including the Arctic and sub-Arctic regions, are commonly excellent situations in which to use a helicopter, because one can land close to prospective showings and move rapidly from one to another. In areas where transportation and supply are difficult, the number of people employed should be kept to a minimum. In terms of follow-up work, an experienced geologist or team of geologists, supported by helicopter, is the cheapest and most effective means of operation. This is apparent from a comparison of the useful work output from lower-salaried personnel, working on foot, with that from higher-paid professional staff, working on the same project, supported by helicopter.

DETAILED WORK

Once a favourable uranium showing has been located, or an area which, from geological considerations, might hide an ore deposit at depth, more detailed exploratory work is undertaken. This stage of the exploration process is commonly the most costly. The detailed work may involve ground geophysics, geochemistry and geological mapping.

Ground control is often achieved by survey grids, which, in the case of forested areas, require cutting lines of sight and the setting of pickets at chained intervals. Such lines cost approximately $75 per line mile and the work is generally contracted (in open country the cost is less). Enlarged aerial photographs may be used for positioning in the place of picket lines and, in many ways, they are preferable. The cost of enlarged photographs is insignificant, but relatively few people have learned to utilize this technique. Black and white aerial photographs are available from governments, but colour photography must be contracted. Contract rates for colour photography vary greatly according to the amount of area to be surveyed and the logistics. A recently completed contract for colour photography at $10 per square mile can be considered low because of advantageous logistics.

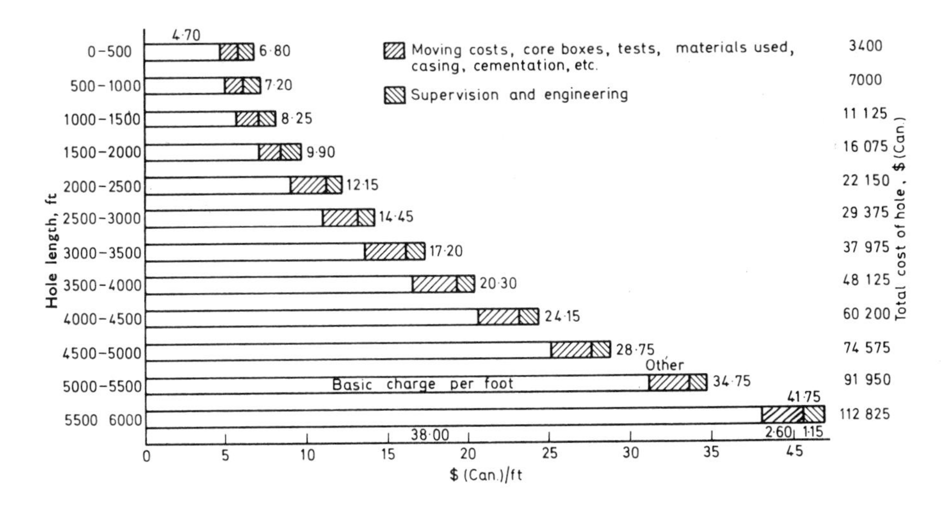

Fig. 1 Average diamond drilling costs in the Huronian of Ontario, Canada

Aside from the stripping of radioactive showings and blasting of the surface for sampling purposes, diamond drilling returns the most significant information per dollar spent. Diamond drilling costs per foot are largely dependent on hole length and the moving and servicing of the drill. In remote areas basic drilling costs may be $10/ft or more. Average drilling costs in relatively accessible areas are given in Fig. 1. Little use of rotary drilling equipment has been made in Canada because of the hard nature of the rocks in the Shield area.

In the detailed work stage ground acquisition is also an important cost factor. Depending on the legislative jurisdiction, the acquisition of mineral rights on Crown lands costs as much as \$1.25 per acre, which is the cost of staking and recording a mineral claim. In Federal Territories, which are generally remote areas, and in some sections of the Provinces, the mineral rights may be obtained for a work commitment.

COSTS

Most areas in Canada that received work in the uranium exploration period 1965–70 could be classed as remote, and the work performed on them would follow the pattern described above from reconnaissance through detailed work. One-half of the total expenditure, however, was in areas where general reconnaissance had already been done and there was little purpose in flying radiometric surveys. In such cases the main work was diamond drilling to obtain stratigraphic information, to test known showings, to extend ore trends or to find improvements along sub-ore trends.

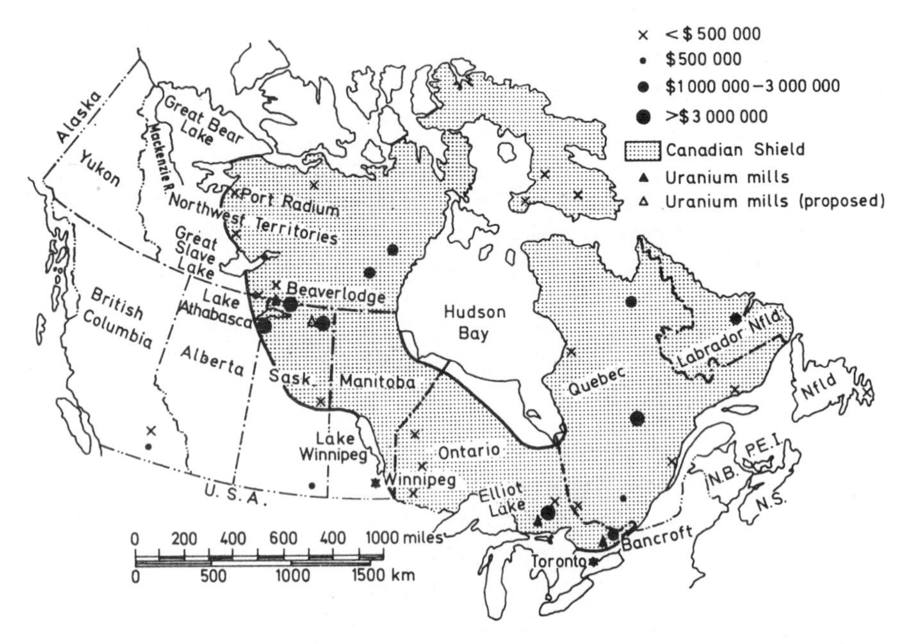

Fig. 2 Areas of Canadian uranium expenditure, 1965–70

The geographical distribution of exploration expenditures in Canada from 1965 to 1970 is shown in Fig. 2, and the percentage breakdown by regions and uranium source types is given in Fig. 3. The breakdown of exploration expenditures for a typical far-ranging programme is presented in Fig. 4. The total cost of the latter programme was about \$750000, mainly for reconnaissance and follow-up, but including some detailed work. In this particular case the programme involved work at various points along the southern and western edges of the Canadian Shield (a length of 5000 miles (8000 km)), and involved work on 26 properties. The programme did not run its full course because of legislative changes regarding the foreign ownership of Canadian uranium reserves.
Costs of detailed work on specific, remote properties are given in Fig. 5.

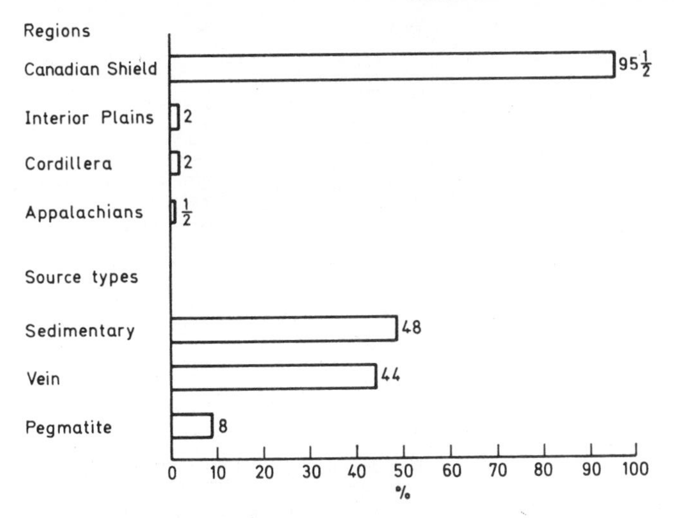

Fig. 3 Canadian uranium expenditure by regions and resource types

Diamond drilling in one case was not warranted after completion of the programme, and, in the other, was planned for a subsequent year. Costs attributable to diamond drilling in this particular case were about $15/ft, holes being limited to about 500 ft in length.

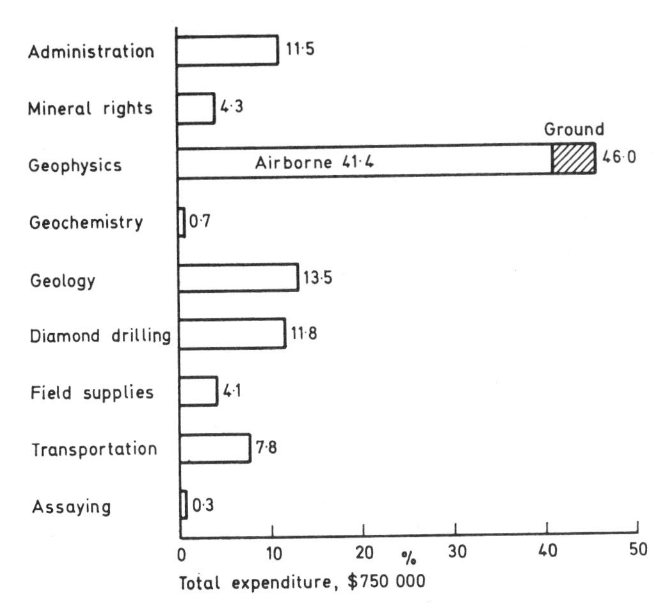

Fig. 4 Cost distribution of a far-ranging uranium exploration programme, 1969–70

Much of Canadian uranium exploration of recent years has been conducted for Elliot Lake type conglomerate deposits in the Huronian sediments of the Province of Ontario, between Sault Ste. Marie, near Lake Superior, and Sudbury. In this region most reconnaissance work had been completed by 1960 and the work performed recently was more of a detailed nature. Fig. 6 gives a breakdown of exploration cost for the Huronian region. Geophysics was mainly

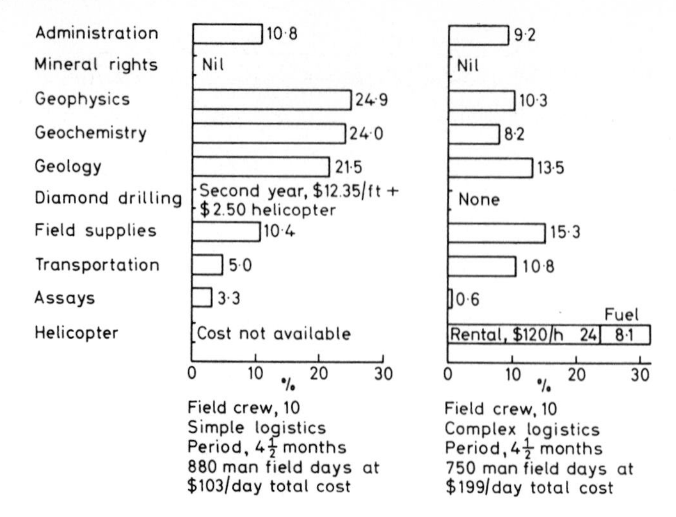

Fig. 5 Expenditure distribution on two far northern uranium properties, 1969–70

airborne. The results had little practical value. The relatively high expenditure on mineral rights and property acquisition reflects the highly competitive nature of the region, which is host to not only 90% or more of Canada's uranium reserves but also the nickel mines of the Sudbury area.

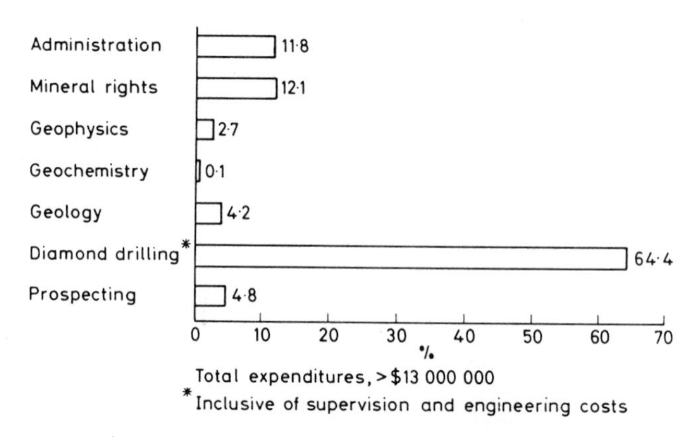

Fig. 6 Distribution of expenditure for Huronian exploration, Sudbury–Sault Ste. Marie, 1969–70

Fig. 7 gives a breakdown of costs for the successful exploration of a uranium property in the Huronian sediments. These expenditures were on the property itself and do not cover investigations in surrounding areas. No geophysical or geochemical work was performed because the nature of the deposit did not require it. The geological work could be described as research in nature with only a small element of routine mapping.

Two new uranium discoveries were made in Canada in the period 1965–70: the Gulf Minerals Company discovery in northern Saskatchewan, with a minimum of 40000000 lb U_3O_8, was a completely new find; the discovery of Kerr Addison Mines, Ltd., now Agnew Lake Mines, 30 miles west of Sudbury, Ontario, was a known occurrence at which additional geological work and diamond drilling indicated a uraniferous conglomerate similar to those of Elliot

Lake. Reserves at Agnew Lake Mines currently stand at 16000000 lb U_3O_8. Additional pounds were found at the mine properties near Elliot Lake, Ontario, during the same period. Notably, Denison Mines added a minimum of 50000000 lb U_3O_8 through development work, and the currently inactive Stanleigh mine added to its reserves of uranium-bearing material by extending the known ore-bearing sheet through surface diamond drilling.

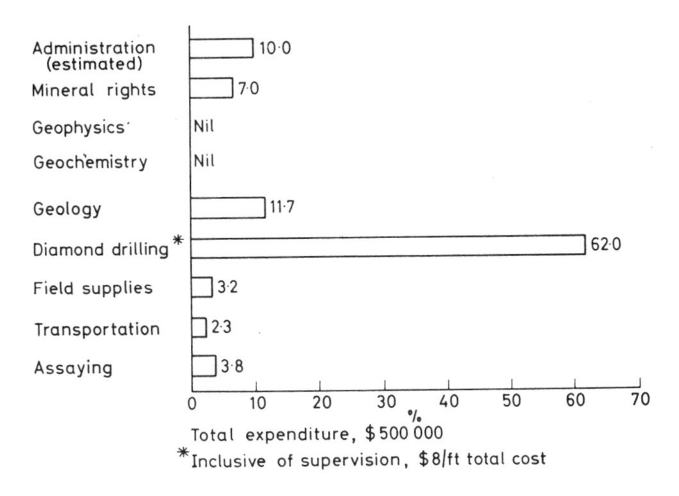

Fig. 7 Distribution of uranium exploration expenditure on a successful property, 1965–70

The total national exploration cost of U_3O_8 reserves added in the period 1965–70 is about 50 cents/lb. If the quantity added at the Elliot Lake mines is considered as part of discoveries made in an earlier period of exploration, the cost in newly discovered orebodies becomes about $1.00/lb (this can be compared with 5 cents/lb for all discoveries prior to the period 1965–70). The low previous figure relates to the large reserves added rapidly at the time of the discovery of the Elliot Lake conglomerate ores.

The United States

Large areas of the United States were flown with scintillometer in the 1950s, and the principal areas of interest were outlined at that time. Except for minor exploration for vein-type deposits in the Front Range, uranium exploration in the United States has been confined to permeable Triassic and younger sandstones derived from Precambrian granitic areas in the western United States and Texas. Recent work has been of a detailed nature in that each specific area chosen for exploration is studied geologically as to favourable stratigraphy and lithology, presence of known radioactive occurrences, alteration and colour phenomena in sandstones, palaeoecology, palaeodrainage, sand–shale ratios, permeability, etc., prior to exploration drilling. Radon measurements in soils and waters are being undertaken in various areas, as well as low-level airborne gamma-ray spectrometer surveying; but the major exploration expenditures are on non-coring rotary drilling, administration, the support of technical staffs and land acquisition.

The information from rotary drilling is frequently important at the start of the exploration process—rather than signalling the end-stage of a specific project, as it frequently does with exploration in Canada. The stratigraphic picture is enlarged with the information obtained about the various rock units.

Core samples may be taken of specific stratigraphic layers for trace-element analysis, and the alteration and colour characteristics of the sands are recorded. All holes are logged with gamma-ray probes and normally with resistivity equipment, which aids both in interpreting radiometric logs and in making geological correlation between drill-holes. A simultaneous record of the probe readings is made with an analogue recorder. Logging costs are generally in addition to drilling charges, and logging is generally carried out by separate contractors by use of truck-mounted equipment. Probing costs are about 15 cents/ft. The rotary drills are also truck-mounted for mobility, and drilling in the poorly lithified, young sediments is fast and relatively inexpensive. In the period 1965–70 more than 90000000 ft of rotary drilling was undertaken in the search for uranium. Drilling costs are summarized in Fig. 8. The truck-mounted drills are unable to drill much deeper than 3000 ft and, for testing beyond this depth, oil-well rigs are required, their hourly rental and ancillary charges averaging about $10/ft over a 4000-ft hole.

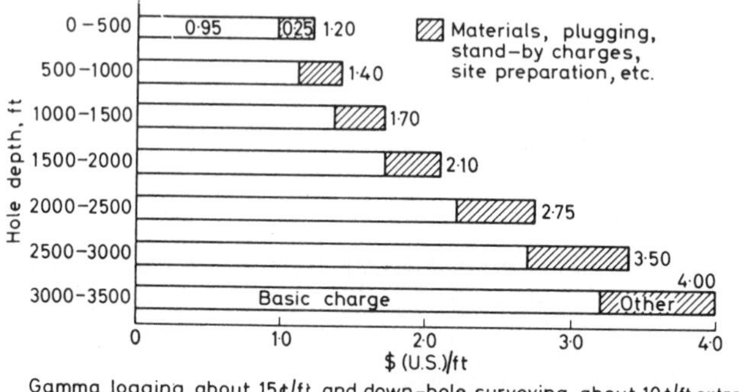

Fig. 8 Rotary drilling cost in Tertiary sandstones, U.S.A.

Exploration costs for the United States are difficult to calculate because of the great number of companies involved and the ambiguity of the information available. The AEC figure of 51 cents/lb U_3O_8 found* for all exploratory and development drilling over the years 1966–69 inclusive infers an average cost of $1.58/ft of drilling. This would appear to be a reasonable figure for non-coring rotary drilling in that the yearly average hole length for the industry varied between 220 and 410 ft in the same period. This is, however, only part of the exploration cost, which also includes administration, support of technical staffs, land acquisition and geochemical and geophysical surveys. One company, which undertakes extensive exploration and whose records could be examined, has a total current exploration cost, covering technical, administrative, land, research and drilling charges, in excess of $3/ft drilled. Over a five-year period its costs total close to $4/ft drilled, and reflect, in part, the high cost of land acquisition early in the programme. An expenditure of $1.50/ft in addition to the inferred drilling charges of $1.58/ft would appear to be reasonable over the past four or five years. The additional expenditure is more or less confirmed by general cost figures supplied by other companies and calculated from AEC statistical data.

The difficulty in calculating an exploration cost per pound for pounds of

*Atomic Industrial Forum, Grand Junction, Colorado, U.S.A., September, 1970.

U_3O_8 found in the United States over the period 1965–70 is partly one of semantics. If all pounds added to reserves in this period, whether found through development or exploration drilling, are termed exploration pounds, then the cost per pound is the average industry cost on a per foot drilled basis, multiplied by the AEC estimate of 94200000 ft drilled in the period and divided by the AEC figure of 330000000 lb found. The cost per pound calculated from these figures comes to 86 cents. The term development drilling, applied to a large percentage of the total footage drilled, represents work on discoveries once a deposit has been outlined and the fill-in drilling begins, and development drilling, from other AEC information, has a relative efficiency, in pounds per foot drilled, about ten times that for exploration drilling. The term also applies, however, to extension drilling of known orebodies. The relative value of the terms may be unimportant as far as pounds added is concerned, but the pounds added to known orebodies do not represent new exploration additions and are not likely to increase productive capability. The same applies to development pounds found at Elliot Lake in Canada, even though the total of the Elliot Lake added pounds is as great as the reserves added from new discoveries.

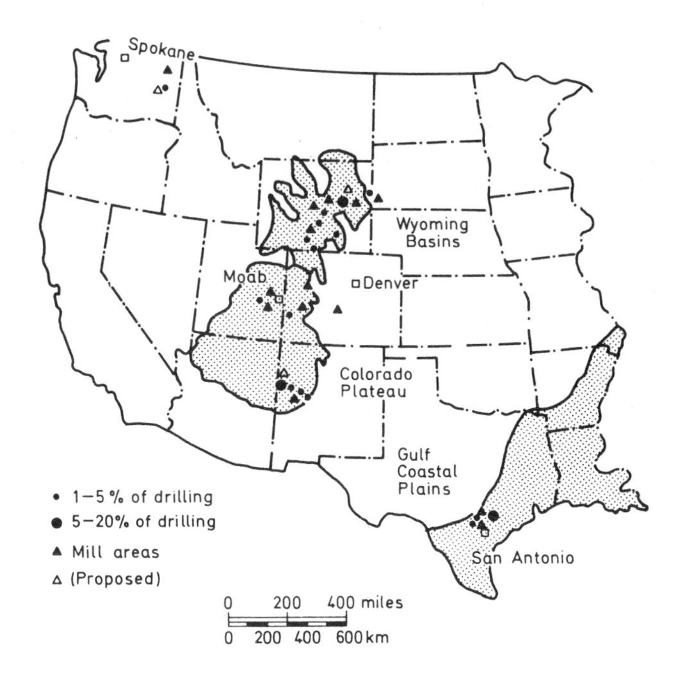

Fig. 9 Areas of United States uranium activity, 1965–70

Ore discoveries in the United States have been classified according to the following types and are in the areas of exploration activity outlined in Fig. 9: (*a*) rank wildcat; (*b*) secondary wildcat; (*c*) stepout; (*d*) deeper find; and (*e*) extension.

A 'rank wildcat' is a discovery in a geologically distinct area in which there has been no previous discovery. A 'secondary wildcat' is an ore find in an area in which a discovery has already been made, but which is off the direct trend of the original discovery. A 'stepout' discovery is made along the recognized geologic trends of a prior discovery, but is located at such distances from it that it cannot be considered an extension of the original find. A 'deeper find' is a discovery in an area of previous discovery, but one where the ore occurs in a zone deeper than the ore zone of the original find. An 'extension' discovery results from exploratory development drilling and, in effect, extends a known orebody.

There have been no 'rank wildcat' discoveries made in the United States since the period of exploration in the 1950s. Of the other four classes of discovery, the percentage distribution of 270000000 lb U_3O_8 added to reserves, for which details are available, was calculated as follows: secondary wildcats, 45; stepouts, 9; deeper finds, 25; and extension, 21.

The balance of about 60000000 lb reported by the AEC is thought to be in extensions to producing orebodies, although there is no breakdown by property or areas in the AEC reports. The current net addition of about 200000000 lb to reserves during the period 1965–70 is an increase of 63% over the 1964 reserve figure, but the addition is calculated to be capable of adding only 30–55% to the productive capability of the uranium industry in the United States. The percentage increase depends upon the level of concentrate prices over the next seven years.

Conclusions

Uranium exploration costs, as calculated by government agencies and the mining industry, are non-interest-bearing finding costs and are, in effect, a measure of exploration efficiency rather than a fully chargeable expense to concentrate. If the full costs were applied to concentrate prices, discovery costs would certainly be one of the main factors influencing exploration philosophy. Exploration incentives, given in the form of pre-tax expenses, tax-free periods, depletion allowances and government-sponsored prospector assistance plans, are continually under review and could disappear. The effect of adverse government legislation in the mineral exporting countries could eventually exert an upward pressure on prices, largely because of exploration expenses.

Uranium exploration in the main producing countries has been geared to the finding of relatively large, high-grade deposits because of the exploration techniques used, high extraction costs and, in recent years, low concentrate prices. Significant changes in either of the latter two factors could ultimately lower the unit finding cost of uranium over the period of forecast high demand.

Exploration for the sedimentary-type uranium deposits is commonly initiated with airborne scintillometer surveys or geochemistry, but, thereafter, geology is the principal means of target selection. For the less predictable vein-type deposits, geophysical methods are applicable at all stages of the search programme, as is geochemistry. Drilling is the method whereby geological ideas are tested and, in part, formulated and, at advanced stages of exploration, not only represents the bulk of the expenditure but is the essential method of assessing occurrences.

The reserves added in Canada and the United States during the period 1965–70 have been both in new deposits and in extensions of known ores. No 'rank wildcat' discoveries were made in the United States, and only one was made in Canada—representing 9% of the North American added reserves.

In the United States discoveries of 330000000 lb U_3O_8 from 1965 to 1970 currently represent a 63% increase over the 1964 reserves, but the additions will increase productive capability by perhaps only 30% of current mill capacity because much of this quantity is ancillary to existing productive units. The cost per pound added in the United States, calculated here at 86 cents, is, from USAEC estimates, approximately the historic average, although the AEC expects a 60% increase in drilling cost per pound as drilling goes to depth. The recent cost per pound added in Canada is about 50 cents—approximately ten times that estimated for the historic period when the reserves at Elliot Lake, Ontario, were found.

The costs of finding uranium are increasing and will continue to increase in the traditional areas of search in North America. The economic discoveries in North America continue to be in formerly proven areas of interest, partly because of concentration of effort and partly because of basic geochemical considerations. Low finding costs in the future are likely to be achieved only in the little prospected favourable areas of the world. Because finding costs are only one factor of several which influence exploration policy, however, the traditional areas of search will continue to receive a major portion of exploration expenditure.

DISCUSSION

Dr. E. E. N. Smith said that it was the practice of the U.S. Atomic Energy Commission, when calculating domestic uranium reserves, to consider exploration and early development costs as sunk costs and, consequently, they were not included in the calculations. Furthermore, there would appear to be no allowance made for profit to the mining companies.

The author said that he would think that exploration costs were taken into account *vis-à-vis* concentrate pricing by mining corporations, although exploration costs of mining companies were hidden. The reserve calculations made by the USAEC, in the $8 class, would possibly not have considered 'sunk' costs in exploration or development but might be based solely on the profitability of a mining operation. The whole question was very controversial during the period of uranium oversupply: certainly, he and his colleagues would not consider that many of the reserves listed by the USAEC were in the $8 class under normal conditions when all costs and a profit had to be returned. Exploration costs were often written off against company profits.

E. W. Grutt said that he was in agreement with Dr. Barnes' figures concerning exploration drilling costs for the future and with his estimates of what it would cost in exploration funds to find a pound of U_3O_8.

Dr. M. Davis said that in referring to the problem of detecting a large number of aeroradiometric anomalies, the author had noted that gamma spectrometry was of great assistance in selecting those anomalies for further examination. He asked if that meant that an anomaly showing up only in the thorium channel would be ignored.

The author replied that, in general, the number of anomalies selected for investigation, on the basis of the ratio of U:Th or U:K, was small, but one must be guided largely by the geology. Favourable anomalies, selected on the above basis, were also found in geological environments which, historically, had not yielded economic uranium deposits. A high total-count anomaly, which also had a high Th:U ratio, might be investigated, but the proportion of current world uranium reserves which would fit that model was indeed small. Experience in the Canadian Shield had shown that the high thorium anomalies were mostly due to pegmatites. Prior to development and mining operations, the economic deposits of uranium at Elliot Lake, discovered through prospecting, would not have been directly detected from the air.

Dr. A. G. Darnley said that the Lower Mississagi Formation (which contained the uranium mineralization of Elliot Lake) provided a recognizable uranium and thorium feature all round the outcrop (and secondary outcrop) which they had covered in experimental airborne surveys. There were associated U:K ratio anomalies, and more localized U:Th ratio anomalies, wherever there were indications of mineralization.

In addition, he would like to question the costs for airborne gamma spectrometry (Table 1, p. 83): he suspected that they were possibly two years old; for the given specification they were perhaps high by a factor of two.

The author agreed that the costs were likely to be half those set out: much depended on plotting charges. (As a result of discussions with Dr. Darnley and checking contract rates for standard data reduction, the cost of fixed-wing spectrometry was reduced by about 50% and by lesser amounts for surveys incorporating magnetometer and electromagnetic equipment. The costs for helicopter surveys were reduced by between 10 and 35%. Amended contract rates were given in Table 1 (p. 83).

R. J. Chico said that at a time when 'money cost reduction' was very important, he wondered what measures should be taken to reduce from $750 000 to $500 000 the budgetary expenditures of an exploration programme, without curtailing quality and the return on investment objectives. He had particularly in mind Fig. 4 (p. 87).

The author said that, primarily, the important feature was to get the money! The major items were costs per man and costs per effective man-day—two important distinctions.

G. R. Ryan said that Geopeko Limited costs for two aerial surveys during 1970, involving about 10000 line miles of flying, were approximately $Aus.12.50 per line mile. That included spectrometer, magnetometer and the production of contour maps. The plane was a fixed-wing twin-engined type. Costs per day for a geologist, field assistant and the logistics (four-wheel-drive vehicle and supplies) were estimated to be about $Aus.200. That included distribution of all overheads on a *pro rata* basis between several parties operating in Northern Australia.

The author commented that for a geologist in a remote area the costs with a helicopter per man-day were $Can.195–199.

Dr. D. A. O. Morgan said that as in most countries in the world expenditure on mineral exploration was allowable against tax (the United Kingdom had become rather a special case under the recent change in legislation), it was legitimate to regard the proportion allowed as having been paid for by the Government. Regarding the balance, accountants were apt to argue that that should earn profits at the average rate achieved by a given company's other investments. In consequence, the accountants could argue further that that proportion of exploration expenditure not borne by the Government should theoretically bear compound interest at the company's average rate of net return (after tax). Without going into details, it was possible to show that, with pertinacity and by carrying out mineral exploration on a sufficiently large scale, there was a good probability of recouping not only the balance not borne by Government but also the compound interest with which it had been notionally loaded.

The author said that the question was largely what the profits would have to be if one were to plough the money back into exploration. Papers on copper by Sir Ronald Prain* were relevant: how exploration for copper went on under proper pricing and the conditions under which it was abandoned. There was a margin, over and above costs, at which exploration would take place and below which it would not take place. Certainly, the investment had to have a return.

*Prain *Sir* R. L. *Selected papers, 1964–1967* (London: The RST Group, 1968), 151 p.

Neutron activation analysis as an aid to geochemical prospecting for uranium

D. Ostle B.Sc., M.I.M.M.
Institute of Geological Sciences, London, England

R. F. Coleman B.Sc., A.R.I.C.
Atomic Weapons Research Establishment, Aldermaston, Berkshire, England

T. K. Ball B.Sc., Ph.D.
Institute of Geological Sciences, London, England

543.53:550.84:553.495.2

Synopsis

A method of neutron activation analysis for uranium is described which is based on the phenomenon of delayed neutron emission from samples containing ^{235}U, following irradiation by thermal neutrons in a nuclear reactor. The technique has been applied to the analysis of a wide range of geological materials collected during the course of a uranium reconnaissance programme in the United Kingdom, including water, stream-sediment, soil, rock and vegetation samples. The technique involves minimal preparation of samples, principally homogenization and weighing, and includes several steps which are automatically controlled. Costs are therefore low, being about £0·20. Analysis is rapid, less than 2 min being required per sample; specific, there being only very slight interference from thorium; and sensitive to about 0·03 μg U—equivalent to 1 μg/l in a 30-ml water sample. Precision is high, with a relative standard deviation of 1% at the 50-μg level, and largely independent of matrix. The accuracy is assessed by reference to analytical data for a number of international rock reference standards.

Examples of application of the method include the results of a regional stream water and alluvium study in northern Scotland, where the potential of such a primary geochemical survey in delineating favourable areas for uranium exploration is demonstrated. In areas where the parent uranium has been partitioned from its decay products, it is shown that neutron activation analysis for uranium in soils may delineate structures more precisely than radiometric gridding.

Geochemical methods were relatively slow to be accepted as an effective aid to the discovery of uranium deposits, and they are still not applied to the same extent as in the search for many other metals. Initially, the reluctance to develop and apply such methods could be attributed to two main factors: the ease of detection of outcropping uranium deposits by direct radiometric means, particularly after the introduction of the scintillation counter with its high degree of sensitivity, and the difficulty of conducting precise determinations of uranium at very low levels of concentration, except under the most sophisticated laboratory conditions.

Even at the early stages of applied geochemistry development, however, which coincided approximately with the post-war boom in uranium prospecting, it was appreciated that the chemical detection of uranium could have useful applications. For instance, Ostle[1] argued that the high solubility of uranium in the oxidized zone would favour its introduction into ground and surface waters at concentrations which might indicate the proximity of uranium mineral deposits but which would not be detectable by instrumental techniques because of the dissociation of the dissolved uranium from its gamma-ray-emitting decay products. An orientation study in the area of the disused uranium–radium property at South Terras, Cornwall, demonstrated the potential of the hydrogeochemical approach, and it has been used extensively elsewhere, with varying success. One of its main advantages lies in the size of the area which can be sampled—a valuable factor in regions which are difficult of access.

More recently, it has become apparent that the establishment of uranium supplies for the future must depend to an increasing extent on the discovery of orebodies which have no surface radiometric expression. Those lying within the zone of oxidation and influenced by groundwater movement, whether hydrostatic or capillary, may produce dispersion anomalies in which partitioning of the uranium and gamma-emitting daughter elements results in only the non-radioactive uranium being detectable at the surface. Although the complete absence of any gamma indication may be rare, the study of the distribution of chemically determined uranium in soils and stream sediments, as well as in water, may serve to indicate the presence of hidden deposits.

Accepting the validity of the geochemical prospecting methods, therefore, the basic requirement for the study of all types of dispersion is a sensitive, rapid, precise, relatively accurate and low-cost method of sample analysis. A number of methods are available (Table 1); but, in view of the variety of materials which

Table 1 Analytical techniques for geochemical analysis of uranium

Method	Field application	General remarks
Fluorimetric		
(*a*) Direct	Yes	Subject to matrix effects; suitable for water
(*b*) With chromatographic separation	Yes	Specific and time-consuming; wide application
Gamma spectrometry	Limited use	Suitable for materials in secular equilibrium; not suitable for alluvium or soils
X-ray fluorescence	No	Subject to matrix effects, but has advantages if multi-element determinations necessary
Colorimetric	Yes	Time-consuming and imprecise at low concentrations
Neutron activation		
(*a*) Radiochemical analysis of fission products	No	Very sensitive and specific, but requires special equipment and staff; slow
(*b*) Delayed neutron measurement	No	Rapid; no matrix effects; suitable for unskilled operators or automation

are analysed in comprehensive geochemical surveys (water, soil, stream sediments, vegetation and rock), it is not surprising that, until comparatively recently, no single method has been acceptable for determinations on all types of sample. Delayed neutron measurement, following neutron activation, adequately meets most of the criteria for an analytical technique applicable over the whole range of geochemical sampling for uranium. A system of analysis employing the Herald reactor facility at the Atomic Weapons Research Establishment (AWRE), Aldermaston, and developed by AWRE in collaboration with the Institute of Geological Sciences (IGS), has been used virtually exclusively for the geochemical aspects of the uranium reconnaissance in the United Kingdom undertaken by the Institute on behalf of the Atomic Energy Authority.

The advantages of the delayed neutron method are discussed later. It is appropriate here, however, to emphasize that its major disadvantage (the need for access to a reactor and the consequent limitation on the location of analytical facilities) requires, at the outset of a programme, an acceptance of the necessity to transport all samples to a single laboratory, which may be remote from the sampling areas. This is not a new problem to uranium exploration geochemistry and opinions differ as to the desirability of there being analytical services immediately accessible to the field geologist. Factors such as cost and rapid turn-around of results are involved, but in the United Kingdom, with good communications, relatively short travel distances and high labour costs for sample preparation, the delayed neutron system of analysis offers advantages which outweigh those of an in-field or regional base laboratory which employs other analytical techniques.

Sampling procedures

Standard field procedures are employed in taking samples for analysis by delayed neutron determinations.

The dimension of the pneumatic sample transfer system at the reactor is such that the size of sample must be limited to that required to fill a 1-oz (30-ml) bottle. In most cases water is collected directly in the 30-ml polyethylene bottles in the field and screw-capped ready for further treatment. On rare occasions filtration is necessary, and this employs a plug of polypropylene or glass wool held in the neck of a polyethylene funnel. Filter paper is avoided because of the ion-exchange properties of uranium with cellulose in soft water. At the field base the screw cap is removed and replaced by a well fitting plug of polyethylene rod, which is then heat-sealed into the neck of the bottle. No further treatment is required prior to analysis. Owing to the very small quantities of uranium being determined, scrupulous cleanliness must be maintained throughout, and there is therefore some advantage in the sealing stage being completed in the field, although this is not essential.

In the event of a scavenging stage being introduced to achieve greater sensitivity, and if the water samples are to be stored for some time prior to treatment, then the pH of the water must be adjusted to $<1{\cdot}5$ to avoid loss of uranium to the walls of the vessel. This precaution is not necessary when the water is hard.

Stream sediments are initially wet-screened through a $\frac{1}{10}$-in mesh nylon sieve. The undersize is then dried at 150°C and sieved on a suitable nylon mesh, normally -100 B.S.S. mesh, a split of 5 g of the undersize being retained for multi-element analysis, where this is required.

Soil samples are taken by auger or spade and dried, sieved and weighed in the same way as the stream-sediment samples. A portable torsion balance is used for weighing, and most of this simple preparation can be carried out in a field laboratory by relatively unskilled assistants.

Since the technique is non-destructive, all samples are available for further investigation following a suitable decay period.

Analysis of uranium by delayed neutron measurement

Uranium analysis by the emission of delayed neutrons, following fission of ^{235}U, has been described by Amiel.[2] The present paper is concerned specifically with the application of the method to the analysis of large numbers of samples arising from a geochemical survey.

^{235}U is the only nuclide found in nature which undergoes fission with thermal neutrons. Some of the resultant fission products—for example, ^{87}Br and ^{137}I—decay by the emission of particles to products which are themselves unstable and immediately decay further by the emission of a neutron. The half-lives of the precursors for delayed neutron emission range from less than 1 sec to about 55 sec. Apart from fission products, ^{9}Li ($t\frac{1}{2}$ = 0·17 sec) and ^{17}N ($t\frac{1}{2}$ = 4·1 sec) are the only nuclides which decay by neutron emission, so for most geological purposes the measurement of delayed neutrons following neutron irradiation can be regarded as specific for uranium.

The conditions for irradiation and counting to obtain maximum sensitivity would require an irradiation about five times the half-life of the longest-lived precursor, that is 5 min, counting beginning immediately after the end of the irradiation. In practice, it is found that an irradiation of 60 sec gives nearly the maximum delayed neutron activity and, at the same time, limits the build-up of excessive gamma-radiation levels arising from activation of the major constituents of the sample. In our system it is not practicable to begin counting until 20 sec has elapsed after the end of irradiation, but it has, in any case, been found to be advantageous to introduce a 25-sec delay for samples of water containing low uranium concentrations in order to reduce the contribution of neutrons due to the presence of ^{17}N. The selected optimum counting period of 60 sec is long enough to record 80% of all neutrons emitted later than 25 sec from the end of irradiation. A longer counting period would result in a larger contribution from counter background activity, such 'noise' being particularly undesirable in the analysis of low uranium concentration.

EXPERIMENTAL PROCEDURE

The samples collected in the field must be in containers suitable for irradiation and free of uranium contamination, because the sample is not separated from the container following irradiation.

For irradiation the sample container is placed in a 'rabbit', made from resin-bonded fabric, which is transferred pneumatically from the laboratory to the side of the Herald reactor core (flux 5×10^{12} n sec^{-1} cm^{-2}). It would simplify operation if it were not necessary to remove the sample from the rabbit before counting, but the possibility of external contamination of the rabbit precludes this modification. As very large numbers of samples are determined, with a wide range of concentrations, it is not considered feasible to avoid slight contamination from time to time and, therefore, a carrier which is separated after irradiation is essential. If the rabbit becomes contaminated, it can be rejected, whereas cleaning of the pneumatic system would be a formidable task.

The loaded rabbit is placed in the pneumatic tube and transferred to the reactor core. On arrival, the timing unit is triggered and all subsequent operations are automatically timed. After 60 sec the sample is returned to the laboratory and the rabbit is opened manually with a tool—to eliminate direct handling and thereby minimize the radiation dose to the hands. The sample is transferred to the neutron detector for a 60-sec count, which automatically commences after a 25-sec delay. At the end of the count the sample is ejected into a shielded store and the neutron count and sample number printed out by an Addo-X printer. The timer has two channels, so a second sample can be irradiated while the first is being counted. With this system samples can be analysed at the rate of 33 per hour.

The neutron counter has been built with a high sensitivity for neutrons but low sensitivity for gamma rays. It consists of six boron trifluoride counters containing enriched ^{10}B (31EB70 from 20th Century Electronics, Ltd., Croydon) which are 2·5 cm in diameter, 31 cm long and form a ring 7·5 cm in diameter in a polythene block 30 cm × 30 cm × 42 cm. The block is surrounded by cadmium foil 0·5 mm thick with an outer layer of polythene 3·5 cm thick to moderate and absorb neutrons from external sources—and thus minimize the background of the detector. The efficiency of the counter is approximately 8% for fission neutrons when biased to reject up to 50 mCi of gamma activity from ^{28}Al (energy, 1·78 MeV). This level of activity is produced by 0·5 g of aluminium under normal irradiation conditions. The background count of the detector is 4–6 counts/min.

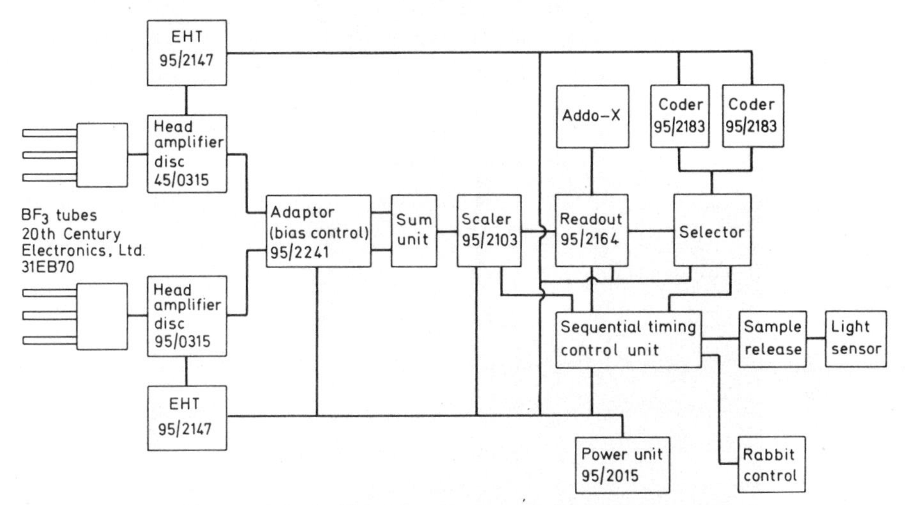

Fig. 1 Block diagram of neutron detection system

The neutron counter was designed and made at AWRE to achieve these parameters; but, whenever possible, commercially available components were used. Most of the associated electronics is from the 2000 Series and the block diagram shown in Fig. 1 indicates the components with the associated model numbers. An overall view of the counting assembly is shown in Fig. 2.

In order to obtain consistently reliable analysis it is essential to perform a few simple checks to ensure the correct functioning of equipment. At the beginning of each day the following procedure is necessary. (*a*) Irradiate and count a standard sample to check the efficiency of the equipment. (*b*) Irradiate and count 0·5 g pure aluminium to check that the gamma rejection of the counter is adequate. (*c*) Irradiate and count either an empty container used for solid samples or a polythene container filled with pure water. The normal background for polythene containers is less than 1 count, so the counter background is recorded. The pure water sample results in a count of about 20 neutrons, mainly due to ^{17}N.

The equipment is then ready for the analysis of routine samples, but throughout the day it is usual to perform about six checks on the background with an empty or water-filled container and to repeat counts on the standard three times.

The uranium content of each sample is simply determined from the relationship

$$\text{Weight of uranium} = \frac{C_S - \bar{C}_B}{\bar{C}_{\text{Std}}} \times \text{weight of U standard}$$

where C_S is the neutron count due to the sample, $\bar{C}_B$ is the mean neutron count for blank samples and $\bar{C}_{\text{Std}}$ is the mean neutron count for standard samples.

Fig. 2 Counting assembly

Reliability of analytical method

INTERFERENCES

The method is subject to very few interferences when used for the analysis of natural materials. ^{235}U is the only nuclide which gives rise to delayed neutrons following irradiation in a thermal neutron flux and a decay period of 25 sec. In most reactors, however, there is a significant flux of fast neutrons, which can cause fission in ^{232}Th, resulting in similar fission products to ^{235}U. Also, the reaction ^{17}O (*np*) ^{17}N gives rise to ^{17}N, which has a half-life of 4·2 sec and decays by neutron emission. In the Herald reactor the sensitivity for thorium is approximately 0·3% of that for natural uranium: thus, only rarely is thorium likely to cause significant interference. The ^{17}N decay results in the emission of 0·7 *n* per gramme of oxygen following a decay period of 25 sec. This is not significant for solid samples, but represents a neutron count of approximately 20 from a 30-ml water sample. Although this blank level can be deducted, it is the limiting sensitivity factor in the direct analysis of natural waters. Other possible interferences due to the absorption of neutrons by the presence of lithium and boron in solution have been discussed by Amiel,[2] but these are unlikely to be of significance in the analysis of geochemical samples.

ACCURACY

The accuracy of the method depends on a number of factors. The method is based upon comparison with a standard uranium solution, and there should be no difficulty in preparing a suitable standard, containing a known quantity of uranium, with an uncertainty in the absolute amount of much less than 1%. For our purpose NBS uranium oxide (U_3O_8) is the starting material. It is important to avoid the use of uranium compounds of unspecified isotopic composition, as many which are available for analytical purposes are depleted in ^{235}U. The sensitivity of the system described is such that about 350 neutron counts are recorded for 1 μg of natural uranium.

For low uranium concentrations in water the accuracy of the method is dominated by the counting statistics (see Table 2). The error quoted is for a single determination based on counting statistics alone and includes in the calculation the error on the mean of six blank determinations. It can be inferred that the limit of detection is approximately 1 μg/l of uranium.

Table 2 Analysis of solution of known concentration of uranium (expressed as μg/l U)

Added	0·7	0·8	1·6	1·7	3·3	3·6	5·7	6·8	8·1
Found	0·9	0·4	1·5	0·8	3·0	4·3	6·8	7·0	8·6
	0·9	1·2	0·3	1·6	3·9	4·7	5·8	7·7	9·6
Error (1σ) calculated	0·6	0·6	0·6	0·6	0·7	0·8	1·0	1·0	1·1

For samples with a uranium content of more than 10 μg the relative standard deviation is approximately 3%. It is possible to reduce this error to less than 1% for samples containing more than 50 μg of uranium if special care is taken over the positioning of the sample in the rabbit so that variations in neutron flux over small distances can be allowed for and standards and samples analysed alternately. A comparison of results by delayed neutron measurement with those obtained by other techniques on rock samples supplied by the U.S. Geological Survey is shown in Table 3.

In order to obtain consistent and reliable results for tens of thousands of samples per year it is important to take stringent precautions against contamination in the laboratory. It has been found desirable to segregate samples so that different sets of rabbits are used for water samples, low-level rock and soil samples and those with a higher concentration of uranium. The most useful indicator of traces of contamination in the system is given by the neutron count of pure water samples used for blank measurements. If these begin to increase, then, almost invariably, it is due to traces of contamination.

ANALYSIS OF SAMPLES CONTAINING LESS THAN 1 μg/l U

Analyses of water samples containing less than 1 μg/l of uranium are of importance in some areas and can be achieved by this method provided that a concentration step is included. This can be most simply effected by first acidifying the sample (say, 250 ml) with dilute sulphuric acid until the pH is less than 3 in order to destroy any carbonate complexes of uranium. This is followed by the addition of ferric sulphate and precipitation of the hydroxide with ammonia. The precipitate is filtered and the uranium content of the hydroxide determined. In Table 4 the results of artificial samples of known concentration are compared, and in Table 5 results by the direct method, by use of a 30-ml sample of natural waters, are compared with those obtained by the ferric hydroxide scavenging procedure.

It is beyond the scope of this paper to discuss the validity of hydrogeochemical sampling based on a limit of detection of 1 μg/l. Experience in a variety of

Table 3 Comparison of uranium analysis of U.S. Geological Survey rock samples*

Ref-erence	Method	Rock sample, uranium in ppm *W1*	*G2*	*GSP-1*	*DTS-1*	*PCC-1*	*AGV-1*	*BCR-1*
3	Average	0·52						
4	γ-spec	0·49	1·32	0·19			1·40	1·20
5	NA		2·16	2·2	0·0032	0·004	2·18	1·81
5	γ-spec		2·1	1·7			1·9	1·6
6	C		1·96	2·2			1·96	1·6
7	DNM	0·62	2·07	2·68	0·004	0·005	2·09	1·80
8	γ-spec		1·96	1·11			2·18	2·17
			1·91	0·27			1·83	1·68
9	I.D.		1·94	2·7	0·004			
10, 11	NA		2·07	2·68		0·005	2·09	1·80
12	F		3·4					
13	γ-spec		2·1	1·7				
14	NA				0·004	0·0049		
15	γ-spec						1·81	2·09
16	I.D.						1·96	1·73
17	F							2·2
18	M.S.	0·46						
Present study	DNM	0·50	2·13	2·87	0·002	0·005	1·94	1·61

*Data from Fleischer[3] and Flanagan.[19]

γ-spec, gamma spectrometry; NA, neutron activation; C, chemical method; DNM, delayed neutron measurement; I.D., isotope dilution; F, fluorimetry; M.S. spark source mass spectrometry.

Table 4 Analysis of uranium solutions after collection of ferric hydroxide (expressed as μg/l U)

Uranium added	0·3	0·3	0·5	0·6	1·1	1·1	2·1	2·1	3·9	4·0
Uranium found	0·3	0·2	0·4	0·6	0·9	0·9	1·9	2·2	3·3	3·9

Table 5 Comparison of direct method with analysis of natural water concentrates (expressed as μg/l U)

Direct method	14	6	3	1	<1	<1
Scavenge	13·4	6·7	2·2	1·6	0·5	0·1

geological environments in the United Kingdom has demonstrated that for reconnaissance purposes such a limit is acceptable since it ensures the delineation of geologically controlled anomalies (e.g. that produced by the Helmsdale granite) and the detection of a high proportion of anomalies arising from discrete concentrations of significant uranium mineralization. In such circumstances the greatest advantage can be taken of the economies of time and manpower resulting from the minimal preparation of samples required for direct delayed neutron measurement. For instance, 70% of those stream-water samples from the north of Scotland reconnaissance area containing uranium in excess of 3 μg/l were shown to lie downstream from a uranium-bearing structure.

Costs

Because of the relative freedom from matrix effects during analysis, the work involved in the preparation of samples for neutron activation is reduced to a minimum. Most of the preparation costs are therefore those common to all analytical systems—the selection of samples in the field, transportation to field base, registration, drying, sieving and weighing. For the latter, tared top-loading balances are used to provide a standardized sample weight of 1 g, which is normally weighed to $\pm$ 2%. The weighing time therefore compares well with that for other analytical methods involving smaller weighings to a similar accuracy.

The true cost of a specific analysis is frequently not appreciated because of the large number of hidden items which have to be incorporated. Several organizations offer activation analysis facilities, although, so far, very few regularly analyse uranium by the delayed neutron method, and there appears to be no regular use of such facilities, outside AWRE, for geochemical purposes. The cost of analysis will be very dependent on the number of analyses carried out. In our own case, in which about 20000 samples per year are analysed, the full cost of a single analysis is about £0·20. If it is necessary to carry out a preliminary separation, as in the analysis of water samples containing less than 1 μg/l of uranium, the cost of a single analysis will rise to 50–75p because of the additional time required for the separation. The operating costs for weighing and sealing are of the order of 5p/sample.

It is apparent even from a comparison between the cost of direct determinations on untreated samples and that of analyses involving only a simple preconcentration stage that the neutron activation technique may represent substantial savings over other systems in an extensive geochemical sampling programme. The very low cost per analysis, amounting to only 5–10% of the cost of occupying a sampling location in many parts of the United Kingdom, enables an economical, but considerable, extension to be made to the coverage which could be justified solely for uranium reconnaissance purposes. Thus, all samples (principally of stream sediments and water) collected by the IGS in regional surveys having prime objectives other than the discovery of uranium are analysed for uranium at little additional cost. The uranium distribution patterns resulting from such routine analysis may contribute the first indications of favourable prospecting areas (see later). Even in circumstances in which analyses for other elements are carried out remote from a reactor facility, the small size of sample required for delayed neutron determinations minimizes the cost of freightage. This analytical technique is therefore well suited to the economical inclusion of uranium in all multi-element regional geochemical surveys.

Examples of application of the method

Advantages in the application of the delayed neutron technique are due principally to its reliability, the lack of interference in geological samples, the speed of throughput and low cost. In the following examples of IGS investigations in which the method has been used, therefore, it must be emphasized that the results could have been obtained by other methods. Within the limits of time and funds set by the programmes involved, however, this could have been at the expense of accuracy and precision. There may be advantages in field or field-laboratory determinations, but it has already been stated that in the opinion of the authors these are far outweighed by the benefits of the neutron activation method.

A very rapid water-sampling programme in the north of Scotland, involving only the collection of samples at road–stream intersections, provided a large amount of valuable data, with the minimum of effort, and served to focus attention on favourable prospecting targets. The results of part of this programme are presented in Fig. 3. The broken line represents the margin of the Newer Caledonian Helmsdale granite mass. All the streams flowing off the granite contain dissolved uranium; with one exception, uranium is undetectable (< 1 μg/l) in drainage from the surrounding schists and gneisses. Subsequent follow-up investigations have confirmed the very high values of uranium concentrations.[20]

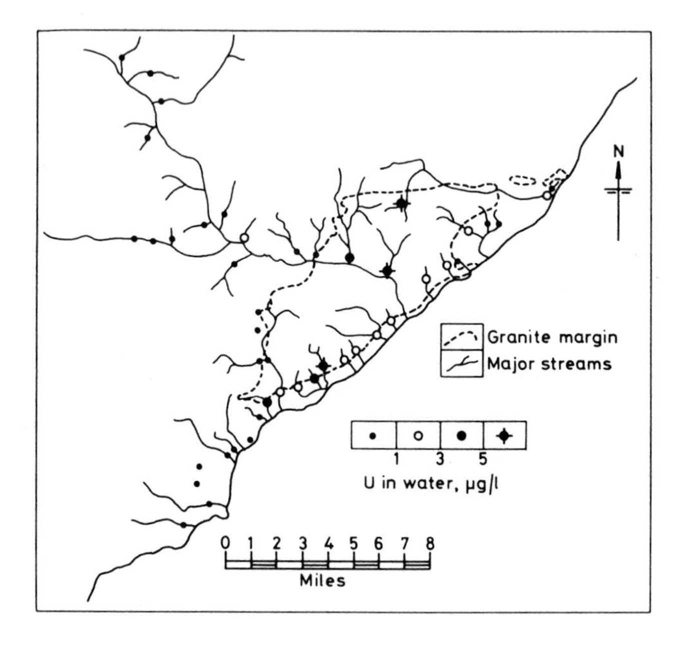

Fig. 3 Rapid roadside hydrogeochemical survey for uranium, Helmsdale, northern Scotland

In the study of the distribution of uranium in the vicinity of surface radiometric anomalies the analysis of soil samples by the delayed neutron method provides adequate resolution of trace values of uranium to show the partitioning of uranium and its decay products resulting from weathering of a uranium mineral occurrence. In the secondary environment radium (and its daughter products, which give rise to the bulk of the radioactivity detected by portable field instruments) behaves differently from uranium, and secondary uranium anomalies may therefore be displaced from secondary radium anomalies. This feature is illustrated by the area shown in Fig. 4, where a thin soil cover of up to 1·5 m overlies a uranium-bearing structure emplaced in slaty hornfelses. The soil was probably originally a forest brown earth, but has been subjected to clearing and ploughing at a later date. The highest radioactivity is located principally in an area in the bottom part of a broad shallow valley, whereas the chemical uranium anomaly is located directly over the vein. In other localities the reverse situation has been observed.

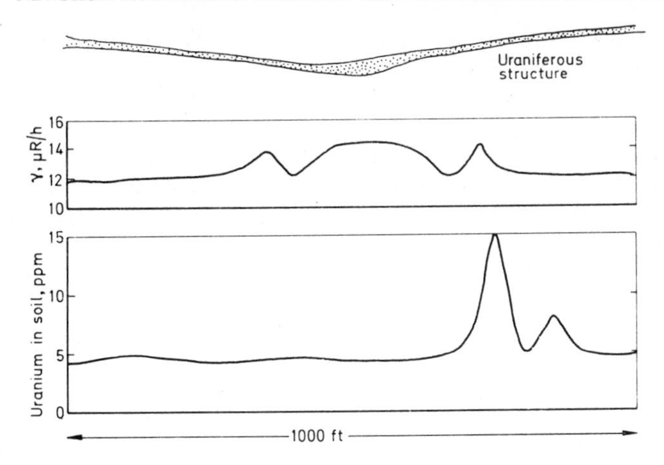

Fig. 4 Profiles of surface radioactivity and uranium in soil showing displacement of radiometric and chemical anomalies. Reverse situation may be observed at other localities

Fig. 5 indicates the use which can be made of stream-sediment and water sampling in the identification of major lithological subdivisions as exploration targets within the geological and climatic environment of northern Scotland. It compares populations comprising about 50 samples representative of areas underlain by Old Red Sandstone sediments, the syenitic complex of Ben Loyal, the Helmsdale granite and metamorphic rocks of the Moine Series.

Despite their inclusion, in part, in the uranium reconnaissance and, in larger part, in a regional multi-element geochemical survey, the Moine Schists have not been shown to contain significant uranium mineralization. Uranium con-

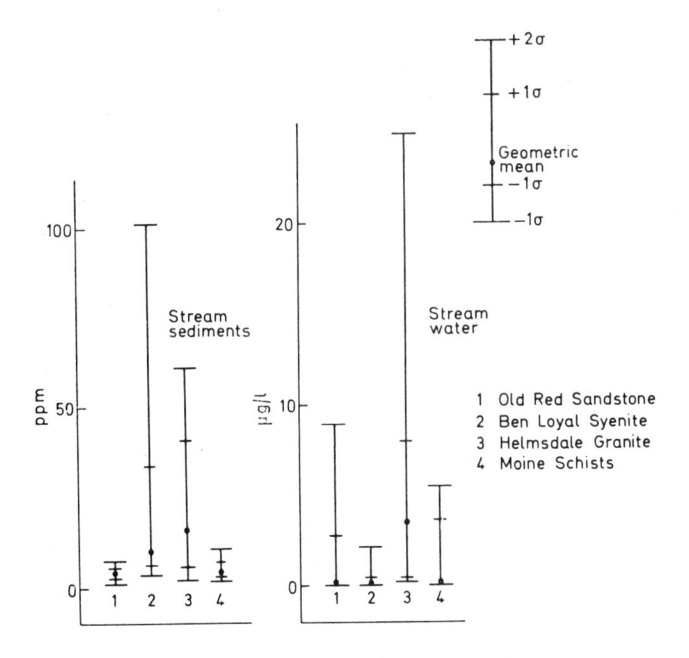

Fig. 5 Comparison of stream-sediment and stream-water uranium populations in four environments in northern Scotland

centrations have, however, been discovered in the Old Red Sandstone, the Ben Loyal massif and the Helmsdale granite. The range of uranium values in samples from the Moine environment may therefore be regarded as background for the region. A comparison of the Moine population with those representing the other major lithologies allows the following conclusions to be drawn: (1) that uranium mineralization associated with the Helmsdale granite is identifiable by reference to the results from either sample type (stream-sediment or water); (2) that the uranium mineralization associated with the Ben Loyal complex is not detectable by water sampling; and (3) that uranium occurrences in the Old Red Sandstone rocks are detectable by water sampling but not by stream-sediment analysis.

The observed distribution can be readily explained by a consideration of the mode of occurrence of uranium in the three lithological subdivisions and by the hydrogeochemistry. Uranium in the Ben Loyal rocks is associated with resistate accessory minerals (thorianite and uranothorianite). In the Old Red Sandstone it is mainly in a readily leachable form and, of prime importance to its detection, it is taken up in water which has a high bicarbonate content: it is, therefore, retained in solution. In the Helmsdale granite area uranium is associated with secondary minerals in addition to resistate minerals. A primary geochemical reconnaissance approach to the region, therefore, would have identified what have, in fact, proved to be the most favourable prospecting target areas, but without the simultaneous coverage by other techniques, which accelerated the discovery of occurrences in the multiple-technique approach adopted.

It is emphasized that the sensitivity of the delayed neutron technique has proved adequate for the purpose of identifying those significant regional variations which can lead to the discovery of uranium deposits.

Conclusions

The discovery of new deposits of uranium will increasingly depend on the detection of highly tenuous dispersions of uranium in superficial materials. Hydrogeochemical surveys, soil and stream-sediment sampling should therefore find wide application to uranium prospecting in the future, and the determination of uranium by delayed neutron measurement following irradiation is a relatively new technique which offers considerable advantages over other analytical systems. These advantages are low cost, specificity, high sensitivity (0·03 μg U), high precision (3% relative standard deviation at the 10-μg level), rapid throughput and independence of matrix for the great majority of geological materials. They far outweigh any disadvantages arising from the need for access to a reactor.

The limit of detection obtainable by direct determinations on untreated samples has been shown to be adequate for effective uranium reconnaissance and follow-up investigation in a variety of environments. Higher sensitivity (with commensurately higher cost) can be achieved by the introduction of a simple pre-concentration stage.

In addition to its value to geochemical programmes aimed specifically at uranium mineral discoveries, the low cost and operational simplicity of the delayed neutron system provide an economical means for the routine analysis of samples collected in broader-based surveys. The resulting data may indicate favourable uranium prospecting areas.

Acknowledgment

D. Ostle and Dr. T. K. Ball wish to thank the Director of the Institute of Geological Sciences for permission to publish this paper.

References

1. Ostle D. Geochemical prospecting for uranium. *Min. Mag., Lond.*, **91**, 1954, 201–8.
2. Amiel S. Analytical applications of delayed neutron emission in fissionable elements. *Analyt. Chem.*, **34**, 1962, 1683–92.
3. Fleischer M. U.S. Geological Survey standards—I. Additional data on rocks G-1 and W-1, 1965–67. *Geochim. cosmochim. Acta*, **33**, 1969, 65–79.
4. Adams J. A. S. Cited in reference 19.
5. Morgan J. W. and Heier K. S. Uranium, thorium and potassium in six U.S. Geological Survey standard rocks. *Earth planet. Sci. Lett.*, **1**, 1966, 158–60.
6. Moore R. Cited in reference 19.
7. Gale N. H. Development of delayed neutron technique as rapid and precise method for determination of uranium and thorium at trace levels in rocks and minerals with applications to isotope geochronology. In *Radioactive dating and methods of low-level counting* (Vienna: IAEA, 1967), 431–52.
8. Wollenberg H. A. Cited in reference 19.
9. Doe B. R. *et al.* Isotope-dilution determination of five elements in G-2 (granite), with a discussion of the analysis of lead. *Prof. Pap. U.S. geol. Surv.* 575-B, 1967, B170-7.
10. Gordon G. E. *et al.* Cited in reference 19.
11. Pleizier G. Cited in reference 19.
13. Lambert I. B. and Heier K. S. The vertical distribution of uranium, thorium, and potassium in the Continental Crust. *Geochim. cosmochim. Acta*, **31**, 1967, 377–90.
14. Hamaguchi H. and Onuma N. Cited in reference 19.
15. Bunker C. Cited in reference 19.
16. Tatsumoto M. Cited in reference 19.
17. McNail E. E. Cited in reference 19.
18. Nicholls G. D. *et al.* Precision and accuracy in trace element analysis of geological materials using solid source spark mass spectrography. *Analyt. Chem.*, **39**, 1967, 584–90.
19. Flanagan F. J. U.S. Geological Survey standards—II. First compilation of data for the new U.S.G.S. rocks. *Geochim. cosmochim. Acta*, **33**, 1969, 81–120.
20. Gallagher M. J. *et al.* New evidence of uranium and other mineralization in Scotland. *Trans. Instn Min. Metall. (Sect. B: Appl. earth sci.)*, **80**, 1971, B150–73.

DISCUSSION

Dr. H. D. Fuchs said that within his organization they had just begun to have some uranium samples analysed by the netron activation method. The results, compared with those obtained by conventional methods, seemed less accurate. He wondered if that were due to a relatively high concentration of uranium (i.e. above 2000 ppm). He asked if the costs of uranium analyses increased drastically if analyses were done for 3–6 other elements, for example, Cu, Zn, Pb, V.

R. F. Coleman replied that high-grade samples could be analysed without loss of accuracy on the Aldermaston equipment. Samples from Elliot Lake containing 0·1% U were analysed regularly; and even much higher grades were sometimes analysed. Suitable control of the analyses was the key to success. Higher-grade samples were simply irradiated for less time. For example, a sample with 0·6% U was irradiated for one-sixth the time for a sample with 0·1%; and that tended to produce approximately the same number of neutrons in each case. Of course, more control and care meant a higher cost per sample and one could take one's choice—low cost and less accuracy or higher cost and accuracy at all grades.

The neutron activation technique could be used for most other metals—some better than others. The most suitable elements in the geological environment were U and Au. He had had little experience of other elements in their geological setting—only in glasses. As to costs, 30 elements in glass would cost about £25 per sample for a large number of samples, but very much more for a small number of samples.

A. Grimbert asked how long it took to obtain the results of the analyses when the work was being carried out in Africa. He also enquired, with regard to the cost of the analysis, if all items were included, or only the labour cost. With neutron activation analysis 'total U' only was determined, but 'mobile U' was of considerable interest in geochemical prospecting.

D. Ostle said that the time taken was largely dependent on communications, i.e. how long it took to get the samples from the field and to send back the results. IGS had no experience in Africa; but, in Britain, the turn-around was very fast. Moreover, the reliability and cheapness of the method must be stressed: the results were much more reliable than those produced in a mobile field laboratory. Also, if one was going to do field analyses, one must pay the salaries and expenses of semi-skilled or skilled staff in the field; and it cost much less to keep a man in London than to send him to Central Africa.

Mr. Coleman said that the speed of analysis depended largely on organizational problems. A single customer could have his analyses 15 min after the sample arrived; but, of course, in practice, one had to wait in a queue, and that was what was time-consuming. The cost of 20p/sample included the labour of one man performing the analysis, the overheads of the laboratory, the cost of the equipment amortized over 5 years and the cost of the laboratory at £9 per square foot. The actual cost of each analysis was only 5p in terms of reactor time.

Mr. Ostle commented that the work was only at the end of the development stage, but they had already found that their results were significant in detecting uraniferous occurrences. Orientation studies were essential to any kind of exploration procedure: their orientation work had shown that to be an effective method of detection.

In reply to questions by L. Løvborg, Mr. Coleman said that Aldermaston electronic equipment could discriminate against gamma rays produced by ^{28}Al at a dose rate of 500 mR/h at 18 in. There was no point in using a lead shield for geological samples of 1 or 2 g in weight: Gale* used a lead shield because he was analysing samples weighing tens of grammes not only for uranium but for thorium. In large samples of rocks with high Al contents the gamma dose rate did become significant. With better electronics one did not even need a lead shield for such samples.

Professor M. Dall'Aglio said that in the trace and ultra-trace analysis of water samples there were two problems, equally important, to the analytical technique adopted—the possibility of contamination and the preservation of the water samples. He enquired the extent of contamination found in water analyses, considering the marked aureoles of uranium in the laboratories. The authors had reported an error, approximately constant, of 0·6–1·0 ppm in a wide range of uranium concentrations, and that might suggest a contamination effect.

Mr. Coleman said that the errors were those arising solely from the counting statistics. At low levels the counting statistics were dominated by the blank from ^{17}N; so they were calculated errors—not measured errors.

No one could ever say that they had never contaminated a sample. They were interested in uranium at the 10^{-10}–10^{-12} g level; and the only thing to do to avoid contamination would be to isolate the laboratory completely and never let any uranium come near it! One had just to be aware of the problem and try to minimize it. The success rate could only be gauged by the number of times a geologist was sent to look for a non-existent anomaly because of contamination of his sample. Trace elements were absorbed on to the sides of containers, but, as the determination was made on the container in which the sample had been collected, that error was eliminated.

Mr. Ostle added that the sample was sealed at an early stage, while still in the field, and so no contamination could be introduced into the sample at Aldermaston.

J. C. Schiltz said that he was working on neutron activation analysis applied to geochemical problems. He thought that among the various methods of neutron activation the delayed neutron method was the best because of its quickness, sensitivity and selectivity. A very short cooling time and the absence of radiochemical separations were important. It was also applicable to other elements, especially in low concentrations, such as rare earths, scandium, gallium, thorium, tantalum, rubidium and caesium.

In reply to a question by R. N. Aitken, Mr. Ostle said that, in the United Kingdom, background values in streams, rivers and lakes were generally below the detection limit of 1 ppb, except under particular conditions of groundwater geochemistry. At that level they became interesting. Results of scavenging work in south Scotland, to see if

*See reference 7 on page 107.

there were natural variation below 1 ppb, had neither been encouraging nor really worth the trouble.

Dr. J. Cameron commented, with regard to the utilization of research reactors in developing countries, that the International Atomic Energy Agency did initiate a small programme a few years ago. It was suggested that research reactors in developing countries having unexplored mineral potential, and generally in tropical areas, could be used for neutron activation analysis in geochemical exploration programmes. Precious metals and uranium were suggested as the principal target metals. Research contracts had been sponsored by the IAEA to develop that work.

Dr. T. F. Schwarzer said that in the hope of using neutron activation as a field method some interesting studies had been made with a ^{252}Cf source to induce neutron activation and render the sample amenable to measurement by gamma-ray spectrometry. In response, Mr. Coleman said that one element which could be analysed by that method was silver. Uranium might have been tried, but it would almost certainly not be successful at low concentrations—for example, at the 1 ppb level in water samples. It might be more successful for soil samples, and a lot of money was being spent to try to make it work.

Dr. H. M. Harsveldt said that in the southeast Netherlands phosphorite nodules occurred at a depth of about 150 m. About 60 m above the nodules water-bearing beds occurred. He asked if it were possible to predict from analyses of water samples the extension of the phosphorite nodules in that region. Water samples could also be taken from the shallower wells, so they could possibly obtain the relevant information more cheaply.

Mr. Ostle said that the method could be tried, but results would be exceedingly difficult to interpret if one were not very familiar with the groundwater regime. Of course, uranium would also have to be present in the phosphate.

W. Dyck said that Atomic Energy of Canada, Ltd., were currently marketing a small reactor called 'Slopoke', with a neutron flux of 10^7–10^8 n/cm^2/sec. He would welcome comment on the relationship between neutron flux and sensitivity.

Mr. Coleman said that the relationship was proportional: if the flux went down by a factor of 10^4 n/cm^2/sec, the sensitivity would go down by the same factor of 10^4.

Dr. Ball, in reply to a question by T. H. Rodger, said that as far as the reproducibility in the long term was concerned, stream waters could be divided very roughly into two components—the groundwater component and that due to rainfall. Rain affected the level of uranium obtained in streams, the more usual effect being to dilute the groundwater component and, thus, to lower the uranium value. The precision of replicate samples taken at the same locality at the same time was related to the counting statistics. With regard to stream sediments, values of about 8% relative standard deviation had been obtained. Better than 6% relative deviation was commonly obtained with soils for 'background' areas, but the variance was somewhat greater in 'anomalous' soils.

Use of geochemical techniques in uranium prospecting

A. Grimbert Ingénieur

Commissariat à l'Energie Atomique, Fontenay-aux-Roses, France

550.84: 553.495.2

Synopsis

Radiometric techniques are not always appropriate to the search for uranium deposits. The use of geochemical prospecting methods can be particularly valuable during the initial stages of exploration, especially in hilly regions and/or in areas difficult of access.

Working methods must include orientation surveys: satisfactory results having been obtained, routine work can then be carried out—followed by interpretation of the results, which is an extremely delicate operation.

The fields of use of geochemical prospecting for uranium vary from the search for and selection of the most promising areas to the location of suspected occurrences. In addition, geochemical prospecting methods may be linked judiciously with other techniques: interpretation, however, must be based on sound geological knowledge.

It is suggested that, with the current interest in the search for concealed deposits, the importance of geochemical prospecting will increase in the future.

It may seem surprising that anyone should try to use geochemical techniques in prospecting for uranium deposits, since such deposits usually show radioactive anomalies easily detectable by radiometric prospecting methods. Moreover, by the *in situ* measurement of gamma radiation from ^{214}Bi(RaC), a daughter product of uranium, it is possible to estimate indirectly the significance of the amount of uranium present without having to collect, prepare and analyse samples, as is necessary in geochemical prospecting.

In practice, however, there are instances when radiometric prospecting does not give completely satisfactory results and for which geochemical prospecting is particularly suitable.

Thanks to the scintillometer, radiometry has succeeded in overcoming the lack of sensitivity which prevented Geiger–Müller tubes from detecting slight changes in radioactivity. Although by the use of portable gamma spectrometers it is now possible to overcome the earlier difficulty of distinguishing ^{214}Bi from ^{208}Tl(ThC″), ^{40}K, etc., disadvantages still remain which are difficult to overcome. Thus, a cover of some tens of centimetres of barren overburden or a thin screen of water can absorb the gamma rays emitted by the uranium series, and uranium separated from its related elements during surface alteration and dispersion is undetectable. Radiometric surveying involves making closely spaced traverses over the area—a long and onerous task to be undertaken during the earlier stages of exploration, particularly in very hilly regions or those difficult of access.

Geochemical prospecting, despite some of its inconveniences (such as the need to take samples and the delays between sampling and obtaining results), has the advantage of being a direct method (a quantitative analysis is made of the element sought) and it allows very widely spaced sampling, appropriate to strategic prospecting, even in humid zones or those masked by barren overburden. It also affords the opportunity of searching simultaneously for several elements, as the samples can be kept indefinitely.

The conditions necessary for the effective use of geochemical prospecting techniques are satisfied in the case of uranium: (1) it is very freely dispersed in hexavalent form in surface layers and gives rise to extensive secondary dispersions, which allow widely spaced investigations of the waters and stream sediments of regions endowed with a sufficiently developed hydrographic network; (2) quantitative analytical techniques for uranium are sufficiently sensitive (0·1 ppb (parts per billion) in water and 0·1 ppm (parts per million) in stream sediments), yield good precision, whatever the conditions or admixtures, and their cost is relatively low; (3) the geochemical behaviour of uranium is sufficiently well known to allow proper interpretation of the phenomena verified.

Geochemical prospecting is thus applicable to different stages of the search for uranium if the climatic conditions are favourable—as has frequently been demonstrated by use of water, stream sediments and soil. Geochemical prospecting in rocks is ignored here, as it represents a different problem, as does that in vegetation, on which there has, until now, been only limited experience.

Working methods

There is no strict method of operating, and there are no precise rules to be followed. The techniques used must be adapted to each particular project to take account of the geology, climate and topography of the environment; the relevant expenditure which will be incurred and the significance of the result will be anticipated. It is, however, necessary to pay attention to the various phases of geochemical prospecting.

Preliminary investigations will confirm that the geochemical techniques are suitable for solving the problem and enable the determination of the best methods to be used to obtain answers in the most effective working conditions and at the lowest cost, known ore deposits serving as a basis. If no uranium deposits exist in the area to be prospected, results obtained in another area, where the same geological and climatic conditions exist can be used as a basis.

Besides the selection of suitable material to be sampled, it is necessary to decide on the best interval between sampling points to detect anomalies of value with the least number of samples, the most attractive way of sampling (surface or depth for soils, active sediments or bank for drainage surveys, etc.), the element or elements for quantitative analysis, as well as the analytical techniques to adopt for the expected composition of the samples (the presence of large amounts of phosphates, chlorides, manganese, etc., may sometimes require the usual analytical techniques to be modified).

The above operation is absolutely necessary because of the risk of undertaking an exploration doomed to failure.

Routine work takes place only when the preliminary results have been judged to be satisfactory and the costs of the operation have been shown to be acceptable: it comprises three phases.

Selected material is *sampled* according to the surveying method decided upon. Samples must be representative of the area and, if possible, collected under identical conditions so that the results obtained are comparable one with another. Each sample must be submitted with full particulars of its source, etc.

Preparation of samples of soils and stream sediments consists of drying them and then disaggregating the indurated clay before they are sifted through a nylon or stainless steel 100-μm mesh to obtain a powder representative of the survey sample. Water samples need no preparation at all, nor filtering or acidifying, which would risk altering their uranium contents. If water samples are not perfectly clear, it is advisable to let them stand for 24 h and only to use the supernate.

Analysis, for which one of several techniques may be used, each having different qualities (neutron activation, fluorimetry, etc.). Fluorimetry is preferred because no large expenditure is involved on equipment or highly skilled personnel. Above all, it provides the opportunity of carrying out analyses near the surveying site, either with the help of a portable laboratory or in a lorry-mounted laboratory, which allows rapid results to be obtained and followed up without delay.

Soils and stream sediments are leached in dilute nitric acid (2·5 N) for 1 h in a water-bath. Uranium and its associated elements are then separated by ascending paper chromatography. After calcinating the part of the paper where the uranium is concentrated, a bead is made up with a mixture of sodium fluoride and sodium carbonate. The fluorescence of this bead is compared under ultraviolet light with that of a standard, with the help of a fluorimeter capable of working from a battery supply. Detection limits are less than 1 ppm with an accuracy of $\pm 15\%$.

For water, the same technique is used after evaporating 10 ml of water on a strip of chromatographic paper. Detection limits are 0·1 ppb with an accuracy of $\pm 30\%$ for weak concentrations.

The daily output per operation is 90 duplicate analyses of soils and stream sediments or 48 duplicate water analyses.

Interpretation of results must allow the recognition of anomalies within the normal distribution of contents, revealing the presence of a deposit and the evaluation of the value of such anomalies, and locating their origin, if possible.

This is a delicate operation, in which personal experience plays a dominant role, and which statistical techniques can help but not replace. Three successive stages must be followed: (1) plotting anomalies on geochemical maps; (2) evaluating their worth; and (3) interpreting in the field, which includes checking and improving on preliminary conclusions. There nearly always follow confirmatory surveys or supplementary sampling, with the aim of specifying certain characteristics of the anomalies.

Inspection of the results obtained will make it possible to decide whether or not the anomaly should be the subject of a follow-up study based on geochemistry or some other technique, e.g. radiometry or geophysics.

These various operations, carried out in conditions appropriate to each project, are effected at the different stages of uranium prospecting, which are examined below to show how they are undertaken in practice.

Fields of use

STRATEGIC PROSPECTING*

The purpose of strategic prospecting is to select the most favourable areas without trying to be precise as to their location.

Orientation surveys have shown that it is better to look for uranium than for associated elements and that the waters and stream sediments of a hydrographic network, which is widely dispersed, are perfectly suitable for this kind of prospecting.

The sampling interval is determined in such a way that the haloes of dispersion in the hydrographic network around an occurrence of average significance are shown by at least three anomalous samples. In France this condition is met by sampling at 50- to 1000-m intervals along the streams, provided that samples are also collected along tributary streams. Active stream sediments are collected as this allows samples to be selected in similar conditions from site to site.

Widely spaced (250- or 500-m) soil surveys are generally only used in addition

*1:200000 to 1:25000 scales.

to drainage surveys in regions with very unfavourable conditions for hydrogeochemical prospecting—for example, areas of very heavy rainfall or flat regions with a poorly developed drainage network.

During *routine work*, on a topographical map of a suitable scale or an aerial photograph, sampling points are selected which, although adhering to the sampling interval to be adopted, allow homogeneous sampling and afford some ease of access. It is the person doing the surveys, however, who finally chooses the sampling point. The travelling required for surveying on a large scale is onerous and it is preferable to take very many more samples than necessary, even if they are not all to be analysed.

In France the network of highways and byways is sufficiently dense for any sampling point selected to be accessible by light vehicle. In areas difficult of access, one can either move along small water courses or regular paths.

Radiometric prospecting, on foot or by vehicle, is carried out on the way to sampling points, where a 60-ml water sample is taken in a polyethylene bag or a 150-g (or so) sample of stream sediment in a Kraft paper bag.

In such conditions the daily output is 15–20 double samples (water and stream sediments) covering 30–50 km^2, according to the density of the drainage system. Samples collected during the day are sent to the field laboratory, if there is one, each evening. If not, they are sent to a central laboratory once or twice a week.

Interpretation of the first stage begins with receipt of results after they have been plotted. This allows the necessary checks and completions to be carried out before climatic conditions have changed significantly. It also allows prospecting to be directed according to the results obtained.

Where the results are clearly abnormal, it is unnecessary to wait before making new surveys in order to remove uncertainties which could affect the value of the conclusions. There must be no doubt when it comes to interpreting the results.

To interpret results obtained from water, use is preferably made of the uranium content of the dry residue calculated from its resistivity. This allows the elimination of variations in metal content, resulting from fluctuations in the flow of water, mixtures of waters of different origins to be taken into account and comparison with waters draining different geological formations. When there is doubt, reference can be made to a complete analysis of the main cations and anions contained in the water so as to compare them only with waters of the same kind.

To interpret quantitative analyses of uranium in stream sediments, despite precautions, heterogeneity of the samples introduces problems because of the varying proportions of the constituents (clay, organic material, oxides, carbonates, etc.) which have different fixation properties in respect of uranium. The results of analyses have a qualitative rather than a quantitative value. Treatment with nitric acid, N/10 sodium fluoride, oxalic and citric acid, etc., which enables the uranium contents of the constituents to be determined, has given encouraging results, but is not appropriate as a routine. Monomineralic phases of the clastic fraction (magnetites, ilmenites, etc.), concentrated by panning, may be examined, but conditions for the use of this technique remain to be defined.

Quantitative analysis of numerous trace elements existing with uranium would certainly allow better interpretation, but would appreciably raise the overall cost of prospecting. It is only in later stages, when the chances of discovering an ore deposit are better, that this step can be justified.

Geochemical prospecting at this stage of the search offers undeniable advantages over other prospecting techniques, including aerial scintillometric prospecting, which involves considerable logistic support and still requires on-site inspection. Thus, it has been possible to discover in an equatorial region of Africa an occurrence which had escaped detection by airborne radiometric prospecting; and, in a part of southwest France, valuable indications which vehicle-borne radiometric prospecting had failed to uncover.

The method is rapid, thanks to a fairly large investigation scale, and economical, since it does not require methods involving significant amounts of material. It can be entrusted to local staff under a specialist.

Finally, it allows harmonious integration of different techniques (geological, radiometric, geochemical) by personnel in contact with the terrain.

DETAILED PROSPECTING*

This is an intermediate stage, which provides, where desired, for the identification of the most attractive zone or zones (of some hectares) within an area of several km² known to be promising and which could become the subject of tactical prospecting.

If anomalies are found as a result of strategic prospecting, it is desirable to know whether it is expedient to use closely spaced radiometry, which, if it provided negative results, would still not rule out the presence of an occurrence. It could, in fact, be that the deposit has no outcrop or is masked by a thick overburden, which makes radiometry ineffective. On the other hand, the movement of underground water can result in the occurrence of geochemical anomalies in surface water or stream sediments. Thus, it is often preferable to try to obtain the maximum information on geochemical anomalies by geochemical techniques before calling on other techniques.

It is necessary, first of all, to delimit the anomaly clearly and to locate its origin within each stream of interest while determining its relations to geology, petrology, stratigraphy and tectonics. The procedure for this is noted below.

Water and stream-sediment samples are taken from all streams in the anomalous zone, progressively reducing the interval (100–50 m) until the desired accuracy is obtained. Soil sampling at the base of the slopes of valleys above high water level is often included in these surveys. Soil is sampled at 50- or 100-m intervals, following the axis of the thalwegs without obvious drainage, where these are considered of interest. In interfluvial zones spring waters and seepages are sampled, in addition to well water, where necessary. If the geology of the region is not well known, geological information on the anomalous zones should be improved.

At this stage of the search a study is made of the distribution of radon or radium, which behave differently from uranium and assist interpretation. Quantitative analysis of radon in spring water has been used successfully in some crystalline areas, but the interpretation of results is very difficult owing to lack of knowledge of the geochemical behaviour of radon. Quantitative analysis of radium in water seems to provide useful indications of the proximity of a deposit because of the low mobility of radium in surface conditions, but this is only applicable in certain circumstances.

If such detailed prospecting only rarely locates uranium-bearing ores, it nevertheless permits the reduction of the zone of interest, the estimation of its value and a better understanding of the problem of the origin of the geochemical anomaly. It is then possible to select the method best suited to the pursuit of further studies, i.e. geochemical prospecting, radiometric prospecting or geophysical prospecting.

TACTICAL PROSPECTING†

The aim of tactical prospecting is to locate a suspected occurrence and to study

*1:20000 to 1:5000 scales. †1:5000 to 1:500 scales.

the extent of an orebody or to find associated bodies.

When alteration is important or barren overburden masks the ores, as well as in marshy zones where radiometry gives unsatisfactory results, geochemical prospecting may provide useful information.

Preliminary investigations have shown that soil sampling is best suited to this phase of searching and that the risk of human pollution is restricted to those occasioned by mining operations and the use of certain phosphate fertilizers. The size of the secondary dispersion pattern which attends a deposit varies from tens to hundreds of metres, according to the significance of the deposit, the geological environment, climate and relief.

A square sampling grid is usually adopted, intervals varying from 10 to 50 m. If there is a prevailing structural or morphological direction, regular sampling can be made along traverses perpendicular to this direction. In general, it is advisable to simplify the geometry of the sampling pattern as much as possible. With the help of a base line it is possible to fix the sampling points, which need not be sited with precision unless they are required for a later check and subsequent exploitation.

Routine work is generally confined to a team of three men, one a technician, who achieve an output of 150 samples a day (covering 6 hectares) if samples are taken at a depth of 40 cm on a 20-m grid, as is most frequently the case. Output falls when the intervals or depth of sampling are increased. The replacement of a manual soil auger by motor-driven equipment has not appreciably improved results, least of all in hilly areas.

Interpretation consists, first, of tracing *iso*-anomalous curves from the results plotted on the map, rendering the geochemical 'relief' with the greatest accuracy possible. The value of the different anomalies detected is then determined after discrimination of non-significant anomalies, and significant locations examined in the field for other characteristics—in particular, their geographic location.

Relief and water circulation have a considerable influence on the displacement of anomalies downstream and it is therefore necessary to guard against hasty conclusions on the origin of anomalies.

It will be possible to effect surveys at different depths of the anomaly and over its extension upstream to obtain indications of the position of the anomalous concentration which has given rise to the anomaly. Increasing uranium content with depth is a favourable sign and may lead to the root of the anomaly. Nevertheless, reduction with depth is not a characteristic which should discourage further investigation since there often exists a strong leaching level in the neighbourhood of the phreatic underground water level, and uranium accumulation on the surface can result from its fixation in surface soil in which organic matter is widely present.

It will, in certain cases, be necessary to add quantitative analyses for radium to the quantitative analyses for uranium to dispel uncertainties about the significance of dispersions. The fast fixation of radium in clays and organic materials causes some shift between dispersions of the radium and uranium, which initially exist together in equilibrium.

Quantitative analysis of radon in soils at the sampling point with the help of an emanometer meets difficulties of interpretation due to variations in results because of atmospheric and morphological conditions. This technique can be used only at certain seasons of the year in special conditions.

In crystalline formations use can also be made of electrical prospecting to identify tectonic features which a geological survey will not have disclosed. At this stage of prospecting, and despite the difficulties encountered, geochemical prospecting of soils can provide useful results in particular cases where other techniques are shown to be deficient.

OTHER USES

There are no limits to the use of geochemical prospecting, provided that preliminary enquiry has demonstrated it to be appropriate. It has so far been possible to use it to resolve problems such as the following. Determination of the uranium-bearing character and value of a radiometric anomaly, whether detected by aerial, vehicle-borne or foot surveys, by sampling stream sediments and soils. The search for blind orebodies through hydrochemical anomalies in underground water levels intersected by wells or boreholes. Evaluation of the worth of mineralized occurrences from the geochemical dispersions they cause in soils. In the absence of significant geochemical anomalies, a spectacular occurrence is very unlikely to indicate an economically valuable deposit; however, the converse is not true and some displaced anomalies may have no nearby ore deposits as their origin.

Other fields will certainly be opened up when knowledge has advanced and techniques have been improved. If it seems difficult to improve surveying techniques, which will always depend on human judgment in the choice of a sampling point and information on its environment, quantitative analytical techniques for uranium can still be made more precise, reliable and accurate without detracting from the method. Quantitative analysis of a dozen trace elements out of each sample will certainly bring about an improvement, but they will not be enough, since each sample has its own individuality, which can only be determined in the field.

It is in the domain of interpretation that improvements are particularly necessary. There also the human factor has an important role to play, and particular attention ought to be paid to the training of geologist–geochemists.

Conclusion

At all stages of prospecting in which it has been used, in very varied geological and climatic conditions, geochemical prospecting for uranium has shown itself to be a useful technique for resolving the most diverse problems when it has been judiciously associated with other techniques. Although considered as a basic method at the strategic prospecting stage, it can also make a contribution to other stages of the search for ore.

It is necessary to stress that geochemical prospecting cannot, by itself, solve all the problems set by prospecting for uranium-bearing deposits, since the interpretation of the results needs the knowledge contributed by other means of investigation—particularly a sound knowledge of the geology.

It has been said, with truth, that geochemical prospecting is more of a way of thinking than a technique. In fact, it is a question of understanding the movement of uranium in superficial and deep zones from a knowledge of its geochemical behaviour, the conditions of the environment and numerous other factors difficult to quantify. Progress will come more from improving the interpretation of results than from the production of sophisticated and burdensome apparatus, which cannot be supplied to each prospecting team. From the latter will follow the centralization of results and interpretation, which will lose all contact with the field. For the moment geochemical prospecting still remains in the realm of the naturalist used to integrating a great number of factors, more than in the domain of the physicist, who is able to go thoroughly into simpler problems. Its importance cannot but grow in the various steps of prospecting to the extent that knowledge improves and interest grows in searching for hidden deposits.

Bibliography

Berthollet P. Method of uranium analysis used by the geochemical section. *France C.E.A. Note* no. 250, 1958. (French text)

Bowie S. H. U., Ball T. K. and Ostle D. Geochemical methods in the detection of hidden uranium deposits. In *Geochemical exploration* (Montreal: Canadian Institute of Mining and Metallurgy, 1971), 103–11. (*Proc. 3rd Int. geochem. Explor. Symp. 1970*) (*CIM Spec. vol.* 11)

Dyck W. Development of uranium exploration methods using radon. *Pap. geol. Surv. Can.* 69–46, 1969, 26 p.

Grimbert A. Les laboratoires de terrain en prospection géochimique. Cas de l'uranium. *Mines Métall.*, 1962, 617–20.

Grimbert A. The application of geochemical prospecting for uranium in forested zones in the tropics. In *Proceedings of the seminar on geochemical prospecting methods and techniques* (New York: U.N., 1963), 81–94, (*ECAFE Miner. Resour. Dev. series* no. 21).

Grimbert A. Evolution et perspectives de la prospection géochimique des gîtes uranifères. In *Geochemical exploration* (Montreal: Canadian Institute of Mining and Metallurgy, 1971), 21–3. (*Proc. 3rd Int. geochem. Explor. Symp. 1970* (*CIM Spec. vol.* 11)

Grimbert A. and Obellianne J. M. Essais de prospection géochimique de l'uranium en pays aride. *France C.E.A.* 2219, 1962, 48 p.

Grimbert A. and Loriod R. La prospection géochimique detaillée de l'uranium. *Pap. geol. Surv. Can.* 66–54, 1967, 167–71.

Morse R. H. Comparison of geochemical prospecting methods using radium with those using radon and uranium. In *Geochemical exploration* (Montreal: Canadian Institute of Mining and Metallurgy, 1971), 215–30. (*CIM Spec. vol. 11*)

Appendix 1

FLUORIMETRIC DETERMINATION OF URANIUM CONTENT IN SOIL

Method

The uranium in the soil sample is dissolved out with nitric acid and the content determined by fluorimetry after chromatographic extraction.

Equipment and materials

Assay balance accurate to 1 mg
Pyrex test tubes, 16 cm × 160 cm
Calibrated dispenser, 5 cm^3
Calibrated dispenser, 4 cm^3
Butane water-bath capable of holding 90 test tubes
Metal carriers holding 30 tubes
Moulded glass tanks
Flat-bottomed jars, 30 mm in diameter, 70 mm high
Micro-pipettes, 0·05 cm^3
Micro-pumps
Epiradiant butane heaters
Butane Meker burner
Metal mesh (Ni,Cr) 130 mm × 130 mm
Fluorimeter–reflector type 'Nucleometre' FPDTU 1, with battery supply
Galvanometer AOIP 10^{-9} A
Pure platinum planchets, 10 mm in diameter, 2 mm deep
Guillotine with stainless steel blade
Chromatographic paper, Whatman no. 1, in 100 mm × 25 mm strips
Nitric acid, 2·5 N
Extraction solvent, TBP + 50% white spirit
Flux capsules
Standard solution containing 1 µg/cm^3 U + diluent

Reagents

Solvent: nitric acid, 2·5 N nitric acid RP (d = 1·33 36 Be), 225 cm^3. Distilled water to make up to 1000 cm^3.

Extraction solvent: mixture of tributyl phosphate + 50% white spirit.

Flux: anhydrous sodium carbonate RP 90%. Sodium fluoride RP 10%.

Standard solution: 1 µg/cm^3 U + diluent. Dissolve 211 mg of uranyl nitrate, $(NO_3)_2$ $UO6H_2O$, analytically pure and accurately weighed, in 100 cm^3 of 2·5 N nitric acid. This stock solution contains 1000 µg/cm^3 U.

Diluent: a solution of 10% ferric nitrate in nitric acid is essential as a diluent for satisfactory chromatographic extraction. The ferric nitrate can be replaced by aluminium nitrate.

Operating procedure

Preparation of samples 1 g of soil sifted through a 150-mesh sieve is placed in a test tube and 5 cm³ of 2·5 N nitric acid is added. The mixture is shaken until the powder is soaked and then boiled for 1 h in a water-bath. When the action has stopped, the tube is again shaken to ensure a homogeneous mixture, which is then decanted and cooled.

Chromatography With a micro-pipette fitted with a micro-pump, 0·05 cm³ of clear solution (there must be no suspended matter in the liquid) is applied to a strip of Whatman paper about 1·5–2 cm from the bottom edge, ensuring that it spreads across the entire width of the paper. The upper limit of the solution must also be as horizontal as possible. The strip of paper impregnated with the reaction solution is then placed in a dish containing 4 cm³ of extraction solution. The same procedure is carried out with a standard solution containing 1 μg/cm³ of uranium for comparison. A strip of unused paper is placed in another dish as a blank. All these operations are duplicated. The solvent is allowed to rise to within about 1·5–2 cm of the top of the paper strip; the part not impregnated by the solvent is removed with a guillotine and a 9-mm strip is then cut off from the top of the impregnated part.

Fluorimetry This strip is placed folded in a platinum cupel and burned under a radiant heater, care being taken to exclude draughts. The ash is covered with a flux capsule and fused on a Meker burner for 5 min (the fusion temperature should be around 800°C). After cooling, the cupel is placed in a fluorimeter and the galvanometer deviation read.

Presentation of results and calculations

For the result to be acceptable, the two readings for the same sample should not differ by more than 20%. The results are given in ug/g, i.e. ppm (by weight). Where V is the mean reading for unused paper, E is the mean reading for standard solution and X is the mean reading for samples

$$\text{U ppm} = \frac{X - V}{E - V} \times 5$$

Appendix 2

DETERMINATION OF URANIUM CONTENT IN NATURAL WATER

Method

Depending on the content of the samples to be analysed, 10 or 200 cm³ is concentrated on a strip of chromatographic paper. The uranium present is separated from the other ions by ascending chromatography and the content is measured by fluorimetry.

Equipment and materials

Support of Whatman no. 1 paper, 25 mm wide × 40 mm high, folded into an 'M'
Whatman no. 1 paper, 25 mm wide, 120 mm high
Butane evaporator
Pyrex glass jars, 30 mm in diameter and 30 mm high
Pipettes, 10 and 20 cm³
Micro-pipettes, 0·1 cm³ with micro-pumps
Glass jars, 30 mm in diameter and 75 mm high
Moulded glass bath, 280 cm × 280 cm × 420 cm
Extraction solvent, TBP, white spirit 50%
Nitric acid, 2·5 N, containing 1% of ferric nitrate
Guillotine
Platinum cupels, 10 mm in diameter
Flux capsules 10% NaF–90% Na_2CO_3
Fusion burner
Fluorimeter with battery supply

Operating procedure

10 ml of the sample to be analysed is evaporated in an evaporating dish, the chromatographic paper being so arranged that, with the top held by a clip, the lower edge of the strip rests on the bottom of the dish. When the paper is completely dry, 0·1 cm³ of the

2·5 N nitric acid solution containing 1% of ferric nitrate is applied to the whole surface previously impregnated. The reimpregnated strip is placed in one jar of the bath containing the extraction solvent, the lower edge of the strip resting on the paper support. When the solvent has risen to about 1 cm from the top of the strip, the advancing edge of the solvent is cut off as for soil chromatography, followed by the same procedure as used for soil samples. For each analytical series the clean papers and standard papers are placed in the bath at the same time as the samples.

Presentation of results and calculations

The results are expressed in ppb. Where X is the fluorimetric reading for the sample, V is the fluorimetric reading for the clean paper and E is the fluorimetric reading for the standard solution

$$\text{U ppb} = \frac{X - V}{E - V}$$

DISCUSSION

D. Ostle noted that the author had referred to analyses of water, stream-sediment and soil samples for radium: he asked (1) if he could indicate the lengths of the dispersion trains for radium in water and stream sediment, relative to those for uranium; (2) how data about the distribution of radium had been applied; and (3) if the analysis for radium was carried out in the field.

The author replied that radium analyses could now be done in a field laboratory; and they had found it useful, particularly in granitic country, in pinpointing the source of a uranium anomaly. In one instance analyses of soils for uranium showed a distribution pattern related to its emission from springs and seepages. Subsequent analyses of the same samples for radium gave, relative to the uranium dispersion, a smaller anomaly displaced upstream, which was related to a small uranium vein. The distance from that vein to the uranium anomaly might be 100 m or more. Clearly, uranium was preferentially leached by waters; but radium stayed nearer the actual source of the anomalies.

Dr. D. A. O. Morgan, with regard to the role that the analysis of associated metals could play in geochemical prospecting for uranium, asked whether any work had been done on mercury. The author replied that none had been done. In the past mercury analysis had been too difficult, but, with the recent rapid progress in developing techniques for mercury analysis, the situation would be reviewed and mercury would probably be used in the future.

In reply to questions by G. R. Ryan, the author said that at −100 mesh there was no dilution due to clastic particles in the sediment, which were generally of coarser grain size. In consequence, the best definition of anomalies was obtained.

They had only analysed other elements where they were of interest in detecting uranium. In France most uraniferous orebodies contained only uranium, so no analyses had been carried out for other elements. An exception was where they wanted to know if uranium in geochemical samples had come from an orebody or had merely been leached from rocks. If other element concentrations were high, it might be concluded that an orebody was present. Analyses for other elements, however, were a matter of judgment related to specific cases—not a general rule.

Dr. R. G. Dodson said that in the Rum Jungle district of Northern Australia geochemical surveys had proved that most radiometric anomalies coincided with Cu–Pb–Zn–Co–Ni anomalies. Surface anomalies of uranium in laterite and ferricrete often had a high manganese content, but such concentrations were believed to be due to enrichment in clay minerals at the surface. A side-benefit of geological survey at Rum Jungle was that the geochemical characteristics of individual lithological units were mostly identifiable, thereby allowing geological boundaries to be drawn in areas mantled by superficial deposits and where drilling had not reached identifiable rock. He wondered what use, if any, was made of geochemical prospecting for uranium at the Arlit deposits in the Niger Republic.

The author said that the applicability of geochemical methods was investigated at a very early stage, but was found to be impracticable. There was no water and very little stream sediment, all of which was coarse-grained. Fossil soils could be used in a small part of the area, but it was much more practicable to use radiometry.

Planning and interpretation criteria in hydrogeochemical prospecting for uranium

M. Dall'Aglio

Laboratorio Geominerario del CNEN, Rome, Italy

550.846:553.495.2

Synopsis

A knowledge of the hydrogeochemical behaviour of uranium is essential to the design of a survey. Environmental factors which control the formation of hydrogeochemical aureoles are considered, and particular attention is given to those climatic conditions in which the detection of uranium aureoles is difficult. The design of a survey involves the definition of criteria for distinguishing anomalous samples, and automatic data processing is an important aid in this respect.

Simple and multiple correlation between uranium and the major elements and the regression analysis of uranium on major elements are particularly useful in discriminating between anomalous and background samples. They are also helpful in demonstrating the geochemical processes responsible for the distribution of uranium in water.

Hydrogeochemistry constitutes an important link between ore genesis and exploration. Case histories related to the detection of negative anomalies are presented and the conditions of formation of some secondary uranium mineral deposits are described for areas in which no primary deposits are known.

Hydrogeochemical prospecting is based on the analysis of natural waters and is aimed at ascertaining the presence, location and quality of orebodies.

The most common application of hydrogeochemistry to mineral exploration is based on the systematic collection and analysis of stream waters. The concentration of uranium in groundwaters tends to reach a value which is roughly proportional to the content of the rocks with which they come into contact. As groundwaters continue on through the hydrological cycle, they tend to converge in the stream waters, and vice versa. In general, systematic sampling of stream waters allows the detection of hydrogeochemical dispersion aureoles.

A complete study of the geochemistry of natural waters should include an examination of the substances which come closely into contact with them. According to Boyle and Garrett,[1] for example, hydrogeochemical methods 'are based on analyses of natural waters and their precipitates, and on stream sediments'. More generally speaking, it is desirable that all the complementary research necessary to understand hydrogeochemical processes be carried out.

The scope of such research should be kept rigidly within economic limits, and the practical benefits must be commensurate with the additional expense involved. On the other hand, it should be stressed that sounder conclusions—and the general economy these imply—can be reached only with sufficient understanding of the natural processes. This is particularly important in hydrogeochemical prospecting.

The circulation of natural waters is the chief cause of the migration of the elements in the supergene environment. This migration is responsible both for ore deposition and the disintegration of orebodies and, hence, for secondary dispersion of elements.

Of all prospecting methods, hydrogeochemistry maintains the most logical link between ore genesis and exploration, because (*a*) it provides information on the circulation of elements in different environmental conditions and (*b*) in the areas explored processes of ore deposition may be occurring. It is principally through hydrogeochemical methods that information is gained on these processes in the region studied.

In order to obtain all this information in a single survey, it is important to understand the geochemical processes, and their intensity in the region studied. In particular, it is necessary (*a*) to gather enough data on the geochemical characteristics and conditions of the water samples examined (for example, the concentration of the major constituents and of some pathfinder elements); (*b*) to employ suitable analytical methods and organizational facilities; (*c*) to have efficient interpretative tools which permit the full utilization of the data collected (i.e. automatic data processing); and (*d*) to obtain exhaustive information on conditions related to the geology, hydrogeology and other environmental factors—especially for the most promising areas.

Hydrogeochemistry of uranium

An overall idea of the geochemical behaviour of the elements in the supergene environment is given by their geochemical balance—that is, by the distribution between rock, hydrosphere and sediments (Fig. 1). The elements considered are clustered around the curve, which represents the exact balance between the igneous rock, the sediments and the hydrosphere. The elements can be subdivided into three principal groups. A few elements pass directly and almost completely from the igneous rock to the sediments (Si, Al, Fe, Ti, Mn, V, Co, Ni, etc.) and these show a very low geochemical mobility. A second group of very mobile elements, which includes Ca, Mg and Na, form ions which are not adsorbed and which are not firmly fixed in minerals. In particular circumstances, e.g. in transitional environments, some metals may pass through an intermediate stage of fixation. Once in aqueous solution, they tend to remain there as stable compounds and a few even tend to become enriched in the hydrosphere (for example, Na, Cl, S and B). In an intermediate position between the geochemically immobile elements of the first group and the mobile ones of the second is a group of minor elements. These have been shown to be suitable for geochemical prospecting. Uranium occurs in the intermediate group of elements along with copper, zinc and arsenic. If an element is non-mobile, only clastic haloes of its minerals are formed. If an element is too mobile, the contrast between water from mineralization and that from non-mineralized areas may not be adequate for prospecting purposes. Mineralogical phases in which the metals are held are also important.

The position of an element in Fig. 1 gives a rough measure of its geochemical mobility in the supergene environment. However, the sea is the major part of the hydrosphere and the average geochemical behaviour of an element indicated in Fig. 1 only reflects in part the dynamics of supergene processes.

Fig. 1 represents the final results of a series of complex processes. The logarithm of the ratio concentration in igneous rock–concentration in sea water (which provides a measure of geochemical mobility) is plotted against the ionic potential of several elements in Fig. 2. Hexavalent uranium has an anomalous position. Its geochemical mobility is much higher than the mobility

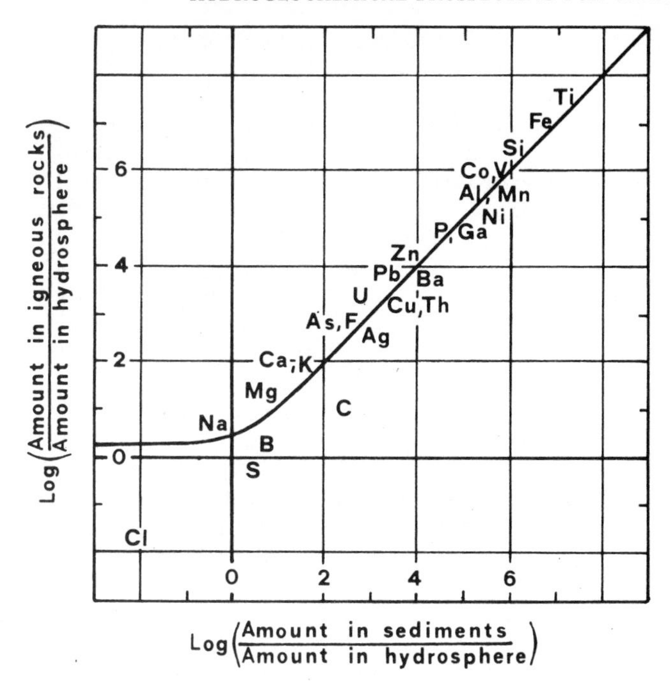

Fig. 1 Geochemical mobility of the elements. Position of an element on the graph gives a measure of its geochemical mobility in the supergene environment. From Dall'Aglio and Tonani[5]

corresponding to the value of the ionic potential. This is because the uranyl, and not U^{6+}, is the stable ion under oxidizing conditions. The uranyl ion has a much lower ionic potential value than U^{6+}. It also forms soluble complexes with the most common anions.

Table 1 shows the concentration of several elements in fresh waters (expressed

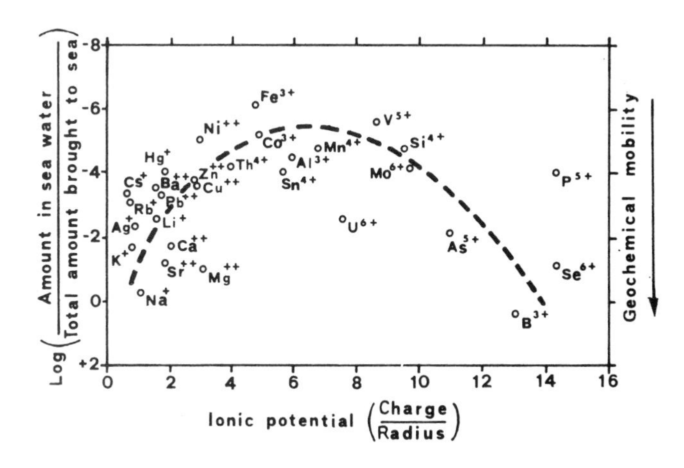

Fig. 2 Geochemical mobility: uranium is in an anomalous position because it shows a higher value of geochemical mobility than that corresponding to its ionic potential. From Garrels and Christ[6]

Table 1 Distribution (ppm) of elements in rock and in fresh water

	SiO_2	Ca	Na	Al	F	B	U	Zn	Cu	Pb	Hg
In rocks	600000	30000	30000	80000	300	10	4	80	30	20	0·02
								150	100		0·1
In dry residue of waters	50000	200000	150000	1000	1000	500	1–10	10	1–5	1	0·1
	100000					1000		30			0·3

as ppm of dry residue) and in rocks. It is apparent that elements such as Ca and Na are highly enriched in water. Uranium is present in water in the same proportion as in rocks, but silica and, to a greater extent, aluminium represent only a small fraction of the salts dissolved in fresh water, even though they are major constituents of rocks.

Thus, the geochemical mobility of uranium is higher than that indicated in Fig. 1 and uranium is considerably more mobile than elements such as Cu and Zn. Hydrogeochemical techniques are therefore extremely useful for uranium prospecting. Uranium is not coprecipitated during the precipitation of common minerals such as carbonates, sulphates and chlorides.

In evaporitic processes uranium is enriched as the precipitation of $CaSO_4$

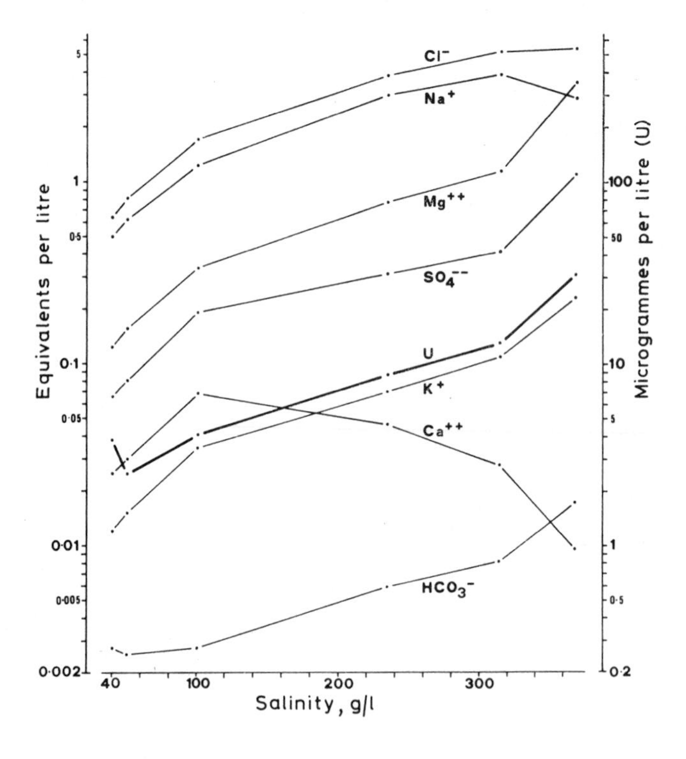

Fig. 3 Uranium and major-element content in brines during solar salt production

and NaCl proceeds. Fig. 3 shows the content of uranium and of the major constituents in brines during solar salt production. The uranium pattern duplicates that of the most soluble elements, such as K and Mg.[4]

In order to apply hydrogeochemical methods, it is necessary to consider the contrast between the element content of the water resulting from leaching of mineralized rock and that of water draining normal rocks. The ratio is very high for uranium, i.e. from 10 000 to 100 000. Normal stream and ground water has a mean content of less than 1 ppb uranium, whereas groundwater, under active exchange conditions and in contact with uranium ores, has a uranium content greater than tens of ppm. Such a high ratio is not reached by other trace metals. For example, the solubility of heavy metals is very limited at pH $\geqslant 7$ by the precipitation of their hydroxides and carbonates. The solubility of fluorine is limited by the precipitation of fluorite. The high solubility of uranium over a considerable range of pH is due, essentially, to the formation of stable and soluble complexes of uranyl ions with the most abundant anions in natural water (carbonate, sulphate and chloride). For uranium to dissolve there must be oxidizing conditions since U^{4+} has an extremely low solubility. Under oxidizing conditions the following processes restrict the formation of hydrogeochemical aureoles of uranium: (*a*) the precipitation of insoluble uranium minerals, such as phosphates, arsenates, vanadates; (*b*) the precipitation of uranium by insoluble hydroxides, such as those of iron and manganese; and (*c*) the precipitation of uranium by organic matter. Thus, precipitation occurs mostly in the first part of the hydrological cycle, when hydrodynamic conditions permit prolonged exchange between water and rocks and soils. Once uranium has dissolved in stream water, the dispersion aureole is stable and persistent for many kilometres.

Despite all these conditions favourable to the application of hydrogeochemical methods to uranium exploration, adoption of these methods has been surprisingly slow and restricted until recently.

Conditions for formation and detection of hydrogeochemical aureoles

CLIMATIC CONDITIONS

A major limitation of hydrogeochemical methods of prospecting for uranium is related to climatic conditions. The formation of hydrogeochemical dispersion aureoles in the vicinity of mineral deposits depends on the leaching of orebodies by oxidizing groundwater.

If hydrodynamic and hydrogeological conditions permit a sufficient water–rock interaction for the formation of dispersion aureoles in the vicinity of ores, stream-water samples can be used to represent parts of the drainage basin. The *temperate climatic* zones favour the development of stable and easily detectable hydrogeochemical aureoles.

In *cold climates* water circulation occurs mainly at the surface. The accumulation of organic matter at the surface limits the formation of aureoles because a cap is formed which is impermeable to oxidation, and the organic matter also tends to precipitate uranium from fresh water.

Desert climates preclude the formation of a regular stream network, the water-table is deep underground, and groundwater movement is dominantly vertical.

In *tropical climates* bedrock is generally deeply altered and leached, and for this reason the aureoles are difficult to detect.

Thus, hydrogeochemical methods can only be applied in areas where the climatic and hydrological conditions are favourable.

SIZE OF AUREOLES

Hydrogeochemical aureoles are characteristically broad features. Best use can be made of hydrogeochemical methods with a sample density related to the dimensions of the haloes. It is wasteful to use a smaller sampling grid. In fact, a direct correlation exists between the mobility of the sampled material and the detail of the information which can be obtained. In general, it is best to use dispersion media less mobile than water for detailed surveys. Stream sediments and soils, for example, are associated with aureoles of smaller dimensions. Thus, hydrogeochemical prospecting should be used mainly for preliminary reconnaissance.

DETECTION OF AUREOLES

Observation indicates that the concentration of many trace and minor elements in natural water varies widely in relation to changes in environmental conditions. Variability, in time and space, is in general greater than that observed, for the same elements, in soils and stream sediments. This has retarded the application of hydrogeochemical methods to uranium exploration. For uranium, the variations found are attributable to well defined geochemical processes. Simple and multiple correlation is carried out between the concentration of uranium and that of other constituents dissolved in water. In this way most of the variability attributable to environmental factors can be eliminated.

ANOMALIES NOT RELATED TO MINERAL DEPOSITS

Environmental conditions may cause anomalies which are unrelated to mineral deposits, and the latter may give rise to only very weak aureoles or none at all. The first possibility is not very serious since the inclusion of one or two additional anomalous areas among those to be subjected to more detailed research would represent only a slight economic burden. Hydrogeochemical anomalies for uranium which are unrelated to mineral deposits are easily identified at the interpretation stage if sufficient information is available on the hydrogeochemistry of the area explored. The second case is more serious and occurs in general when an orebody, although close to the surface, is not subjected to erosion. Special cases in which the regional approach will not be effective can be identified from morphological and environmental factors and must be investigated in greater detail. Experience, however, has proved that this is not always possible.

Even if conditions are generally favourable for the application of hydrogeochemical methods, there may be limited areas in which a regular drainage system has not been developed. This may be the case for hills near the coast.

Planning of hydrogeochemical surveys

GENERAL

The planning of hydrogeochemical surveys is not readily described in general terms. It depends on the facilities available, and on logistic and environmental conditions. Success largely depends on the experience and skill of the project manager. Planning should try to anticipate all other stages, and, in particular, define the interpretative possibilities of the data to be collected. Some of the characteristics of hydrogeochemical haloes of uranium which are important in planning have already been described.

To optimize the exploration, the limits and advantages of the technique should be considered by (*a*) comparing it to the characteristics of other exploration methods and (*b*) relating it to the environmental and geological conditions of the area studied.

In uranium exploration the possibilities of using radiometric methods must be considered. The choice between geochemical or radiometric methods depends on the aim of the exploration and on environmental conditions in the exploration area.

The scale of exploration is a very important factor. Detailed research is under the control of the geologist, who has at his disposal prospecting techniques, such as radiometry, which are equally or even more efficient than the geochemistry.

Hydrogeochemical prospecting is best applied to surveys on a large or intermediate scale, that is from 10 to 2 km on the stream network. For the purpose of preliminary reconnaissance surveys, hydrogeochemical prospecting is the quickest and the cheapest method.

Geochemistry is very useful in studying particular problems, such as, for example, the detection of the negative anomalies which can indicate that ore-bodies are now forming, the study of primary haloes, the study of the causes and conditions for the formation of secondary minerals, etc.

PRELIMINARY RECONNAISSANCE GEOCHEMICAL SURVEYS

Prospecting on a regional scale is based on the systematic collection and analysis of stream-water and stream-sediment samples. A mean density of one sample per 5–50 km^2, which corresponds to one sampling site every 2–10 km, is maintained. The aim is to determine anomalies related directly to mineralization, thus establishing new favourable areas. In this type of survey the heterogeneity of the local and environmental conditions is reflected in the variation of the uranium content of the water samples.

Interpretation criteria in hydrogeochemical exploration

GENERAL

In hydrogeochemical prospecting the frequency distribution of an element is not a suitable criterion for the selection of anomalous samples. The term 'anomalous' indicates a content which differs significantly from the norm. The frequency distributions for elements dissolved in waters are of a polymodal type, and valid criteria are lacking for assigning samples to the statistical populations which make up the observed frequency distribution. A sample may be normal or anomalous according to the statistical population it belongs to.

This difficulty arises for the following reasons: (*a*) the composite character of natural water—which comprises a mixture of waters which have undergone natural processes of different kinds and intensity; (*b*) the existence of natural processes which can cause an increase in the content of the element sought, even in the absence of mineral deposits. These processes include the leaching of geological formations which contain different amounts of the metal sought and differences in environmental conditions. The content of the mobile trace elements increases as the water proceeds along the hydrological cycle. This increase is accompanied by a corresponding increase in major-element content. Moreover, the presence of some elements may indicate that geochemically important processes are taking place; for instance, a high boron content may indicate contribution from hydrothermal fluids.

Simple and multiple correlations between uranium and other elements in solution provide objective criteria for the definition of anomalous samples. For a correlation to exist in surface waters between the content of one minor element and some of the major constituents characterized by high geochemical

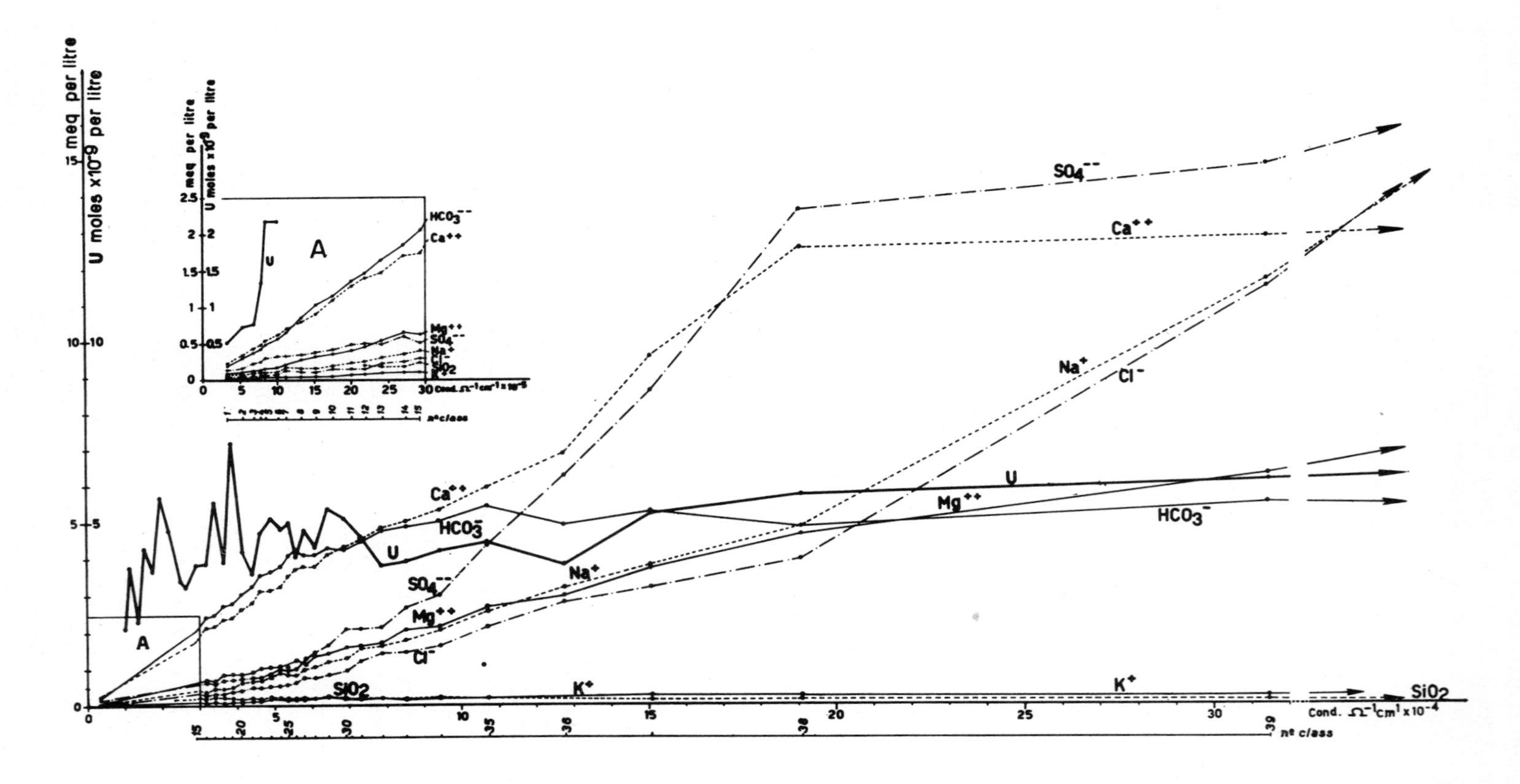

Fig. 4 Pattern of uranium and major elements as a function of electrical conductance value. Data obtained from automatic processing of approximately 4000 stream water samples in Italy

mobility, two conditions must be met: (*a*) in the leached rocks the content of the element considered must be related to a particular chemical or mineralogical composition; and (*b*) the geochemical mobilities of the elements considered must be of the same order. In this respect uranium is a suitable element because solution proceeds with that of other constituents. Also, its geochemical mobility is high enough to make the correlations with the other constituents of the waters valid for long distances and/or periods. The distribution of uranium in water is regular in comparison with the distribution of the major and some of the minor constituents.

Fig. 4 shows the content of uranium as a function of electrical conductivity and of composition. It is based on results for about 4000 samples of stream water collected on a systematic basis in many areas of Italy. All samples are arranged in order of increasing electrical conductivity and then subdivided into groups consisting of 100 terms. For each of these groups the mean and the standard deviation of the elements dissolved in the water are calculated. Each point on the chart therefore represents the mean value of 100 determinations. The concentrations of all elements are expressed in milliequivalents per litre, except that of silica, which is expressed in millimoles per litre. Uranium is in moles $\times 10^{-9}$ per litre.

At low concentrations the uranium content increases rapidly with increase in salinity. The first part of the diagram refers to the water collected mainly in the Alpine Range. Examination of the proportions of uranium in relation to that of the other constituents indicates that under such conditions both increase together as the water proceeds along the hydrological cycle. Calcium bicarbonate is the major constituent.

With further increase of salinity the uranium content diminishes slightly, whereas the concentration of calcium sulphate increases. The decrease depends on the fact that in this range of salinity numerous water samples derived from the leaching of marine formations, containing gypsum and salt, are included. The solution of these minerals, which are poor in uranium, causes a considerable increase of the salinity with practically no increase in uranium content. Salinity increases further as greater amounts of gypsum and salt are dissolved, but the uranium content remains practically constant.

The curve of the uranium concentration varies very regularly as a function of the electrical conductivity, and, thus, of the composition of the waters.

In order to classify each sample of water, the concentration of all the major constituents dissolved in the waters, namely Ca, Mg, Na, K, HCO_3^-, SO_4^{2-}, Cl^- and SiO_2, is determined. All the processes to which a water sample has been subjected, including those due to the various lithological, environmental and hydrodynamic conditions, leave an indication of their effect on the concentration that each element achieves.

AUTOMATIC PROCESSING OF HYDROGEOCHEMICAL DATA

Considering the major constituents as independent parameters and the uranium as the dependent parameter, random variations of the latter can be considerably reduced. The data processing involved in the interpretation of data obtained in the course of a survey of part of the Peloritani Mountains (eastern Sicily) is considered as an example of such statistical treatment. A suite of 242 samples of stream water was collected on a regular grid over an area of about 3000 km². Table 2 shows the values of the linear correlation coefficient between uranium and each of the major constituents of the samples. On the basis of these results, which indicate a positive correlation between uranium and the other constituents, a linear function in the multiple-regression analysis of the uranium on

Table 2 Value of correlation coefficients between uranium and major constituents

	Ca	Mg	Na	K	HCO_3	SO_4	Cl
Uranium	0·644	0·209	0·510	0·558	0·243	0·644	0·450

the major constituents was adopted. The type of equation and the results of the calculation are given in Table 3.

Table 3 Results of regression analysis

$$U = A_0 + A_1[\mathrm{Ca}] + A_2[\mathrm{Mg}] + A_3[\mathrm{Na}] + A_4[\mathrm{K}] + A_5[\mathrm{HCO_3}] + A_6[\mathrm{SO_4}] + A_7[\mathrm{Cl}] \qquad (1)$$

$A_0 = 0{\cdot}00000074$	$A_4 = 0{\cdot}0189 \pm 0{\cdot}000027$
$A_1 = 0{\cdot}0038 \pm 0{\cdot}0000010$	$A_5 = -0{\cdot}0031 \pm 0{\cdot}0000010$
$A_2 = 0{\cdot}0027 \pm 0{\cdot}0000010$	$A_6 = -0{\cdot}0018 \pm 0{\cdot}00000096$
$A_3 = 0{\cdot}0026 \pm 0{\cdot}0000017$	$A_7 = -0{\cdot}0022 \pm 0{\cdot}0000013$

		DEGREES OF FREEDOM
SUMS OF SQUARES DUE TO REGRESSION	$= 22{\cdot}4 \times 10^{-10}$	7
SUMS OF SQUARES RESIDUAL	$= 21{\cdot}2 \times 10^{-10}$	234

F RATIO $= 35{\cdot}25$

LOSS IN SUMS OF SQUARES FOR DELETED VARIABLES

$A_1 = 13{\cdot}4 \times 10^{-11}$; $A_2 = 66{\cdot}3 \times 10^{-12}$; $A_3 = 36{\cdot}1 \times 10^{-12}$;
$A_4 = 12{\cdot}0 \times 10^{-11}$; $A_5 = 88{\cdot}6 \times 10^{-12}$; $A_6 = 32{\cdot}6 \times 10^{-12}$;
$A_7 = 35{\cdot}0 \times 10^{-12}$

R SQUARES $= 0{\cdot}513$ $\qquad R = 0{\cdot}715$

Fig. 5 shows the final results in a diagram displaying the observed uranium values versus the values computed on the basis of the equation (see Table 3).

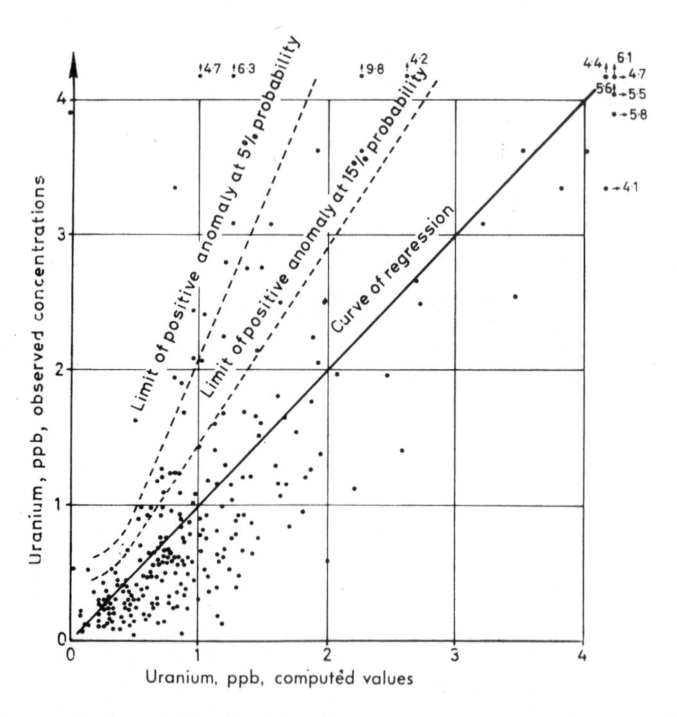

Fig. 5 Multiple-regression analysis (see Table 3) of uranium on the major constituents. Stream-water samples from Sicily

Samples which lie outside ± one standard deviation of the regression curve are shown as anomalous. A highly significant value of the multiple correlation coefficient, i.e. 0·716, was obtained.

Recently, more efficient calculation procedures have been developed.[3] Other methods of data processing which are of great help in drawing objective conclusions from hydrogeochemical surveys include trend and factor analysis.

In the writer's opinion, correlation and regression analysis are the most effective means to a better understanding of the geochemical processes which are the basis of the experimental observations.

URANIUM ANOMALIES UNRELATED TO MINERAL DEPOSITS

The conditions of formation of hydrogeochemical aureoles unrelated to mineral deposits and which occur in the metamorphic basement of the Alpine Range have been described elsewhere.[2]

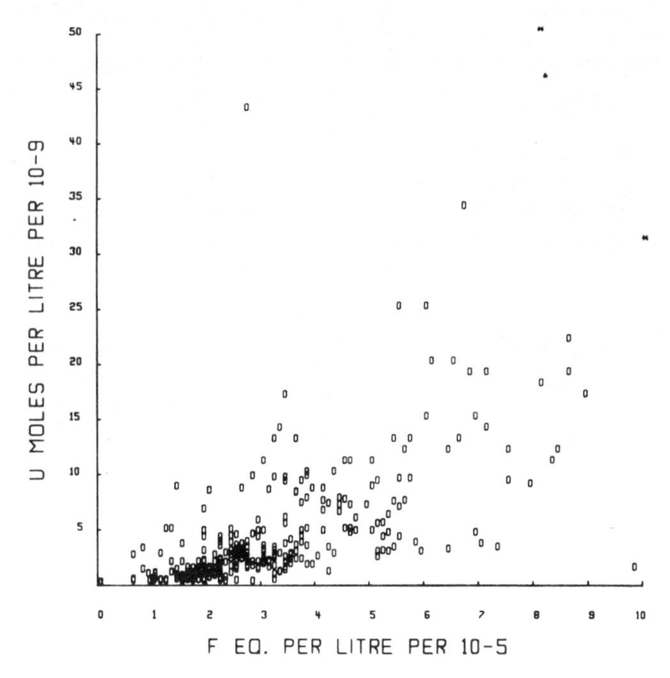

Fig. 6 Uranium–fluorine diagram plotted directly by computer. Stream-water samples from eastern Sicily

Fig. 6 shows the correlation between uranium and fluorine in water samples collected from several geochemically anomalous areas of the Peloritani Mountains. The existence of a significant positive correlation ($R=0{\cdot}36$) between the two elements indicates that particular environmental conditions, namely the presence of mylonitized pegmatite, and not that of uranium ores, is responsible for the anomalies in the waters. More detailed field and laboratory studies, including leach tests on rock samples from various geological formations, substantially confirmed this hypothesis.

NEGATIVE ANOMALIES

Knowledge of uranium circulation in regions studied has shown that minerogenetic considerations can be an aid in field exploration. In the Latium Region of Italy, for example, hydrogeochemical studies showed that the uranium concentration in circulating water may reach 10 ppb and that in these conditions secondary uranium mineral concentrations may occur in porous formations where reducing conditions exist. Follow-up investigations may also be aided by the identification of negative anomalies.

For some springs, which leach uranium-rich volcanics, the concentration of uranium has been observed to be very low. This suggests that they may have deposited, and may still be depositing, uranium underground. The application of this concept has led to the discovery of new uranium orebodies.

HYDROGEOCHEMICAL STUDIES ON THE FORMATION OF SECONDARY MINERALS

Hydrogeochemical studies elucidated the processes of autunite formation on the Sila Plateau in Calabria. Autunite mineralization had been known to exist in Calabria for many years and it had formed the object of several investigations designed to determine the existence of a primary deposit. Hydrogeochemical studies indicate that autunite is present only in granite where water has a very low carbonate and sulphate content (approximately 0·2 meq/l). If the content of phosphates, calcium and carbonates and the activity of the uranyl ion are taken into account, water of this type may be considered to be saturated with respect to autunite. This applies even if the uranium contents are as low (less than 1 ppb) as those found in the waters in question. The formation of autunite is therefore caused by the local hydrogeochemical conditions and not by exceptionally high uranium contents. In such circumstances assumptions of the existence of a primary deposit may be erroneous.

Conclusions

The general principles which form a basis for the application of hydrogeochemical methods of uranium exploration have been considered and an attempt has been made to provide a brief review of the problems connected with the planning of hydrogeochemical surveys for uranium. The limits imposed on the methods by environmental and geological conditions are emphasized, since rational planning must take these into consideration, as well as the possible advantages offered by other methods. In exploration for uranium it is necessary to keep constantly in mind the effectiveness of radiometric methods.

In considering the interpretation of hydrogeochemical data, with particular reference to criteria and procedures adopted in Italy, emphasis is placed on the value of examining simple and multiple correlations between uranium and other constituents dissolved in the water. Automatic data processing is not a substitute for the geochemist's interpretation; rather, it constitutes assistance in applying more rigorous criteria to a study of the geochemical processes.

With regard to the respective roles of the geochemist and the geologist in mineral exploration in general, and in hydrogeochemical exploration in particular, it is worth expressing the view that, in the planning stage, and to a greater extent in interpretation, the geologist and the geochemist must operate in close collaboration. Only in this way can the results of the exploration be optimized. Although there is general agreement with this view, it is still common practice for surveys to be conducted either by the geochemist, without an adequate appreciation of the geological factors, or by the geologist lacking full knowledge of the chemical processes involved.

References

1. Boyle R. W. and Garrett R. G. Geochemical prospecting—a review of its status and future. *Earth Sci. Rev.*, **6**, 1970, 51–75.

2. Dall'Aglio M. A study of the circulation of uranium in the supergene environment in the Italian Alpine Range. *Geochim. cosmochim. Acta*, **35**, 1971, 47–59.
3. Dall'Aglio M. L'applicazione della statistica alla prospezione geochimica: la selezione dei campioni anomali. In *Studi di probabilità, statistica e ricerca operativa in onore di G. Pompilj* (Gubbio: Oderisi, 1971), 278–87
4. Dall'Aglio M. and Casentini E. The distribution of uranium between precipitates and brines in the solar salt plant of Margherita di Savoia. *Boll. Soc. geol. ital.*, **89**, 1970, 475–84.
5. Dall'Aglio M. and Tonani F. Metodi di prospezione idro-geochimica. *Studi Ric. Div. geominer. CNRN*, **3**, 1960, 32 p.
6. Garrels R. M. and Christ C. L. *Solutions, minerals, and equilibria* (New York: Harper and Row, 1965), 463 p.

DISCUSSION

E. W. Grutt enquired about the presence of phosphate or arsenic minerals in the area of investigation. Concentration or retention of uranium in the zone of oxidation as phosphates and/or arsenates could be an intermediate stage in the process of leaching from source rocks and eventual concentration by reduction in deeper parts of the aquifer beds. Such oxidized uranium minerals were known to exist in the Spokane, Washington and Gas Hills areas.

The author replied that no work had been done on the arsenate content of water. With regard to phosphate, there was no significant variation in the phosphate content of the waters analysed. The formation of secondary minerals, for example, autunite, occurred where fresh waters had a low content of complexing compounds of uranium.

Dr. L. Haynes said that in southwest England, where soluble arsenic in the form of löllingite ($FeAs_2$) was present in lodes, and uranium occurred in the Hercynian granites, correlation of high uranium (up to approximately 50 ppm) and As (up to 1500 ppm) in stream sediments could occur. That might be due to complexing of the elements and coprecipitation in the secondary environment. Manganese and iron were not in particularly high concentrations, which would seem to rule out scavenging as the reason for the association.

P. H. Dodd asked if correlative studies had been made to determine the isotopic ($^{238}U/^{234}U$) composition of uranium in the hydrogeological environment. For instance, Russian work* had shown that the proportion U^{234}/U^{238} was useful in arid regions to indicate the existence of ore deposits as against high enrichments from normal background sources. J. N. Rosholt (personal communication) had found high ^{234}U concentrations in waters 'leaching' from the Sweetwater Granite and from Tertiary sediments derived from the granite.

The author replied that there had been no studies based on that ratio in Italy.

W. Dyck said that a spring in the park north of Ottawa contained 10 000 pCi/l of radon, but no detectable amounts of uranium: yet, within a few hundred feet, uranium concentrations of up to 0·1% had been found. He wondered if the author also found high radon contents in his springs.

The author replied that there were high radon contents in the waters from Latium, but interpretation of the radon contents in springs with high CO_2 partial pressures (and other gases) was not easy.

R. J. Chico noted that in the Wyoming Basins not only was U_3O_8 being determined from the borehole cores but also V_2O_5 and selenium in ppm (see Fig. 1, p. 77). In addition, sulphate contents within and outside the 'roll fronts' had been measured. Initial data showed that the mineralized uranium-bearing sandstones in the 'roll fronts' contained about twice the amount of sulphate (ppm) compared with sulphate contents outside them. (Maximum sulphate reached 15 000 ppm, as against a minimum of 7000 ppm.) It might be of interest, as Mr. Grutt had suggested, to follow up that type of sulphate and carbonate mineralization in the roll fronts. Uranium 'rolls'

*Syromyatnikov N. G., Ibraev R. A. and Mukashev F. A. Interpretation of uranometric anomalies in arid regions by means of the U-234/U-238 isotopic ratio. *Geokhimiya*, 1967, 834–41; *Geochemistry intn.*, 1967, 697. (Abstract) possibly, but not certainly, formed at the groundwater level where bicarbonate and sulphate waters mixed. Theoretical considerations concerning that argument had appeared elsewhere.*

The author said that he would stress that the precipitation of uranium was not explained by a decrease in the content of bicarbonate ions because, at low pH, there was a sulphate complex of uranyl ions which was sufficiently soluble. In the case of the 'roll' formations, as in the case of uranium precipitation, the principal agent was always a change in Eh conditions.

*Hagmaier J. L. The relation of uranium occurrences to groundwater flow systems. *Earth Sci. Bull., Wyoming geol. Ass.* **4**, June 1971, 19–24.

Instrumental techniques for uranium prospecting

J. M. Miller B.Sc., F.G.S.

Institute of Geological Sciences, London, England

W. R. Loosemore B.Sc., A.R.C.S.

Atomic Energy Research Establishment, Harwell, Didcot, Berkshire, England

539.16.08:550.83:553.495.2

Synopsis

A review is given of uranium exploration field instrumental techniques, discussion of the underlying physical principles and descriptions of currently available instrumentation being included.

Examination of the natural radioactivity emitted by rocks precedes consideration of the problems of estimating their radioelement content and the advantages and limitations of the various types of detector. The techniques described are total radioactivity and gamma-spectrometry survey by foot, car and aircraft, borehole logging and radon determination in soil and water. Descriptions of techniques include their field of application in different climatic regimes, the optimum sizes of detector and the results to be expected from their use, and details of facilities which it is desirable that the instruments should incorporate. Attention is drawn to the different ways in which instrumental problems have been solved in various countries.

An attempt is made to foresee future development in this field, stress being laid upon the probable increasing importance of neutron activation, X-ray fluorescence and cryogenic semiconductor detectors.

This review is restricted to radiometric techniques, geochemical methods being covered elsewhere in this *Handbook*; it is appropriate to begin with a brief consideration of the radiation emitted by uranium and thorium.

Radiation from uranium and thorium

The naturally occurring radioisotopes ^{238}U, ^{235}U (99·3 and 0·7% of natural uranium) and ^{232}Th (100% of natural thorium) are all long-lived alpha emitters, which, with their decay products, contribute the greater part of all the naturally occurring radiation of terrestrial origin—the remainder being due to ^{40}K.

These decay products include elements with atomic numbers ranging between 92 (U) and 81 (Tl), having chemical properties which may be very different from those of the parent uranium and thorium, and half-lives ranging from $>10^5$ years to $<10^{-6}$ sec. Each of the three decay chains includes an intermediate nuclide of atomic number 86, which is a gas under normal conditions, decaying by alpha emission to a solid daughter. These are ^{222}Rn in the ^{238}U series, ^{219}Rn in the ^{235}U series and ^{220}Rn in the ^{232}Th series; the first is commonly known as radon, with a half-life of 3·8 days; the second, actinon, has a half-life of only 3·9 sec; and the third, thoron, is of 51·5 sec. Radon and thoron provide the basis for a valuable prospecting technique, which will be considered later.

The most common instrumental detection techniques, however, make use of gamma emissions from the decay process because of the comparatively low absorption coefficient for gamma radiation above 0·5 MeV in most materials, allowing detection through a metre or so of rock and much greater distances in air.

When all decay products are in secular equilibrium, the dominant contributor to the gamma spectrum from the ^{238}U chain is ^{214}Bi, which exhibits a specially characteristic group at about 1·76 MeV, having a total abundance of 22–23 per 100 disintegrations of ^{238}U. Decay of ^{232}Th is chiefly characterized by a 2·62-MeV radiation from ^{208}Tl (100% abundance); other strong emissions are present at 0·86–0·97 MeV from ^{208}Tl and ^{228}Ac, and at 0·58 MeV from ^{208}Tl.

Gamma radiation from ^{235}U and its daughters is relatively unimportant because of the low abundance of the parent in natural uranium. Estimation of uranium from measurements made on radiations from its daughters is liable to large error if differential leaching or other processes have disturbed the radioactive equilibrium between members of the decay series. The need to determine and correct for the effects of disequilibrium could be avoided if it were possible to utilize radiations characteristic of uranium itself or of daughters with which it is always in equilibrium. These are shown in Fig. 1. The most promising emission is that of ^{234}Pa, with an energy of 1·001 MeV, but this is present at a very low intensity; a 150-g sample containing 1000 ppm U_3O_8 will have an emission rate for this radiation of only 2–3/sec. Even with a large-volume Ge(Li) detector, counting times of the order of 1 h are needed for an acceptable statistical accuracy

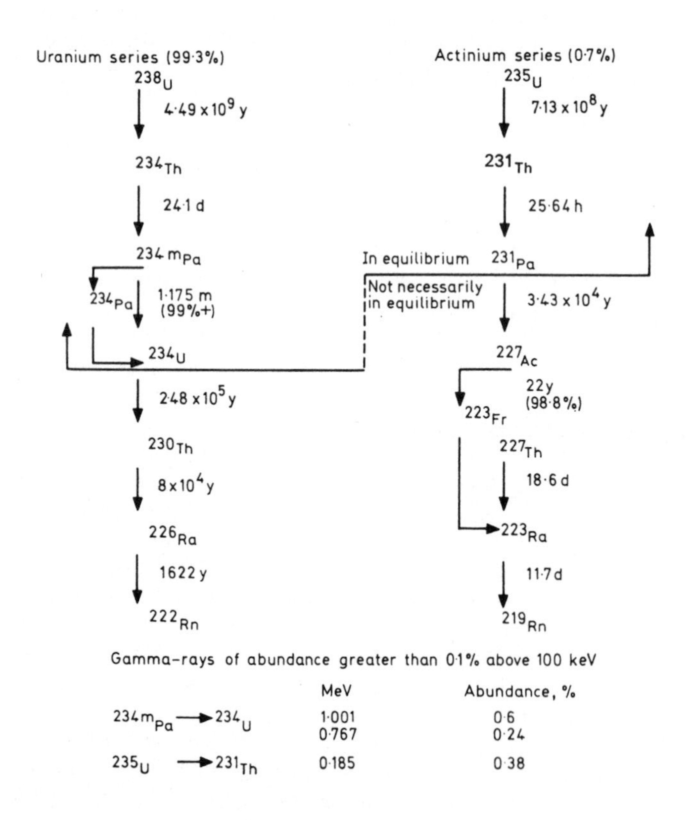

Fig. 1 Nuclides in equilibrium with ^{238}U or ^{235}U

to be achieved at this concentration in the presence of Compton-scattered background due to higher-energy radiations from ^{214}Bi, ^{228}Ac or ^{208}Tl. The method is unlikely to prove useful for routine analysis of low-activity samples.

It is not proposed to consider the beta activity produced as part of the decay process, although in the past the ratios of beta/gamma or beta/alpha activity have been used to obtain estimates of uranium disequilibrium or uranium/thorium abundances.

Detectors

Many large ore deposits still being worked were originally located by simple survey instruments containing a Geiger–Müller counter detector. Indeed, similar instruments are still in use in many parts of the world, especially for borehole logging where the uranium concentration reaches ore grade, or for control of mining operations. But the thallium-activated sodium iodide scintillation detector viewed by a high-gain photomultiplier tube is now almost universally preferred for gamma survey instruments, both because of its improved sensitivity as a total-count detector and for its property of producing output signals containing information about the energy spectrum of incident radiation.

The detection efficiency of this type of detector varies markedly with energy, leading to ambiguities in specifying the overall sensitivity of total-count instruments unless the calibration source and geometry are themselves specified exactly. Further, changes in ambient temperature produce significant variations in light output from a sodium iodide crystal for a fixed photon energy, and also alter the photomultiplier tube gain. As a result, the photomultiplier output signal amplitude usually exhibits a temperature coefficient which is typically negative above 0°C, and of magnitude 0·3–2% per °C; at low temperatures the sign of the coefficient may reverse. The effect is serious in gamma-spectrometry applications, as temperature changes of 20°C may be expected in the field. In elaborate systems it may be largely eliminated by mounting the detector(s) in a temperature-controlled enclosure, but where limited space or power supplies are available it is necessary to incorporate facilities for energy calibration by use of a radioactive reference source of known energy. This may be achieved by periodic manual adjustment of the system gain, but a more effective method is to arrange for an automatic electronic adjustment by continuously comparing the signal outputs from two auxiliary channels set one on either side of the reference peak. Modern low-power circuit techniques allow the design of a portable field spectrometer incorporating fully automatic stabilization against temperature changes, at the same time giving an acceptable battery life from 8 'D'-type cells.

A typical NaI(Tl) scintillation detector has a limiting photo peak energy resolution of about 8–9% (fwhm) for a single mono-energetic radiation of 1 MeV (equivalent to 80–90 keV). Two main reasons for this spread are the large amount of energy (300 eV) required to create one photo-electron at the photo cathode, which leads to statistical fluctuations in the signal amplitude, and the effects of Compton scattering in the crystal. Considerable difficulties are therefore experienced with this type of detector in resolving the complex spectra obtained from mixed uranium–radium and thorium sources.

The use of semiconductor gamma detectors (GeLi or SiLi) gives a very great improvement in resolution down to 2–2·5 keV (fwhm) for a 30-cm^3 GeLi detector. In order to achieve these values, however, the germanium crystal and amplifier input circuit must be maintained continuously at the temperature of liquid nitrogen, which limits their application to field laboratories. Silicon detectors can be stored at room temperature and need be cooled only when operating; but the low atomic number of silicon restricts their application to energies below 100–150 keV. With these detectors a complete analysis for uranium, radium and thorium can be achieved on a single sample, together with measurements of disequilibrium, if the additional expense and complexity can be justified. It is possible that during the next few years this technique will be introduced in field operations.

Prospecting techniques

Uranium prospecting techniques may be divided into primary and secondary categories: the primary are suitable for covering large areas rapidly, and the secondary techniques are used to evaluate the discoveries made in the primary phase.

In order to achieve rapid, yet reasonably full, coverage of the terrain, primary prospecting instruments must be capable of providing a continuous flow of statistically significant data from sources which extend over a large area. This enables them to be operated along widely separated traverses—if necessary, at high speeds. Because of these requirements most primary uranium prospecting instruments measure natural gamma radiation, which has a long range in air, whereas the less penetrative alpha and beta radiations are only used in detailed secondary prospecting operations.

AERORADIOMETRY

The primary prospecting technique which exploits the characteristics of natural gamma radiation to the full is aeroradiometry. Surface gamma radioactivity can be detected up to heights of nearly 1000 m, but useful measurements are best made at 50–200 m, where the surface flux is reduced to 75–25 %, respectively, of its original intensity. The detector receives the bulk of its signal from a circle on the ground beneath the aircraft with a radius equal to the flying height, which implies that, for full coverage of the terrain, traverses should be separated by about twice the flying height.

To produce information which may be correlated between traverses the survey aircraft must, ideally, maintain a constant ground clearance, but the need for this is reduced where the equipment incorporates a height compensator, which corrects the detector output according to the absorption of gamma radiation in air by reference to a radar altimeter. This type of compensation, however, assumes that the ground is flat and of uniform radioactivity. It therefore applies the wrong correction to radiation from small area sources in which the intensity falls off with distance approximately according to the inverse square law, and also takes no account of variations in solid angle due to changes in the topography. The consequent errors are greater in areas of high relief where compensation is most needed, but such systems are quite effective over relatively flat country.

The need for high sensitivity and light weight in this type of equipment requires the use of scintillation rather than Geiger–Müller detectors. These are generally NaI(Tl), but organic scintillators are also used.

Until recently, most instruments recorded the total count above a minimum energy threshold (of about 50 keV), and detector sizes were chosen which gave 500–1000 counts/sec above the residual count (cosmic and contamination) when flown over unmineralized rock types at a ground clearance of 150 m. In the United Kingdom this was achieved with three unshielded NaI(Tl) crystals 117 mm in diameter and 25 mm thick, whereas the U.S. Geological Survey obtained the same count rate with six partly shielded crystals 102 mm in diameter and 51 mm thick.[1] With such systems it is possible to resolve differences in broad-source radioactivity of 3 ppm eU_3O_8 and in favourable circumstances to detect small area anomalous sources 40 m across averaging 30 ppm eU_3O_8 above their surroundings.

During the last four years energy discrimination techniques have been applied to aeroradiometry.[2] In addition to total radioactivity, the count rates are recorded from three channels, 2–400 keV in width, centred upon 1·46, 1·76 and 2·62 MeV, to include important gamma emissions from the potassium, uranium and thorium decay series, respectively. Given an adequate count rate within the channels, it is possible to estimate the surface concentration of the three parent radioelements, assuming secular equilibrium.

The restricted energy ranges covered by the channels, however, reduce their count rates to less than 1% of the total above 50 keV, and necessitate the use of large and expensive detectors. Because of the high costs involved, airborne gamma spectrometers have tended to develop along two lines: those which employ the smallest detector size able to give any useful results and much larger equipments designed to produce the maximum amount of information.

The smaller systems have detector volumes of 4–8 l of NaI contained in one large crystal or up to four small crystals. When flown over areas of normal radioactivity (20–40 ppm eU_3O_8) net count rates in the uranium and thorium channels are commonly 4–5/sec, which is too low to reflect reliably the small concentrations of radioelements occurring in ordinary rock types. The position is improved over strongly anomalous sources, but, even here, qualitative determinations only can be made of the U/Th/K ratio. The smallest anomalous surface radioelement concentration which can be resolved by an 8-l detector is of the order of 100 ppm eU_3O_8 evenly distributed over 30000 m^2.

Although the channel counts are low, the crystal volumes are such that the total-count rate averages several thousands per second, which allows recognition of broad-source differences down to 1 ppm eU_3O_8. These small spectrometers are therefore best regarded as very efficient total radioactivity survey equipments able to indicate the predominant radioelement in medium to large surface anomalies.

Automatic stabilization of the channel boundaries against changes in temperature is usually provided by use of a reference source. If the reference energy lies within the signal energy range, however, the total-count measuring sensitivity will be reduced and changes in the signal spectrum may affect operation of the stabilization circuits. This may be avoided by raising the total-count energy threshold above that of the reference source, but this will reduce the count rate produced by activity on the ground. A commonly used reference source is ^{137}Cs, which has a mono-energetic gamma emission of 662 keV, when the signal threshold must be raised to about 1 MeV. The total count rate from the ground is then reduced to less than 5% of its 50-keV threshold value, at which level its sensitivity is little better than that of the potassium channel.

There is clearly a need for lower-energy stabilization sources: two which have recently been used successfully are ^{57}Co (half-life 270 days), with a main peak at 120 keV, and the 59·6-keV gamma emission from ^{241}Am (half-life 458 years). These allow the threshold to be lowered to levels at which the total count made has more acceptable sensitivity.

Large airborne spectrometer systems with sodium iodide detector volumes up to 50 l have been discussed by Darnley,[2] but it should be noted that their operating costs are at least three times those of the smaller equipments.

CAR-BORNE RADIOMETRIC SURVEY

If the area to be prospected is covered by a network of roads, much useful primary exploration can be done by traversing with a car-mounted scintillation counter. The equipment consists of detector, preferably mounted upon a short mast projecting above the vehicle, connected to a rate meter and recorder situated within the vehicle near to the front passenger seat. There should be a facility for making positional marks upon the recorder trace at road intersections, etc., and an alarm unit which operates when the radioactivity exceeds a preset level.

Such a survey can be carried out by the driver alone, but its value is enhanced greatly if two additional crew members are available—one to navigate while the other interprets changes in count rate by reference to his surroundings and the geological map. His observations are best recorded on a pocket tape recorder fitted with a throat microphone.

It is impracticable to traverse with the detector mounted at a height greater than about 3 m, at which level over flat ground 80% of the gamma radioactivity detected comes from a circle 25 m in diameter, of which about one-quarter of the area will be road. The contribution from road metal can be reduced by positioning the detector horizontally towards the nearside, but this attenuates the offside signal and necessitates covering the road in both directions. The short effective range of the detector results in the signal being affected by variations in solid angle due to roadside buildings and road earthworks. A roadside granite-built house or cutting in mica-schist may produce a larger peak on the recorder trace than a uranium vein—hence the need for the interpretative observer.

At traverse speeds of 50–70 km/h adequate sensitivity for total radioactivity surveys can be obtained with a NaI(Tl) crystal of 286-ml volume, which has a count rate sensitivity of about 60 counts/sec per μr/h when calibrated with ^{226}Ra. With a rate meter time-constant of 1 sec such a system can detect a change of 1 μR/h persisting over a distance of 30 m, which is more than adequate for survey purposes.

Gamma-spectrometric vehicle-borne surveys require larger detectors. Crystal sizes of 1·5–2·0 l will give background count rates in the uranium, thorium and potassium channels of about 10–30 counts/sec, which are sufficient to resolve

Fig. 2 Carborne gamma spectrometer (Harwell type 3027) with 76 mm × 76 mm NaI(Tl) detector mounted on 6-m telescopic mast

qualitatively any anomalous peaks in the total-count channel. The total-count energy threshold is usually increased to about 100 keV to reduce the maximum count rate produced by a large detector volume.

In the United Kingdom a compromise has been reached between total-count rate and spectrometer surveying by use of a 76 mm × 76 mm detector coupled to a sweep spectrometer with a total count channel (Fig. 2). Road traverses are made with the total-count facility and anomalies are qualitatively resolved by the sweep spectrometer with the vehicle stationary. Satisfactory results have been obtained over anomalies where the count rate is 3 × background, and sensitivity can be improved by detaching the detecting head from the mast and operating it on a cable up to 100 m in length.

LIGHTWEIGHT PORTABLE RATE METERS

Although the airborne and vehicle-borne systems can cover large areas quickly, there are many places, particularly in mountainous terrain, where the lightweight portable scintillation counter is the only means of making effective ground radiometric surveys.

The portable instrument must combine adequate sensitivity with light weight and the ability to resist shock and moisture. Adequate sensitivity is generally considered to be the ability to recognize a 20% change in background radioactivity, and this can be achieved with a NaI(Tl) scintillation crystal 25–38 mm in diameter and thickness. Rate meters currently in production on both sides of the Atlantic incorporate these in instruments weighing less than 2 kg, but at this weight only the AERE type 1597A (Fig. 3) is housed in a hermetically sealed waterproof case. This sealing has proved to be essential in Scandinavian forest operations and will also resist the ingress of mould in tropical areas, although it is less important in semi-arid and arid conditions.

The small size of this type of rate meter generally restricts them to total radioactivity measurements, with switched count ranges capable of measuring

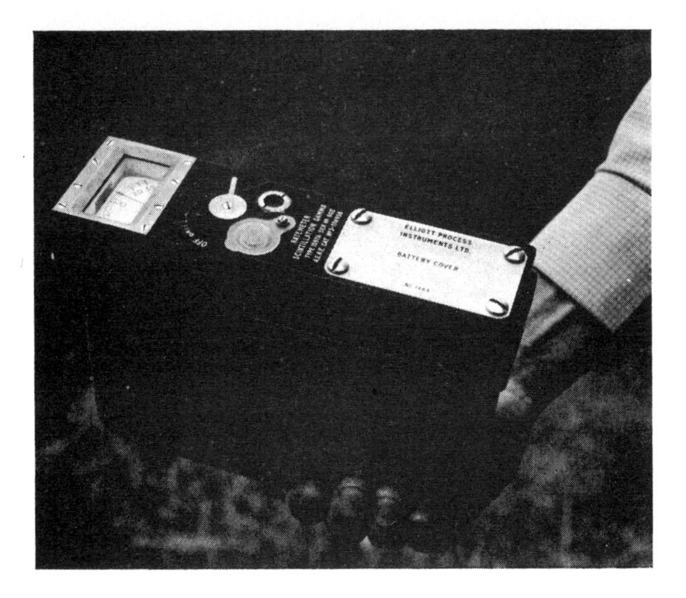

Fig. 3 Portable scintillation counter (Harwell type 1597A) employing 25 mm × 37 mm NaI(Tl) detector

radioactivity up to about 1·0% eU_3O_8, and a meter display. All these instruments incorporate some form of audio presentation—ranging from a simple earphone or loudspeaker to an adjustable-threshold continuous-tone generator or 'squealer', the frequency of which varies with count rate. The 'squealer' type is most useful in densely forested and precipitous areas, where the operator cannot afford to look at the meter. It can also locate quickly the most active parts of a mineralized outcrop. There is less demand for it in open rolling terrain, such as the western U.S.A.

All modern portable scintillation counters are powered from 'D'-type torch cells and have operating lives of 50–400 h, the shorter times being due frequently to squealer power requirements.

The detector is mounted internally in smaller rate meters, but some of the larger equipments are divided into a gun-type detector/rate meter, connected by cable to a separate battery box on which the 'gun' may be clipped, if required. These instruments are heavier and vulnerable to cable faults and snags, but the 'gun' does have some advantages in handling.

PORTABLE GAMMA SCINTILLATION SPECTROMETERS

In recent years several types of instruments incorporating gamma energy discrimination have been developed for field use. In general, these units are larger and heavier than simple rate meters and are usually employed to determine the radioelement content of gamma anomalies located with total-count rate-meters or for detailed gridding. They are of two main types—those which measure the total-count rate above one or more adjustable energy thresholds and the true spectrometers, which allow measurement of the count rate in one or more energy regions or channels, each having adjustable upper and lower boundaries. The variable-threshold varieties are usually lighter and cheaper than the channel spectrometers, but, inevitably, the information which they provide is less precise. The size of NaI(Tl) detector is usually limited to about 76 mm × 76 mm for reasons of weight, cost and fragility, in consequence of which count rates in the higher-energy channels are very low.

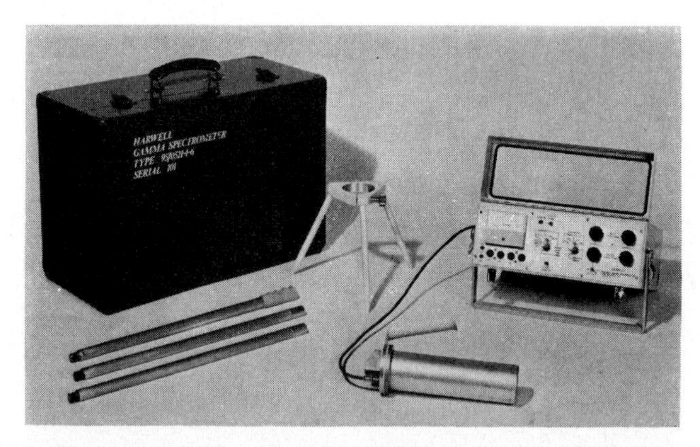

Fig. 4 Portable two-channel gamma spectrometer (Harwell type 3086) incorporating automatic gain stabilization on ^{241}Am source

In order to obtain useful thorium channel measurements in low to medium radioactivity levels it is necessary to replace the simple rate meter circuit by an integrating rate meter or scaler, controlled by a timer. A thorium anomaly of 3 × normal background will produce a count rate of about 75 counts/min in a

400-keV channel centred on 2·62 MeV, with a 51 mm × 46 mm detector. One example of this type of instrument is the portable gamma spectrometer type 3086 (Fig. 4), designed by the Atomic Energy Research Establishment, Harwell, and the Institute of Geological Sciences, which represents a considerable advance on earlier equipments by the incorporation of automatic gain stabilization.

The instrument contains two separate counting channels, each of which may be independently adjusted in position and width within the gamma energy range 0·25–3·5 MeV by calibrated front panel controls. The width adjustments (0–500 keV) operate symmetrically about their respective channel centres as set by the position controls. The position of one channel may also be controlled externally for swept-channel operation. An additional 'total-count' discriminator is set at a fixed low-energy level.

Counts in both channels are accumulated simultaneously in two four-decade stores with 10^5 overflow indication, and the contents of either store may be selected at any time for display on a single four-decade numerical display without affecting information already stored. In addition to manual count control by 'reset/start' and 'stop' push-buttons, the channel counting time may be preset in the range 1–8 min by a selector switch and a built-in electronic timer. A 'bright-up' facility incorporated in the 'stop' button enhances the numerical display at the end of a counting period, if required.

Alternatively, the counting rate above the single 'total-count' discriminator or in either channel can be fed to a five-range linear scale rate meter with meter display (0–100, 300, 1000, 3000, 10^4 counts/sec). Provision is made for connecting the selected count rate signal to an external 100-mV strip-chart recorder.

The spectrometer is intended for use with a NaI(Tl) crystal and photomultiplier mounted as a separate probe, but no manual adjustment of EHT to the photomultiplier is provided. EHT is set automatically to the correct value when the instrument is switched on by locking on to the 59·6-keV gamma emission from a built-in ^{241}Am reference source. This facility also provides automatic compensation for gain changes due to the effects of temperature on the crystal, photomultiplier and other circuit components over the range -10°C to $+60$°C. A low-voltage dc output signal is also provided for the similar control of a detector assembly containing its own EHT direct to the cable.

Any size of crystal may be used, though 51 mm × 46 mm has been chosen as standard. A ^{241}Am source of about 0·1 μCi is attached to the inside of a cup-shaped graded filter assembly, which fits closely over the crystal to attenuate incident radiation below 200 keV. This avoids interaction with the reference radiation. The filter is constructed as a sandwich from three materials of different atomic number and has a total thickness of 1·5 mm. At 360 keV (^{131}I) the effect of the filter is to reduce the count rate sensitivity to about 70% of its unfiltered value.

Attachments for the probe enable it to be held (*a*) in the hand, (*b*) attached to a rigid rod to give an extended reach, (*c*) supported on a collapsible stand facing upward for assaying prepared samples placed on the crystal or (*d*) facing downwards close to a surface for *in situ* measurements of ground activity.

The probe may be separated from the spectrometer by up to at least 100 m of cable.

The spectrometer is contained in a weatherproof light alloy case with overall dimensions about 30 cm × 11 cm × 25 cm high, all controls being mounted on the front panel under a transparent hinged protective cover. Fixing links for a shoulder strap are provided, together with a carrying handle, which also supports the instrument at a convenient angle for bench operation.

The instrument is powered by eight 'D'-type cells mounted in a sealed compartment, giving an operating life of 40–60 h, or by rechargeable cells, which can be recharged without being removed, or from an external 12-V dc supply.

The total weight, including batteries, but excluding a probe, is approximately 7 kg.

RADON MEASUREMENT

Determination of the radon content of near-surface overburden and water may be used either as a primary or secondary technique, the mode of operation adopted depending upon the type of deposit sought and the geological and climatic environment. Thus, in high rainfall areas the many lakes and streams may be sampled on a systematic basis with the type of equipment developed by the Geological Survey of Canada, whereas on the more arid plains of the western U.S.A. radon in soil determinations have been used successfully in both the primary and secondary role.

Where radon is being measured in soils the sampling interval is determined by the size of the expected orebody. Thus, in France, where most of the uranium ore occurs in steeply dipping narrow veins, a 5-m interval is often adopted, whereas in Colorado a disseminated sandstone orebody was located recently in the Denver basin by sampling on one-mile centres.[3]

The principal advantage of radon over gamma prospecting is that the alpha-radioactive gas can diffuse upwards from uranium mineralization through barren overburden sufficiently thick to absorb all gamma radiation. Tanner[4] estimated that ^{222}Rn derived from ^{238}U can migrate by diffusion alone through 7 m of dry sand before the concentration is reduced by $\times$ 100 by radioactive decay, and this distance can be extended considerably by ground air movements due to variations in barometric pressure or other atmospheric disturbances. Such migration clearly depends on the degree of permeability of the overburden, which is determined by its grain size and moisture content. Radon in soil measurements are most successful in dry arenaceous and rudaceous overburdens, but they are of limited value where the orebodies are covered with wet clayey or peaty deposits.

Limited rainfall on sandy overburden can assist the radon prospector by sealing the surface pore spaces and building up the subsurface radon concentration. In these circumstances the radon may be trapped for a sufficient length of time to decay into its gamma-emitting daughter isotopes, giving rise to surface gamma anomalies which disappear as the ground dries.

The short half-life of the thorium-derived ^{220}Rn restricts its diffusion length to a few centimetres. It only affects the alpha activity of the ground air when thorium is present in the surface layers. The concentration of ^{220}Rn present may be determined by taking the difference between a count made immediately after sampling and one made about 10 min later on the same sample. During this period virtually all the ^{220}Rn will decay, leaving only the activity due to ^{222}Rn.

Radon prospecting instruments may be divided into two types—the alpha probe and the pump monitor. Both obtain the soil air sample by making a hole in the ground up to 38 mm in diameter and 0·5–2·0 m deep with a metal bar. If the bar is solid and withdrawn in one clean motion, its removal draws soil air into the hole, in which alpha activity can be measured by inserting the alpha probe. The detector in this type of instrument (Fig. 5) consists of a perspex rod coated with silver-activated zinc sulphide phosphor, and alpha scintillations on this are internally reflected up the rod to a photomultiplier optically coupled to the upper end.

Alternatively, the hole may be lined with a pipe and the air in the hole and surrounding soil pumped via a flexible connexion into a measuring chamber. The walls of the chamber are coated with zinc sulphide, and alpha scintillations within the chamber are viewed by a suitably situated photomultiplier.

An alternative method of detection used in one pump monitor involves

Fig. 5 Portable radon probe monitor (type M8560) taking soil radon measurement in sampling hole made with special bar (left)

drawing air into a mean current ionization chamber, which measures the ionization current produced in the air due to alpha activity. This has the advantages that alpha contamination of the detector can easily be compensated by an adjustment of the amplifier zero to offset any residual current, and the measuring period can be very short once the chamber is filled with a representative air sample.

Probe monitors are simple, with no moving parts, and cause minimum disturbance to soil air in the vicinity of the measuring hole. Pump monitors have greater sensitivity than the probe and are more easily adapted to the measurement of radon in water. All scintillation monitors are liable to alpha contamination, which necessitates a periodic change of phosphor assembly.

BOREHOLE LOGGING

Borehole logging is a secondary technique employed when promising surface indications of uranium mineralization have been located and mapped by the other techniques.

The need to enclose the detector in a watertight steel probe precludes the possibility of measuring other than gamma radiation, and both G–M tubes and NaI(Tl) crystals have been used for this purpose. The small diameter of most exploration drill-holes restricts the diameter of the probes to less than 40 mm, which imposes a maximum detector diameter of about 25 mm. At this size the signal from a G–M tube contains a greater proportion of counts from high-energy primary radiation than a scintillation detector, although the total-count rate of the former is less. The G–M signal is thus more directly related to the radioelement content of the surrounding rock, and for this reason is still used on a limited scale for uranium assessment work. Generally speaking, however, the greater sensitivity of scintillation probes has led to their adoption on an ever-increasing scale.

At the present time virtually all logging of uranium exploration boreholes is for total radioactivity, and ore reserves are frequently calculated directly from the gamma profiles. This can lead to serious miscalculation of grades and tonnages on two counts: first, where groundwater leaching has caused a partial separation of the uranium from the gamma-emitting daughter products, and, secondly, a reduction in gamma activity of high-grade uranium ores due to internal self-absorption of their soft gamma emissions.

Straight gamma spectrometry by use of channels centred on primary emission energies is not a complete solution to either of these phenomena, as the emissions come from the daughter products. Recent work suggests, however, that these problems may be resolved by more sophisticated energy discrimination techniques, discussed at greater length by Dodd and Eschliman.[5]

The ultimate solution undoubtedly lies in neutron-activation techniques which employ the delayed neutron effect of ^{235}U, but these are not likely to become available for several years.

Borehole logging equipment may be either reasonably lightweight and transportable by two or three people or more substantial and mounted in a vehicle. Transportable equipments have been produced in the U.S.A., France and the United Kingdom, and usually measure total radioactivity only down to depths of 3–600 m. Car-mounted loggers frequently incorporate other sensors, such as self-potential electrodes, within their probes and can be built to accommodate the depth requirements of their users. Below 1000 m, however, the rise in temperature due to the geothermal gradient may cause electronic problems.

Conclusion

In this brief survey it has not been possible to discuss the operating characteristics of the instruments mentioned, or the measuring techniques which employ them, in any detail. The principles have been illustrated by reference to instruments largely designed in the United Kingdom, since the authors have had greater experience in their use.

One outstanding requirement of uranium geologists is the need to measure directly, *in situ*, the uranium content of surface and subsurface rocks. Development of the X-ray fluorescence technique can be expected to prove increasingly useful in attaining this objective. The availability of high-resolution intrinsic and lithium-drifted silicon semiconductor gamma detectors of relatively large volume, together with intense radiation sources from portable X-ray generators, should allow the estimation of uranium in rock samples down to very low concentrations, even in the presence of thorium, once the associated cryogenic problems have been satisfactorily overcome for field operation.

References

1. Davis F. J. and Reinhardt P. W. Instrumentation in aircraft for radiation measurements. *Nucl. Sci. Engng.*, **2**, 1957, 713–27.
2. Darnley A. G. Airborne gamma-ray survey techniques. This volume, 173–207.
3. Stevens D. N., Rouse G. E. and De Voto R. H. Radon-222 in soil gas: three uranium exploration case histories in the Western United States. In *Geochemical exploration* (Montreal: CIM, 1971), 258–64. (*CIM Spec. vol.* 11)
4. Tanner A. B. Radon migration in the ground: a review. In *The natural radiation environment* Adams J. A. S. and Lowder W. M. eds. (Chicago: University of Chicago Press, 1964), 161–90.
5. Dodd P. H. and Eschliman D. H. Borehole logging techniques for uranium exploration and evaluation. This volume, 243–70.

Portable equipment for uranium prospecting

Y. Puibaraud

Commissariat à l'Energie Atomique, Fontenay-aux-Roses, France

550.8.002.5–182.3:553.495.2

Synopsis

Uranium prospecting is very similar to exploration for other metals. Gamma rays emitted by the daughter products of uraniferous minerals are used, however, to detect new occurrences and even to make indirect estimates of the uranium content, depending on the equilibrium state of the ores. Alpha radiation, emitted by radon, can also be used.

Instruments for field purpose have two types of detector: (1) Geiger–Müller tubes, whose mechanical robustness, electronic stability and modest price compensate for their weak response to gamma rays; and (2) scintillators, which are more fragile, unstable and rather expensive, but their greater gamma-ray efficiency allows easier measurements with lower statistical fluctuations.

Ideally, portable equipment should be (1) light and easy to handle; (2) robust, watertight and able to work over a wide temperature range; (3) a low power consumer, with cell or battery supply; (4) accurate, reliable, highly sensitive, and able to measure radioactivity from low to high count rates and to function as a spectrometer; and (5) reasonably priced.

Types of portable equipment and areas of usage go hand in hand. Scintillometers are especially appropriate for reconnaissance and general prospecting along lines as well as for detailed studies requiring more systematic measurements, usually following a grid. Aerial or car-borne surveys demand more sensitive equipment. Once uranium ore has been discovered, the use of Geiger–Müller counters may be advantageous for drilling or mining operations.

Broadly speaking, prospecting for uranium does not differ from prospecting for other metallic substances. As in the search for lead or copper, knowledge of the physical and chemical properties and the types of uranium deposits will enable the prospector to seek this metal in the geologically favourable areas of the region under investigation. In addition to geochemical prospecting, conventional geophysical techniques are used to select and pinpoint formations—with the aim of locating uranium-bearing structures: for example, portable instruments for the measurement of gravity, magnetism and resistivity can be used to assist the geologist in his search for mineralization which lacks any obvious surface indication.

Nevertheless, most conventional instruments in use in uranium exploration are still based on the detection of ionizing radiation—indeed, this particular radioactive property of some minerals and uranium ores has been widely used for their detection in the field and has been responsible for many of the advances in their exploitation during the past 20–25 years. A large variety of techniques and types of apparatus designed to detect and measure this radiation has been developed, and it is this equipment which is examined in greater detail in this paper, following a few general comments on natural radioactivity and the means by which it is detected.

Radioactivity of uranium ores

The elements ^{238}U, ^{235}U and ^{232}Th and their daughter products form three radioactive series, each having 11–15 natural radioactive isotopes. The latter emit alpha or beta rays, together with gamma rays. Under normal prospecting conditions radiation given off by the elements ^{238}U, ^{235}U and ^{232}Th is not easily measured, and it is their daughter products which serve as natural tracers, the most important being radium C (^{214}Bi). Uranium prospecting based on instruments for measuring radioactivity is therefore an indirect method.

The radioactivity measured is a function of the equilibrium state of the ores, i.e. a function of the ratio between the content of uranium (or thorium) and the amount of daughter elements resulting from disintegration in a confined environment. The radioactivity is thus not strictly or directly proportional to the amount of contained uranium. Its order of magnitude is, however, very largely sufficient for estimation of the value of a radioactive anomaly encountered in prospecting.

This natural radioactivity of ores is, moreover, very complex. Different radiations are emitted, the energy spectra are very wide and the half-lives vary. Gamma radiations are the most easily and directly measurable in the field. Variations in the characteristics of the energy spectrum can also be used to determine the presence of uranium, thorium or even potassium (by its radioactive isotope ^{40}K). Some field spectrometers—and particularly those used in airborne radiometric surveys—theoretically allow detection of anomalies due to these elements.

Alpha and beta particles, the detection of which is more difficult, are used less often in uranium prospecting. Nevertheless, alpha radiation is used in the field to detect the presence of radon, and beta radiation can be measured by special instruments designed to measure the uranium contents of ores. There is a relationship between the uranium content of an ore and the amount of beta and gamma radiation emitted.

Radioactivity detectors

Radioactivity detectors for use in the field, in particular, gamma-ray detectors, have benefited from the advances made on similar apparatus designed for laboratory use. The ionizing chambers or gold-leaf electroscopes used by pioneering prospectors have been replaced by more practical instruments. First of all came the Geiger–Müller counter, followed by the scintillation detector, which generally consists of a sodium iodide crystal activated by thallium. Other types of instruments, which may be mentioned are those which employ plastic or liquid scintillators, but the practical applications of these have been very limited. On the other hand, it is possible that new types of detectors, the so-called 'junction' types, which employ germanium–lithium and are at present used only in the laboratory, could be of value in the field. It will, however, doubtless be necessary to await significant technological progress before the sort of equipment which could provide an improved spectrometry service can be made available in portable form.

Each of the two detectors most commonly used, the Geiger–Müller tube and the NaI(Tl) scintillator, has its advantages and disadvantages. The Geiger–Müller counter has the advantages of mechanical robustness, relatively simple circuits, electronic stability and modest price. Its weak response to gamma rays, which is often less than 1 %, is certainly inconvenient, but this is generally of little importance and may even be an advantage in use in highly radioactive surroundings such as in a mine. The scintillometer (and its scintillator) is a more fragile instrument and its price is rather high. It shows instability due to changes in temperature of around 0·5% per degree Centigrade, which is important when it is used as a spectrometer. Its efficiency, even to very soft gamma rays, however, allows accurate measurement of the smallest fluctuations; moreover, by providing a means of measuring the energy of the rays, it fully justifies its use in mineral prospecting and radiogeology.

Also worthy of mention is a detector which consists of a compartment coated with zinc sulphide. It is fairly widely used to measure alpha rays emitted by radon. When exposed to such rays zinc sulphide has the property of emitting photons, which can be detected by a photomultiplier.

All these detectors are fitted with batteries and electronic circuits, and can amplify impulses which may then be measured by a galvanometer. The technological revolution, which has in the past ten years caused the change from vacuum electronics to solid state, has also brought about significant improvements in the characteristics and reliability of circuits.

The highly spectacular advances achieved at the laboratory level have been no less important for field equipment. Even if size has not changed significantly, the reliability and possible applications have been considerably increased. Measurements have thus been greatly refined, and phenomena formerly lost in errors and deviations can now be detected and call for further improvements for their exploitation. It is, nevertheless, necessary to point out that this has not been achieved without some inconvenience to the user, who, from a financial point of view, tends to suffer because of the rapid obsolescence of his tools.

Points looked for in portable equipment

Users hope to find a number of characteristics in a piece of equipment: (1) it should be as light and easily handled as possible; (2) it should be robust, not affected by vibration or humidity, and watertight, and it must be able to be worked accurately in a wide range of temperatures; (3) it must be powered by cells or batteries which can easily be replaced or recharged in the most remote areas, and power consumption must be as low as possible; (4) it must be accurate, reliable, very sensitive and capable of measuring radioactivity of high count rates, and it should also be capable of being adapted to function as a spectrometer; and (5) its cost should not be too high.

These requirements are certainly extensive and demanding, and some of them are even contradictory. For most of the time the manufacturer will have to find a compromise in order to cover the user's wishes in the best possible way, e.g. between lightness and sturdiness, reliability despite electronic complexity, performance and cost. Fortunately, present electronic techniques allow these demands to be met more and more successfully.

For prospecting equipment some improvements, however, need to be made—especially in ways of increasing reliability under very severe conditions.

The equipment must be carefully protected against humidity, including, where possible, the electronic circuits themselves. It should be able to withstand drenching or even immersion in some tens of centimetres of water. The probes must remain completely watertight under pressures several times higher than atmospheric. The circuits must function normally over temperature ranges much greater than those for laboratory equipment. The International Electro-Technique Commission studied the usual working temperatures for this type of activity in order to be in a position to advise manufacturers. In temperate to hot climates the normal working temperatures range from −10 to +40°C, with extremes of −25 and +55°C. In cool to temperate climates the range is −25 to +30°C, with extremes of −40 to +40°C.

The Commission also advised the use of 1·5-V cells (type R.20), which are readily available in all countries, to power portable prospecting apparatus. This kind of recommendation cannot but be endorsed as it helps the production of such equipment.

Likewise, the electronic circuits of the equipment must function with great reliability so that distortion of readings and complete breakdown are avoided: either or both would have serious and expensive consequences for the user. The user will, in fact, have a choice in his work between robust, relatively simple equipment and that which is more sophisticated and able to yield considerable information but possibly be less reliable.

Over and above such reliability in operation, which will be one of the major assets in prospecting equipment, the equipment will also have to be capable of taking physical measurements. The geologist who is searching for radioactive anomalies, which may often be very weak, has to be able to take reproducible measurements for subsequent analysis and interpretation. It is also necessary for him to identify meaningfully very weak variations in radioactive values. Going further in measurement analysis, he will want to identify the emitters of the radioactivity by use of a spectrometer. The latter is normally more complex than a simple count rate meter for the measurement of overall radioactivity, but here also electronic improvements allow good equipment to be made available.

In the end, all the properties demanded by the conditions under which prospecting equipment is used, and by the users themselves, come down to a question of cost. Instead of merely looking at the purchase price of equipment it is, however, necessary to consider its operational cost price, which will consist of amortization, care, maintenance of components and estimated cost of immobilization due to a breakdown. It will be this cost and the quality of the measurements which have to be taken into account (where it is possible to do so) in order to judge the contribution of a piece of equipment to a given task.

Types of portable equipment and areas of usage

The methods and techniques of uranium prospecting will vary according to the type of equipment used. A geologist will, for example, use a very sensitive scintillometer to discover weak anomalies on the surface during reconnaissance work, but he will need a reliable Geiger–Müller counter capable of measuring high radioactivity when working in a mine. Different types of equipment and their methods of use are noted below.

During reconnaissance or general prospecting on foot it is necessary to have a light piece of equipment, strong and very sensitive—the use of scintillation detectors is especially appropriate to this last requirement. The NaI(Tl) crystals 1 in (2·5 cm) to 1½ in (3·75 cm) in size in such equipment allow measurement of weak anomalies ranging from variations of some tenths of ppm of uranium in the composition of rocks to the radioactivity emanating from obvious mineralization. The count rate in such conditions varies from some tens of impulses per second to ten or twenty thousand. The weight of the equipment is 3–4 kg at the most.

An interesting refinement consists of fitting to the equipment an audible warning device with an adjustable threshold and frequency which varies with radioactivity. This relieves the user from constantly watching the galvanometer while he is traversing the terrain, and may enable him to reduce the lag in measurement, due to the electronic circuits of the count rate meter.

Basic prospecting equipment can be used in several ways. During a general survey the user takes continuous measurements over the terrain, measuring radioactivity along his route. The variations observed in radioactivity allow a geological survey to be undertaken and completed by giving the contrasts between the different formations encountered, radioactivity being one of their characteristics. Thus, there will be a difference of background between various granites, and between granites and metamorphic rocks; or again, there may be a variation in the natural radioactivity within a sequence of sedimentary rocks, continental pelitic rocks, for example, being more radioactive than evaporites. When prospecting in sedimentary terrain, such measurements allow, in particular, abnormally radioactive series to be recognized which provide favourable indications of the presence of mineralization.

For detailed prospecting, when looking for extensions of known ore deposits, measurements made in the field will be more systematic. They can, for example, be taken at intervals of a few paces on a grid; they can then be plotted on a map, and curves may be traced through points of equal radioactivity to ascertain the alignment of the radioactive structure. As overburden can, to a great extent, mask radioactive anomalies, the sensitivity and reliability of the equipment will be of major importance in obtaining good results. It is also important, if isorads are to be drawn, that measurements should be stable enough to be compared and interpreted with those which have been taken with different apparatus or at different times. Unfortunately, a scintillometer will often lack the precision and reliability desired for this kind of systematic survey work.

Whenever prospecting is done on foot, the most refined equipment is always used to identify the various radiation emitters and to estimate the proportion due to uranium, thorium or potassium. The advances made on electronic circuits allow spectrometers to be manufactured of a size and weight which are compatible with use in the field. In this type of equipment the detector is usually separate from the measuring apparatus, being connected to it by a cable. In addition, it is often of a greater size than that in portable scintillometers to enable the count rate recorded to be increased. This type of apparatus can have a variable threshold or a variable channel. In the first case, radioactivity above an energy level selected by the user, generally between 50 keV and 2 MeV, can be measured. For example, in examining the spectrum higher than 2 MeV, the amount due to thorium can be estimated. In the second case, radioactivity is measured through a channel adjusted to the energy, its size being varied either absolutely or as a function of that energy; the channel can also be displaced over the whole range of the spectrum in order to measure the value of certain characteristic peaks. For uranium the ^{214}Bi (radium C) peak at 1·76 MeV is normally used; for thorium the ^{208}Tl (thorium C″) peak at 2·62 MeV; and for potassium the ^{40}K peak at 1·46 MeV. In more refined equipment the channel can automatically explore the whole spectrum, which is then recorded on a graph or directly observed from the characteristic peaks.

The count rates recorded are relatively low, so as well as increasing the volume of the detector crystal it is necessary to count for periods of up to 15 min in order to obtain a significant total count. It must also be noted that the observed spectra for natural ores are rather complex, thus presenting a difficult problem. Nevertheless, the geologist in the field will very quickly be able to distinguish between thorium and potassium in the observed anomaly.

For aerial or vehicle-borne prospecting the user will need to have still more sensitive equipment, as radiation is very rapidly absorbed by the air the farther one moves from the radioactive source. To offset this attenuation the size of the detector has to be increased considerably. The crystals used often exceed 4 in in size and can even reach 11 in. The associated electronics become more complicated and either total count only or total count in combination with discrimination is used. Frequently, data such as altitude or positioning by radar

Doppler effect are recorded graphically or as numerical data suitable as input for computer analysis. Equipment of this kind is becoming larger and larger in size. It is generally used to make continuous regular traverses 500 m to 2 km apart, according to the types of reconnaissance.

All the types of equipment outlined above are designed for reconnaissance prospecting, and their main property is that of sensitivity. When indications of ores have been found and it is required to discover their extent, either from the surface or by probes, as in a mine, the radioactivity encountered will be much stronger and the geologist will want to have precise and reliable measurements. A scintillation detector would still be of value, but in conditions of strong radioactivity a Geiger–Müller tube has distinct advantages, as has already been mentioned. Equipment applicable to such situations exists, generally comprising a Geiger–Müller tube linked to a robust electronic counter and power supply. It is often possible to fit different kinds of probes and to make measurements in boreholes normally filled with water—hence the need for the equipment to be watertight under pressures of 100 kg/cm^2 or even higher.

For accurate investigations of boreholes rate meters must have properties such that dead time is reduced to a minimum and measurements can be made without significant count losses in conditions of high radioactivity, such as are found in very rich ores with uranium contents of several per cent. Indeed, it is most important that a geologist have accurate measurements available in order to establish the correlation between radioactivity and grade of ore, at the very least, when thorium is not associated with uranium.

Here also equipment becomes more complex, and besides that for measuring radioactivity there are types of apparatus with variable-speed winches, recorders, devices for geophysical measurement, etc. As regards equipment for use in mines, some types are fitted with special items to permit precise measurements in situations in which ambient radioactivity is very high by eliminating background influence.

Reference should be made to devices which, in mines or open-pits, will allow measurement of the radioactivity of excavated material and, from this, the estimation of the grade of ore by means of statistical correlations. In such cases it is important to make measurements as rigorously as possible by use of ore detectors in geometrical patterns. Scintillation detectors are used, or batteries of Geiger–Müller tubes associated with electronic circuits with digital recorders. Taken together, the results obtained are excellent when large numbers of measurements are involved, provided that good and representative samples are taken. Again, however, it is not, strictly speaking, a question of portable equipment, although in small exploratory mine workings prospecting equipment can be used.

Finally, mention should be made of prospecting equipment which is used to discover the presence of radon in the field (this gas can be a useful tracer in the search for uranium-bearing ores near the surface). Radon is an alpha emitter and it is necessary to have special detectors to pick it up. An ionization chamber can be used, but the most practical detector consists of a glass receptacle coated inside with zinc sulphide. When alpha particles reach this, photons are given off which can be detected by a photomultiplier tube, which fulfils the same function as in a sodium iodide crystal scintillometer. The electronic counter and power supply portion will be identical in both instruments, apart from an adaptation for the much weaker count rate in the radon monitor. Air is usually pumped out of the ground and circulated through the scintillating flask. This equipment is usually heavier than conventional scintillometers as it includes a pumping device and soil sampler. The operator normally moves along lines some tens of metres apart, readings being taken at intervals of a few metres. This method is mainly suited to terrain with extensive overburden.

Conclusion

The spectacular developments in the uranium mining industry during the past 20 or 25 years certainly owe much to the use of portable detectors of radioactivity, which have allowed the rapid detection of ores at a time when the metallogeny of uranium and the conditions of its deposition were still little known.

The progress achieved in the field of electronics has permitted the production of portable prospecting equipment more and more suited to arduous working conditions in the field. This improvement in the reliability of equipment has been attended by an equal improvement in the sensitivity, reliability and accuracy of measurement. Their use has added a great deal to the efficiency of conventional prospecting methods—to the detriment of other techniques, such as those which make use of the geochemistry of soils.

Users have available a range of equipment of very wide characteristics, from which they can select the type best adapted to the conditions of the work they propose to undertake. Further advances will undoubtedly be made, thereby improving the tools which are indispensable to the exploration geologist.

DISCUSSION

G. R. Ryan said that field geologists were concerned mainly with robustness and simplicity in field instruments and, as geologists made very poor electronics technicians, a spectrometer was felt to be unnecessary at the reconnaissance stage. Once an area of anomalous radiation had been detected, sophisticated instruments and different techniques could be introduced.

The author said that the deposits in which he had been working did not contain thorium with uranium and, therefore, they needed only a Geiger counter at an early stage to determine the grade of the ore. They did, however, use scintillometers to detect very small variations of radioactivity (10% background) as those had proved significant in finding orebodies. In the Vendée district of France, for example, where the area had been surveyed twice before, a new orebody was discovered by that instrument. For work in mines and for borehole logging Geiger counters were used where there was some variation in radioactivity because they give more reliable measurements.

P. H. Dodd said that a 3-G–M tube differential face scanner had been used since 1958 in the western United States for in-place assays of ore in per cent equivalent U_3O_8. A filter, consisting of $\frac{1}{8}$-in lead, sandwiched between $\frac{1}{16}$-in steel, might be moved between the G–M tubes and the sample. Two readings were taken at the same location and, by subtracting, a difference was obtained which eliminated ambient background. The filter absorbed a fixed percentage of gamma radiation from the 'sample', which was the cone formed by the solid angle from detector through filter to the rock. The geometry was therefore controlled as well. Recent tests indicated that with a scintillator of 1 in $\times$ $1\frac{1}{2}$ in NaI(Tl), 4 in $\times$ 4 in filter of $\frac{1}{8}$-in thick Pb, 8 or 16 sec total count, there was sufficient sensitivity to measure 10 ppm $\pm 2eU$ (1·28Δ counts/sec/ppm).

Magnetic susceptibility measurements seemed to be useful in the sandstone-type deposits, where it appeared that small amounts of magnetic minerals were partially altered by mineralizing fluids and, thus, small negative anomalies indicated 'alteration'.*

*Ellis J. R., Austin S. R. and Droullard R. F. Magnetic susceptibility and geochemical relationships as uranium prospecting guides. *U.S. AEC Rep*. RID-4, Nov. 1968, 1–21.

The author said that that scanning technique had not been used in France, but he believed that it had been used in Canada.

In reply to a question by H. Müller, the author said that when they started work on radon detection in soil in 1959 flask detectors were used. Samples of soil air were collected inside the flasks and analysed later in the laboratory, but the results were unacceptable. The present method of pumping out soil air was then begun: it was best to pump at least 4 l of air from the soil. Radon measurement was only one method of prospecting, and the geologist might also use all the other methods discussed at the meeting. It was a question of selecting the best method for the area concerned.

The author then explained the mechanics of the pump used and why pumping was continued until stability was obtained in the zinc sulphide coated measuring chamber.

Dr. U. Klinge said that, several years ago, a boom in emanometry began in France: he wondered why the method was abandoned so soon afterwards. The author replied that emanometry was not exactly abandoned. The area of field operations determined the choice of techniques, and until a few years ago they were working mainly on vein deposits in France, for which the radon technique was useful. Since then they had expanded operations to other countries, where other techniques could be applied more successfully—especially the use of more sensitive scintillometers, giving precise results and good discrimination of anomalies.

B. O. Kvendbo said that he wished to comment on the importance of a careful study of the accessory minerals associated with the uranium mineralization so that a selection of deeper penetrating instruments could be used in the search for uranium together with radiometric instruments. For example, where magnetite was associated in sufficient and consistent amounts with the uranium, as at the Scandia Mining and Exploration, Ltd., uranium prospect at Lac Forestier in Quebec Province, Canada, a magnetometer survey proved to be very successful besides the several radiometric and geochemical methods used. Additional anomalies associated with uranium were discovered by the magnetometer survey at depths far beyond the reach of any present known radiometric, radon and geochemical methods. A simple rugged fluxgate magnetometer was used at low cost, as the survey lines had already been cut for the radiometric work.

Since other uranium deposits were also associated with considerable amounts of pyrite and other types of sulphides, the additional use of an electromagnetic instrument with considerable depth penetration might prove to be very useful, together with the other radiometric methods selected as most suitable for each particular situation. When Scandia Mining and Exploration, Ltd., conducted an airborne radiometric survey, an airborne magnetometer survey was always made at the same time; and they would even add an electromagnetic unit to the survey if it were found to be justified. If magnetic minerals or sulphide minerals were associated with uranium mineralization at an approximately constant ratio, that extra survey equipment would help in the search for deeply buried uranium deposits at very low cost.

The author said that he agreed with that: he and his colleagues used resistivity and magnetometer surveys in their work, although the latter had not been very effective.

In reply to a question by Dr. P. Johnson regarding the depth of overburden below which radiometric measurement became inefficient, the author said that that depended on the nature of the orebody. In Vendée, where the orebody was covered by 80 cm of soil, there was no surface radiometric anomaly. Generally, more than 50 cm of *barren* overburden would mask a moderately intense radioactive anomaly. Quite often, however, overburden was not barren and 'in-soil' techniques could be used.

Assessment of uranium by gamma-ray spectrometry

Leif Løvborg M.Sc.

Danish Atomic Energy Commission, Research Establishment Risø, Denmark

543.422.8:553.495.2

Synopsis

Gamma-ray spectrometry is a simple and versatile method of assaying uranium, thorium and potassium, provided that secular equilibrium in the U and Th decay series can be established or inferred. The technique is equally applicable in the laboratory and the field. At Risø, Denmark, laboratory gamma spectrometers have been developed which permit automatic analysis of crushed rock specimens and raw drill core. A portable gamma spectrometer specially designed to meet the requirements of field work in rough terrain proved capable of measuring the contents of U and Th in the surface of exposed rocks. The laboratory and field instruments were used for an evaluation of low-grade uranium resources in the Ilímaussaq alkaline intrusion, south Greenland.

On condition that ^{238}U and ^{232}Th are both in secular equilibrium with their respective daughters, and assuming that the ratio $^{40}K/^{39}K$ is constant in nature, gamma-ray spectrometry can be used for the direct determination of uranium, thorium and potassium in rocks. This method of analysis has been used extensively at various geological laboratories concerned with basic research on the distribution of the elements in the earth's crust. In the field of uranium prospecting airborne gamma spectrometry plays an important part as a remote-sensing technique for the detection of uranium anomalies. Assaying for uranium by gamma spectrometry is more rarely used—perhaps because the possibility of disequilibrium in the ^{238}U decay series is considered a strongly prohibitive factor. Once equilibrium has been inferred or established, gamma spectrometry is, however, a versatile method of assaying for uranium, combining simplicity of analysis with adaptability to a variety of tasks in the evaluation of uranium resources.

In the first place, gamma spectrometry is an obvious substitute for gross gamma-ray counting in the evaluation of potential uranium ores which contain abundant thorium. Most radiometric procedures for the evaluation of uranium resources can be directly replaced by gamma spectrometry. In the laboratory automatically operating gamma spectrometers are suitable for assaying surface samples and drill cores; in the field portable gamma spectrometers permit *in situ* assaying of exposed rock surfaces. Logging of drill-holes by gamma spectrometry is possible in principle, but ordinary exploration holes are too narrow to permit spectral gamma-ray logging.

Gamma spectrometry may also be used as an analytical aid in the tracing of uranium anomalies by systematic collection of rock specimens. Laboratory gamma spectrometers can be designed to provide the high sensitivity required in this kind of work. Again, simplicity and suitability for automation are factors which may promote the use of gamma spectrometry. Since weathering processes promote disequilibrium in the ^{238}U decay series, soils and stream sediments should not be analysed for uranium by gamma spectrometry.

The general aspects of gamma-spectrometric analysis for uranium, thorium and potassium have been thoroughly reviewed and discussed by Adams and Gasparini.[1] Another useful source of information on the subject can be found in a collection of papers edited by Adams and Lowder.[2] The present paper deals mainly with gamma-spectrometric techniques developed and used by the author and his colleagues in connexion with their studies of the uraniferous rocks of the Ilímaussaq alkaline intrusion in south Greenland.

Laboratory gamma spectrometry

One gramme of average crustal material containing 3 ppm U, 10 ppm Th and 2·5% K emits only about 0·3 gamma rays per second ($E\gamma > 0{\cdot}1$ MeV). For satisfactory gamma-spectrometric analysis of ordinary rocks, samples weighing several hundred grammes must be available, and strict precautions should be taken against the influence of extraneous sources of gamma radiation. The usable gamma radiation emitted by the sample materials has an energy range covering the decade from about 0·3 to 3 MeV. Gamma-ray detectors in which 3-MeV gamma rays are readily absorbed must therefore be used. In practice, only scintillation detectors equipped with large, thallium-activated sodium iodide crystals (NaI(Tl)) will be considered. Basic information on scintillation gamma spectrometry can be found elsewhere.[3]

BACKGROUND SUPPRESSION

Carefully planned shielding is the most important means of suppressing background gamma rays having their origin in the permanent surroundings of the gamma spectrometer. The concrete of the laboratory structure may be a major source of background radiation owing to its content of natural radioelements. Wollenberg and Smith[4] have shown that serpentinized ultramafic rock is a suitable aggregate for making low-radioactivity concrete. The terrain surrounding the laboratory also contributes to the background radiation, especially if the building is located on granitic bedrock. A shield equivalent to 15 cm of iron is about the optimum for reduction of these external radiations. Steel and lead are commonly used shielding materials.

An ordinary background shield has a limited effect on background arising from cosmic radiation and atmospheric radioactivity. Cosmic showers give rise to the production of gamma rays in the shield itself, and radon possibly emanating from the ground inevitably penetrates into the interior of the shield. The last-mentioned phenomenon can be very troublesome, a variable background being introduced by fluctuations in the radon content of the local atmosphere. High and variable radon contents must be expected in the vicinity of porous, uraniferous rocks. Residual background also occurs as a result of weak radioactivity in the shielding material and in the components of the detector unit(s). Steel and lead may contain traces of ^{60}Co and ^{210}Pb, respectively. Furthermore, a crystal–photomultiplier assembly usually contains minute amounts of ^{40}K, the glass face of the photomultiplier being the main contributor of ^{40}K background gamma rays.

Fig. 1 shows two background gamma spectra recorded with a multi-channel analyser in the gamma-spectrometry laboratory at Risø. This laboratory is located in a concrete-structured basement surrounded by glacial, alluvial deposits. One background spectrum (no. 1) applies to a standard assembly of a 15 cm diameter × 10 cm high crystal and a 5-in photomultiplier; the other (no. 2) applies to a low-background assembly of a 12·5 cm diameter × 12·5 cm high crystal, a 7·5 cm thick light guide consisting of pure NaI, and a 5-in photomultiplier equipped with a low-radioactivity base. The background measurements were made with the detector units mounted in a shield equivalent to 10 cm of lead. According to Courbois and co-workers,[5] the externally generated

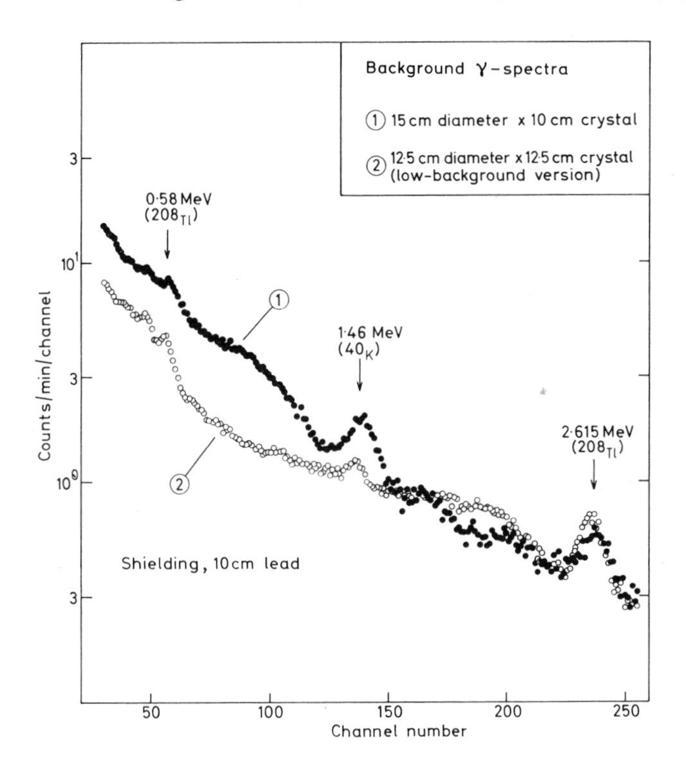

Fig. 1 Background gamma spectra recorded in basement of Electronics Department building, Risø

background is approximately proportional to the overall crystal geometric cross-section, so the two crystals used were considered comparable with regard to their responses to environmental and cosmic background sources. Distinct gamma-ray peaks with energies of 0·58 and 2·62 MeV in the spectra show that thorium is present in the concrete structure of the laboratory. The absence of peaks with an energy of 1·76 MeV indicates that the room was essentially free from radon when the spectra were recorded. Significant radon levels have never, in fact, been observed in this laboratory. A comparison between the two background spectra in Fig. 1 clearly demonstrates the substantial reduction in ^{40}K background which can be obtained by use of a specially designed crystal–photomultiplier assembly.

DATA RECORDING AND DATA REDUCTION

Gamma-spectrometric rock analysis can either be based on multi-channel spectrum recording or on three-channel gamma-ray counting. In both cases concentration levels for U, Th and K are determined from a comparison of a set of sample spectral data with three sets of standard spectral data. Suitable uranium and thorium standard samples can be prepared on the basis of analysed, powdered material available for sale at the New Brunswick Laboratory of the

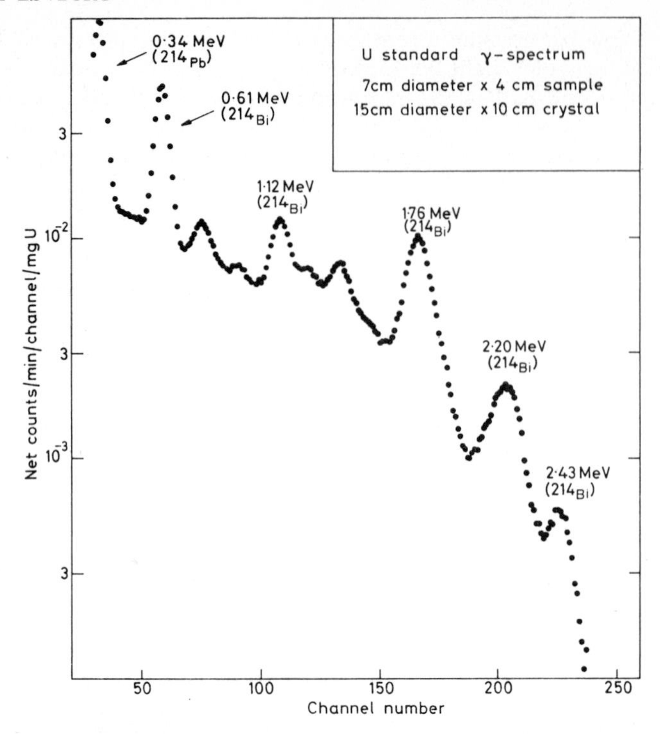

Fig. 2 Normalized gamma spectrum of ^{238}U in secular equilibrium

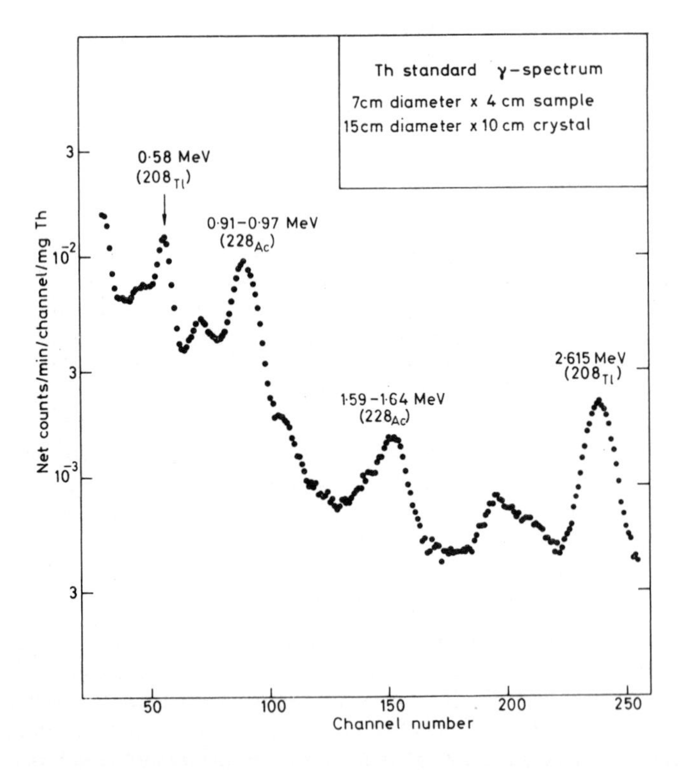

Fig. 3 Normalized gamma spectrum of ^{232}Th in secular equilibrium

USAEC. A potassium standard sample may simply consist of KCl or another potassium salt.

Typical multi-channel standard gamma spectra normalized to net counts per minute per unit weight of U, Th and K, respectively, are shown in Figs. 2, 3 and 4. Provided that all samples, inclusive of the standard samples, fill up exactly the same volume in the gamma spectrometer, the spectrum of a sample is given by a linear combination of the background spectrum and the three standard

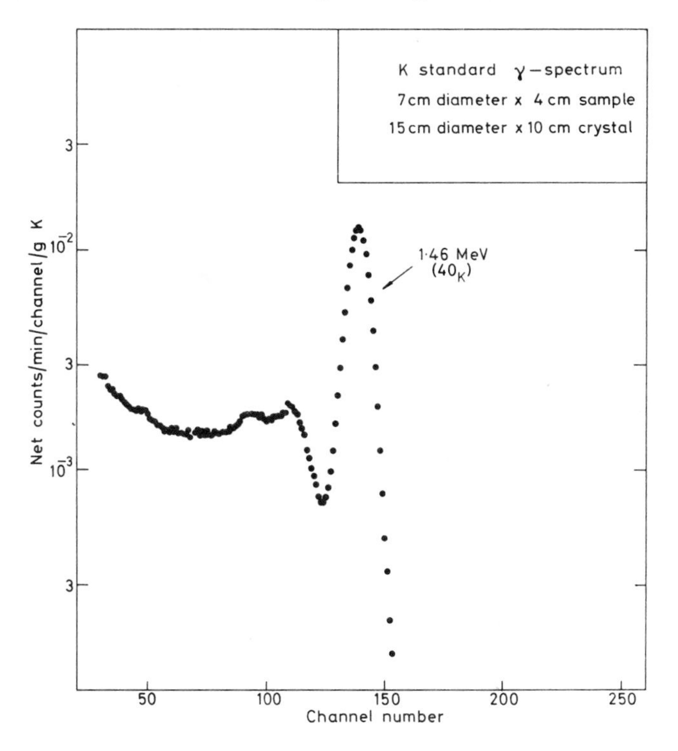

Fig. 4 Normalized gamma spectrum of ^{40}K, assuming ^{39}K/^{40}K = 8500

spectra. Thus, the sample contents of U, Th and K can be determined through application of the method of least squares to the background-corrected sample spectrum, as described by Rybach and co-workers.[6] This kind of data reduction presupposes that a digital computer is available. Spectrum read-out on punched or magnetic tape is commonly used for transfer of data from multi-channel analyser to computer. An example illustrating the performance of the least-squares method is given in Fig. 5, in which the solid curve fitting the sample spectrum represents the spectrum generated by the computer.

Analysis by three-channel gamma-ray counting is based on the solution of three linear equations, which relate the counting rates in three energy channels corresponding to the contents of U, Th and K of the sample. The instrumentation required comprises a timer and three single-channel analysers, each connected to a scaler. One energy channel should cover the 1·46 MeV total-absorption peak in the gamma spectrum of ^{40}K (Fig. 4), whereas the two others can be more freely adjusted. Settings at the 1·76 and 2·62 MeV peaks in the gamma spectra of U and Th, respectively (Figs. 2 and 3), are often preferred as the corresponding equations then permit uranium and thorium to be determined independently of potassium. Energy intervals located between 0·3 and 1·3 MeV can, however, also be used. Although three equations with three unknowns are

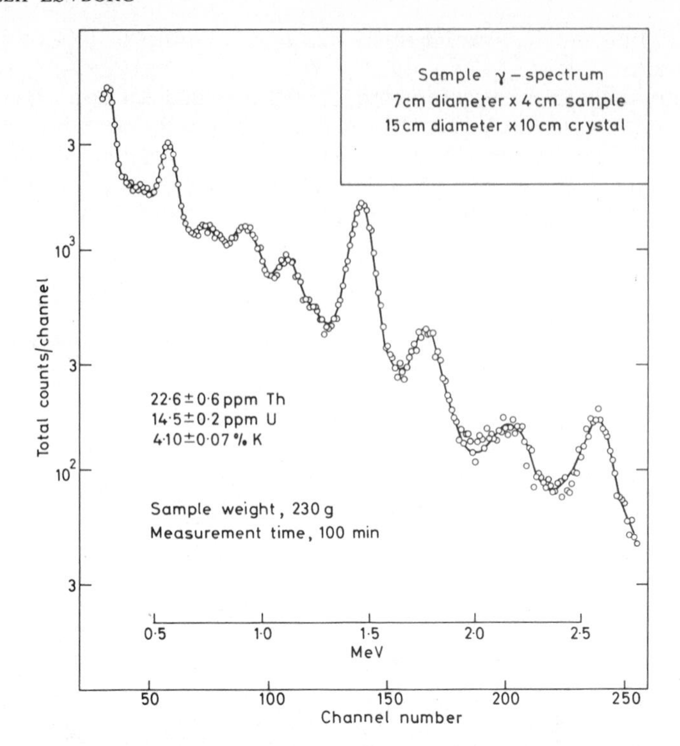

Fig. 5 Measured sample gamma spectrum (circles) and gamma spectrum generated with a least-squares fit (solid curve) of sample spectrum to standard spectra of U, Th and K

relatively easy to solve on a desk calculator, an estimation of the experimental standard errors arising from counting statistics involves extensive calculations. Data read-out on punched tape—for instance, by means of a Teletype printer—followed by data reduction on a digital computer is therefore recommended.

In the present author's experience the two gamma-spectrometric techniques outlined above are quite comparable in regard to their precision of measurement: given the time of measurement, three-channel counting on the 1·46-, 1·76- and 2·62-MeV energy peaks has about the same instrumental precision as multi-channel spectrum recording with subsequent use of the least-squares method. Both techniques presume a long-term gain stability of $\pm 0{\cdot}5\%$ or better of the spectrometer system. As gain drift is a common phenomenon in scintillation gamma spectrometry, mostly owing to the temperature sensitivity of the photo-multiplier(s), it may often be necessary to incorporate an analogue or digital gain stabilizer in the system. Possible gain drift is immediately revealed in the form of abnormally high standard errors on results obtained by the least-squares method, but simultaneous equation calculations provide no means of detecting instrumental instability.

ASSAYING OF HAND SPECIMENS

Rock samples in the form of hand specimens have to be crushed and homogenized before they can be analysed by gamma spectrometry. Coarse crushing by means of a jaw crusher, followed by grinding to a grain size of about 2 mm, is suitable. Sample containers in the form of regular plastic cannisters are commonly used. The filled containers are weighed and then carefully sealed in order to prevent contamination of the gamma spectrometer. A storage time of about two weeks before measurements are made secures the replacement of radon that may have been lost from the sample materials in the grinding process.

Fig. 6 shows an automatically operating gamma spectrometer developed by the Electronics Department at Risø. Cylindrical sample containers (volume, 150 ml) are placed in the magazine on the right, from which they are successively passed through the spectrometer by means of a motor-driven wheel which makes one-quarter of a revolution at a time. The background shield is 12 cm thick and has overall dimensions of 60 cm in diameter × 85 cm long. The shielding material consists of an easily workable alloy of 84% Pb, 12% Sb and 4% Sn, which is also attached to the edge of the sample-changer wheel so that complete shielding is obtained in each of the four positions of the wheel. The scintillation crystal faces the bottom of each sample at a distance of about 2 mm. Until now, a standard assembly of a 15 cm diameter × 10 cm high NaI(Tl) crystal and a 5-in photomultiplier has been used.

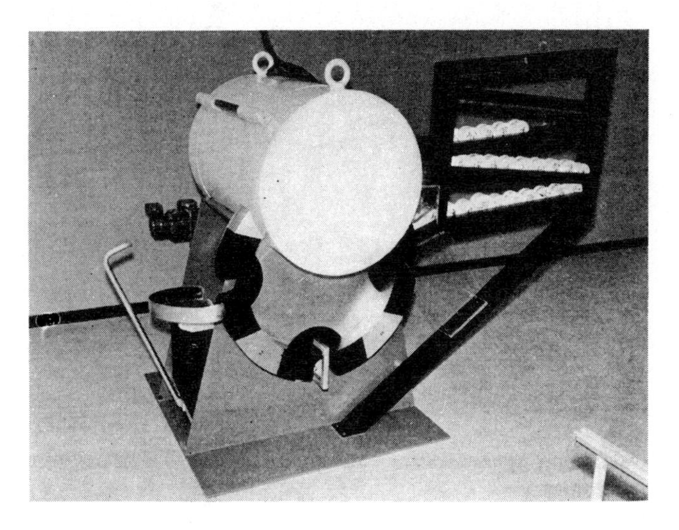

Fig. 6 Automatically operating gamma spectrometer for assessment of crushed material. Sample containers used are 7 cm in diameter and 4 cm in thickness (corresponding to sample weight of about 230 g)

The spectrometer described above is normally operated in the three-channel mode of analysis with the channels set on the 1·46-, 1·76- and 2·62-MeV peaks, respectively. The three standard samples contain 1000 ppm U (NBL no. 74), 1010 ppm Th and 40 ppm U (NBL no. 80), and 52·44% K (pure KCl), respectively. A calibration run which also includes a background measurement is usually made at least once a week. In spite of the heavy shielding, the background count rates are fairly high (45, 24 and 12 counts/min), but stable. As was mentioned earlier, thorium in the concrete of the laboratory structure is the major background source. The measurement time normally used is 100 min/sample, giving a measurement capacity of 14 samples per 24 hours. The corresponding detection limits for assaying about 230 g of material are roughly 0·2 ppm U, 0·5 ppm Th and 0·1% K. Ordinary igneous rocks can be analysed with a precision of about 10%.

ASSAYING OF DRILL CORE

Because of its regular shape, solid drill core can be assayed non-destructively by gamma spectrometry: this kind of rock analysis has been described by Rybach and Adams.[7] Their procedure consisted of splitting the core into 15-cm long sections followed by successive measurements of the core sections by means of a gamma spectrometer equipped with an automatic sample changer. A further development of this technique has been made at Risø, where gamma-spectrometric core scanning came into use in 1969.

The core-scanning apparatus designed and constructed at Risø is shown in Fig. 7. The table in the centre of the photograph carries a background shield made of lead. The centre part of the shield forms a 60 cm long, 4·7 cm wide tunnel through which 3·2- or 4·2-cm drill core can be guided by means of a motor-driven conveyor belt. The two side parts of the shield each form a

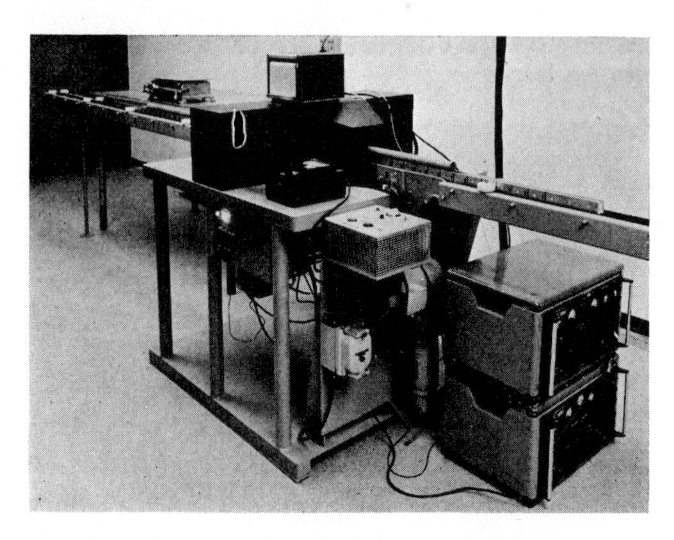

Fig. 7 Core-scanning apparatus for continuous or stepwise scanning of drill core by gamma spectrometry

housing for a scintillation detector equipped with a 5 cm diameter × 7·5 cm long crystal. Each crystal is aligned so that it views the drill core at a distance of 2·5 cm between the crystal face and the core axis. Shielding in the downward direction is provided by a lead bar fastened to the underside of the table. The U-shaped carriage for the drill core is 2·5 m long; it rests on the conveyor belt, from which it can slide on to two rollways on either side of the table.

Having passed a summing amplifier, the output signals from the two scintillation detectors are fed to a multi-channel analyser, the operation of which is controlled by a photoelectric cell located at the entrance of the background shield. A control signal is obtained each time a pair of opposing holes in the sides of the carriage allows a beam of light to shine on the photo-cell. The pairs of holes are located at 5-cm intervals, permitting a control signal to be generated for each 5 cm of core movement. The functioning of the control system depends on a choice between one or other of the two scanning modes described below.

For *continuous scanning* the conveyor belt moves at a constant speed (typically, 1–5 cm/min). Only two pairs of holes separated by several tens of centimetres are left free to produce control signals, the rest being made inoperative by means of black tape covering one side of the carriage. The first control signal switches the multi-channel analyser into the analysis mode of operation; the second stops the spectrum recording. In this way a gamma spectrum is obtained from which can be derived information on the average contents of U, Th and K in the scanned section of core. The scanning interval normally used is 1 m, which corresponds to the assaying of about 2 kg of drill core per scan. Core material missing in the joints between core pieces is accounted for by weighing the core sections before they are scanned. Variations in total radioelement concentration along the drill core are registered by a strip-chart recorder connected to a rate meter which monitors the gross gamma-ray counting rate.

Step scanning consists of an automatically controlled sequence of point measurements at preselected locations along a length of drill core. Measurement locations are made with respect to the homogeneity and the quality of the core, and of the time available for the scanning sequence. Locations characterized by poor core recovery must be left out. The minimum spacing between measurement locations is 5 cm, as determined by the spacing between hole pairs in the carriage; greater spacing can be obtained by covering holes with strips of black tape. The time for a single measurement is normally 20 or 40 min. Each gamma spectrum is automatically read out on punched tape, after which the core rapidly advances to the next measurement location. The scanning resolution is two to three times better than the resolution which can be obtained in gamma-ray logging of drill-holes.

Both continuous and step scanning modes depend on calibration measurements of standards which consist of sealed aluminium tubes filled with closely packed powders of known radioelement contents. Assaying by continuous scanning of 1-m long core sections has been found to be accurate and precise to within $\pm 10\%$ or better. Assaying by step scanning is less accurate and precise; peak values in core contents of U and Th arising from strongly localized mineralization can be estimated with an error margin of about $\pm 15\%$.

Field gamma spectrometry

The gamma-ray flux at the surface of a rock outcrop has an energy distribution which is strongly influenced by the interaction of the gamma rays emitted by the radioelements in the rock with the rock material. Since most rocks are characterized by average atomic numbers between 10 and 13, Compton scattering is the predominant process of interaction, resulting in a build-up of a flux component which shows a continuous energy distribution. A field gamma spectrum is therefore different from a laboratory gamma spectrum of the same material. This difference is very pronounced in the energy region below 1 MeV, where the gamma-ray peaks in the field spectrum are hardly recognizable against the Compton continuum. Determination of U and Th by field gamma spectrometry must therefore be based on measurements of their high-energy gamma rays. Simultaneous or successive counting in channels centred on the 1·76-, 2·62- and 1·46-MeV energy peaks is the preferred method of data recording in complete field assessment of U, Th and K.

Battery-operated, portable field gamma spectrometers designed on the basis of NaI(Tl) scintillation detectors and three-channel scalers or rate meters are now commercially available. Their successful use depends on the exploration geologist's proper understanding of the peculiar characteristics of *in situ* assaying by gamma spectrometry. Thus, it must be realized that the counting rate observed in each of the counting channels is governed by the solid angle of detection and the linear attenuation coefficient for gamma rays of the corresponding energy. It is therefore important to maintain a constant angle of detection at all measurement localities. Since linear attenuation coefficient equals density times mass absorption coefficient, the latter being almost independent of chemical composition, a field gamma spectrometer has essentially the same calibration in all geologic environments.

An isotropic gamma-ray detector views the terrain at a solid angle which depends on the local topography. In flat terrain a constant solid angle of detection equal to 2π can be obtained by elevation of the detector unit to a height exceeding the maximum level of the surface irregularities. If the terrain is very rough, it is generally necessary to use a collimated gamma-ray detector. A suitable collimator which suppresses 2·62-MeV gamma rays coming from the sides by a factor of ten may simply consist of a 5-cm thick lead ring surrounding the scintillation crystal (the extra weight of this lead ring adds significantly to the total weight of the spectrometer).

If absorption and scattering of gamma rays in the air are disregarded, the counting rates observed above an exposed, infinite rock layer are independent of detector elevation. The elevation of the detector is, however, the factor which controls the amount of material being assayed. As was suggested by Løvborg *et al.*,[8] it is adequate to introduce the concept of an effective sample, which is defined as the volume of rock in which the variance of a particular radioelement (U, Th or K) is equal to the estimation variance of a gamma-spectrometric determination of this radioelement. It follows from this definition that the effective sample has the same statistical significance as a real sample of the same size. Based on the exponential and inverse-square laws of gamma-ray attenuation, it is possible to calculate the effective sample viewed when field assaying in flat terrain. Assessment of U and Th with an isotropic detector with a centre elevation of 5 cm is characterized by a bowl-shaped effective sample of thickness $\sim$ 15 cm, surface diameter $\sim$ 80 cm and mass $\sim$ 50 kg. Therefore, much more material than that covered by a normal hand specimen can be assayed by a single field measurement. The diameter of the effective sample increases almost in proportion to the elevation of the detector, whereas its thickness remains essentially constant. A configuration in which an uncollimated scintillation crystal views the terrain at a distance of a few tens of centimetres is recommended as a reasonable compromise between the contradictory requirements of constant solid angle of detection and high horizontal resolution.

Calibration of portable gamma spectrometers presents a rather difficult problem. One method of calibration is based on a collection of rock samples at a number of field-measurement localities. After laboratory assessment of the rock samples, linear regression analysis is applied to the sets of field counting rates with the sample contents of U, Th and K taken as the independent variables. This technique has been used by Løvborg and co-workers[9] and by Killeen and Carmichael.[10] Because of the great difference in the weight of material assayed in the laboratory and in the field, the statistical errors on the calibration constants obtained are often unacceptably high. Moreover, it is very time-consuming to collect and analyse the great number ($\approx$ 100 or more) of samples required to obtain usable calibration constants.

A more direct calibration can be made on the basis of artificial calibration standards. These may conveniently have the form of concrete slabs containing known, homogeneously distributed amounts of U, Th and K. Uranium and thorium ore concentrates and crushed potassium feldspar are suitable materials for making radioactive concrete as they can be mixed with sand and cement. The three standards belonging to a generally applicable calibration facility should each be at least about 3 m in diameter and about 0·5 m in thickness.

Fig. 8 shows in operation a field gamma spectrometer developed by Løvborg.[11] After adjustment, this spectrometer permits consecutive counting in the 1·76- and 2·62-MeV energy channels with the transparent lid of the scaler unit in its permanent position. Measurements can therefore be made even in wet weather. Because of the rough terrain seen on the photograph, a collimated detector is used. The crystal size is 7·5 cm in diameter $\times$ 7·5 cm high, and the collimator consists of a stainless steel pot lined with 5 cm of lead. The total weight of the detector unit is 36 kg, which could be accepted as the spacing between measurement localities was no more than 1 m. Compared with an isotropic detector, the collimated detector in Fig. 8 reduces the diameter of the effective sample by a

Fig. 8 ***In situ* assessment of U and Th in Ilímaussaq intrusion, south Greenland. Portable gamma spectrometer operates on basis of collimated detector unit providing a constant solid angle of detection in rough terrain**

factor of 2·5. Calibration of the spectrometer was obtained through measurements on two concrete standards containing 250 ppm U and 140 ppm Th, respectively. The accuracy of measurement resulting from calibration uncertainties is about 2·5% in assaying Th, and between 2·5 and 7·5% in assaying U, depending on the Th/U ratio. Important factors governing the uncertainty of measurement are the topographical features of the terrain, the distribution of U and Th in the rocks, instrumental stability and counting statistics. Under favourable circumstances an accuracy and precision of about 10% may be obtained in field assaying of rocks containing several hundred ppm U and Th.

Applications

The gamma-spectrometric instrumentation developed at Risø was used for an evaluation of uranium resources in the Ilímaussaq alkaline intrusion, south Greenland. As described by Sørensen,[12] these resources are mainly in a complex body of uraniferous rocks in the northernmost part of the intrusion. The highest contents of both uranium and thorium are found in fine-grained lujavrite, essentially arfvedsonite–aegirine/acmite nepheline syenite, and in a migmatite consisting of deformed and metasomatically altered country rocks and fine-grained lujavrite cut by veins of medium- to coarse-grained lujavrite. The U and Th mineralization is characterized by the occurrence of steenstrupine, a uranium-

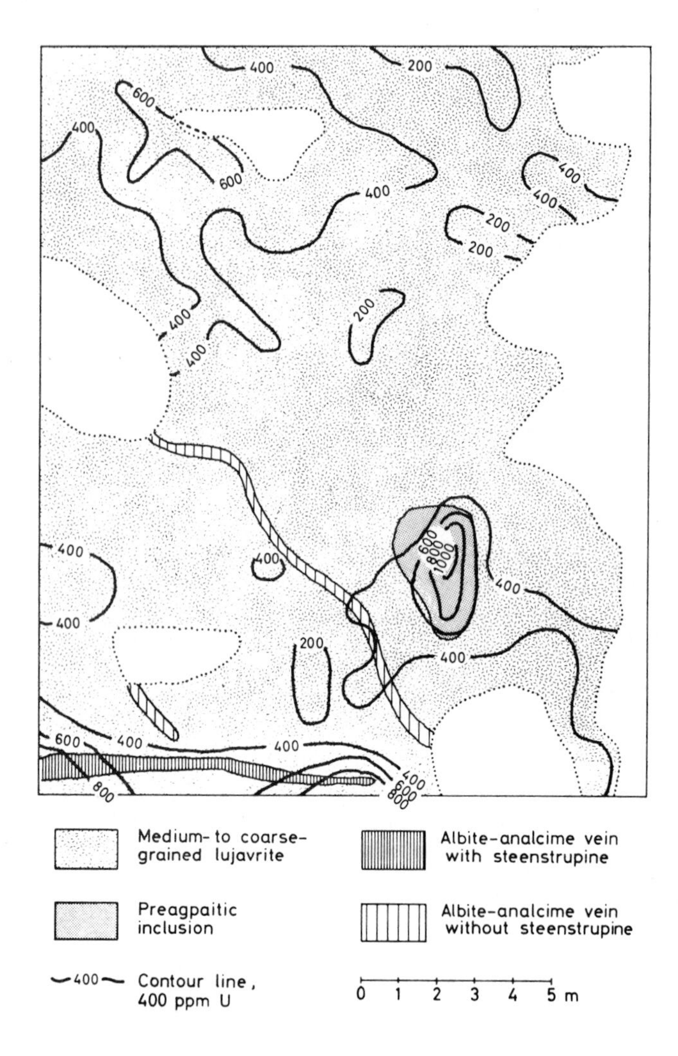

Fig. 9 Contour lines for concentration of uranium superimposed on geologic map of intensively mineralized area in Ilímaussaq intrusion. Contour lines were drawn on basis of 225 single measurements with instrumentation shown in Fig. 8

rich variety of monazite, eudialyte and thorite. Uranium also occurs adsorbed on pigmentary material (an oxide). Whole-rock contents of Th range from several hundred to several thousand ppm, and those of U from a few hundred to more than 1000 ppm. Consistent with the fact that the Ilímaussaq rocks are about 1000 m.y. old, and that they have been subjected to moderate chemical weathering in recent times, no significant deviation from secular equilibrium in the ^{238}U decay series has ever been observed in rock specimens from the Ilímaussaq intrusion.

The field work forming the basis for an evaluation of uranium resources comprised detailed geologic mapping followed by collection of hand specimens and grid measurements with the portable gamma spectrometer shown in Fig. 8, and the sinking of about 50 diamond drill boreholes. The hand specimens and the drill cores were brought to the laboratories at Risø for U and Th analysis by gamma spectrometry.

Gamma spectrometry of crushed hand specimens proved useful, particularly for obtaining data on U and Th concentration levels in the areas surrounding

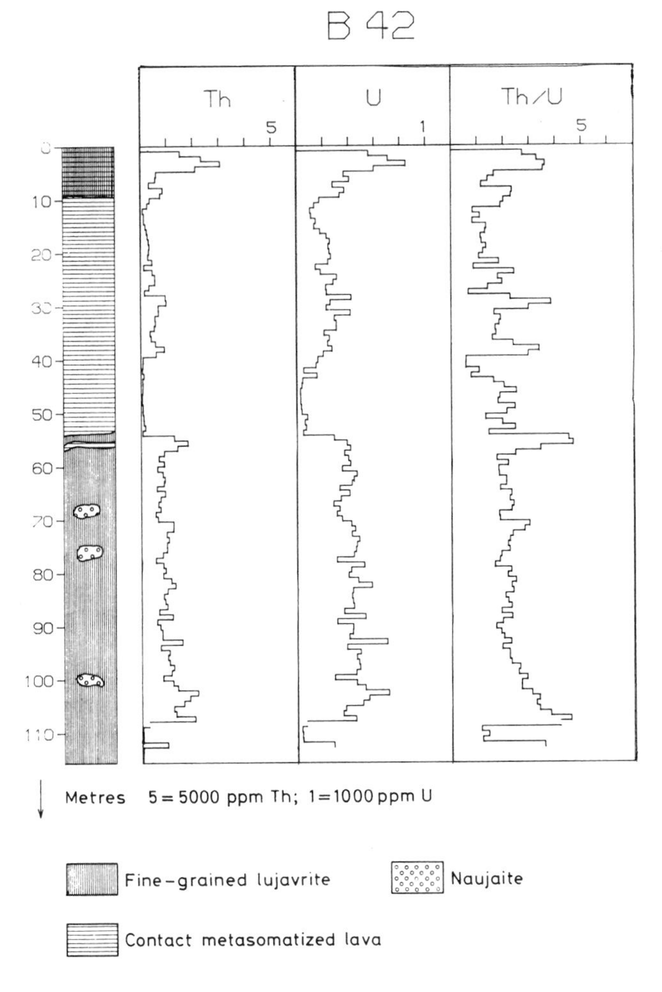

Fig. 10 Depth distributions of U, Th and Th/U as determined from continuous scanning of 1-m long core sections. Histograms were generated by computer-controlled digital plotter; drawing of core lithology was done by hand

the strongly mineralized zone. Thus, rather than carrying the portable gamma spectrometer over long distances in mountainous terrain, it was considered more expedient to radiometrically grid the mineralized outcrop with lightweight Geiger counters and collect rock samples at regular intervals.

As was described earlier,[8] field gamma spectrometry was successfully used for the determination of U and Th surface concentration levels in selected areas of fairly small size ($\sim$ 50 m $\times$ 50 m). The example given in Fig. 9 shows surface *iso*-concentration lines for uranium, drawn on the basis of 225 one-metre-spaced, single measurements. There is a close correlation between contour lines and lithology. With a total of more than 4000 field assays reliable frequency distributions of U and Th were obtained which showed that specific rock types have characteristic contents of U and Th.

A calculation of the grade and tonnage of the potential uranium ore was made by a division of the strongly mineralized rock body into about 50 triangular prisms, each defined by three neighbouring drill-holes. A total of 3500 m of core recovered from these holes was analysed for uranium and thorium by a progressive, continuous scanning of 1-m long core sections carried out with the core-scanning apparatus. After the core had been scanned from each drill-hole, the vertical distributions of U and Th and the ratio Th/U were represented in the form of a computer-generated plot (Fig. 10). From an inspection of these plots an 'ore zone' could be assigned to each drill-hole, and the tonnage of uranium in each prism was then calculated by means of standard procedures of estimation. The task of scanning the core lasted seven months.

Core scanning by gamma spectrometry was also used for geochemical studies of the Ilímaussaq rocks. Again, continuous scanning of 1-m long core sections proved useful for whole-rock assaying, and step scanning carried out at intervals of 5 cm was found to be an expedient technique for localization and estimation of peak contents of U and Th.

Conclusion

Gamma spectrometry is a radiometric method of assessment which can be used for the determination of uranium, thorium and potassium. The method is an obvious substitute for gross gamma-ray counting in the exploration and evaluation of uranium ores containing abundant thorium. Automatically operating gamma spectrometers can be designed which permit efficient laboratory assay of crushed material and raw drill core. Furthermore, *in situ* assessment of exposed rock surfaces can be made with commercially available, portable gamma spectrometers.

Analysis for uranium by gamma spectrometry depends on the assumption of secular equilibrium in the ^{238}U decay series. Where this assumption is doubtful, other, generally more complicated, techniques must be used.

Acknowledgment

The author wishes to express his gratitude to staff of the Electronics Department, Risø, the Institute of Petrology, Copenhagen University, and the Geological Survey of Greenland for their cooperation on the development and use of gamma-spectrometric instrumentation. Special thanks are due to Harold Wollenberg of the Lawrence Radiation Laboratory, Berkeley, for his creative work in the field of gamma-spectrometric rock analysis during his two-year stay as a guest scientist at Risø.

References

1. Adams J. A. S. and Gasparini P. *Gamma-ray spectrometry of rocks* (Amsterdam: Elsevier, 1970), 295 p.
2. Adams J. A. S. and Lowder W. M. eds. *The natural radiation environment* (Chicago: The University of Chicago Press, 1964), 1069 p.
3. Crouthamel C. E. ed. *Applied gamma-ray spectrometry* (Oxford, etc.: Pergamon, 1960), 443 p.
4. Wollenberg H. A. and Smith A. R. A concrete low-background counting enclosure. *Health Phys.*, **12**, 1966, 53–60.
5. Courbois Th., Van Gelderen L. and Leutz H. Background, peak efficiency and dimensions of NaI(Tl)-crystals. *Nucl. Instrum. Meth.*, **69**, 1969, 93–100.
6. Rybach L., von Raumer J. and Adams J. A. S. A gamma spectrometric study of Mont-Blanc granite samples. *Pure appl. Geophys.*, **63**, 1966, 153–60.
7. Rybach L. and Adams J. A. S. Automatic analysis of the elements uranium, thorium and potassium in solid rock samples by nondestructive gamma spectrometry. In *Proc. Conf. appl. phys.-chem. Methods chem. Anal., Budapest*, **2**, 1966, 323–30; *Chem. Abstr.*, **69**, 1968, 8212.
8. Løvborg L. *et al.* Field determination of uranium and thorium by gamma-ray spectrometry, exemplified by measurements in the Ilímaussaq alkaline intrusion, south Greenland. *Econ. Geol.*, **66**, 1971, 368–84.
9. Lövborg L., Kunzendorf H. and Hansen J. Use of field gamma-spectrometry in the exploration of uranium and thorium deposits in south Greenland. In *Nuclear techniques and mineral resources* (Vienna: IAEA, 1969), 197–211.
10. Killeen P. G. and Carmichael C. M. Gamma-ray spectrometer calibration for field analysis of thorium, uranium and potassium. *Can. J. Earth Sci.*, **7**, 1970, 1093–8.
11. Lövborg L. A portable γ-spectrometer for field use. *Risø Rep.* no. 168, 1967, 36 p.
12. Sørensen H. Low-grade uranium deposits in agpaitic nepheline syenites, south Greenland. In *Uranium exploration geology* (Vienna: IAEA, 1970), 151–9.

DISCUSSION

Dr. H. D. Fuchs said, with regard to the spectrograms of uranium, thorium and potassium, that the uranium spectrogram had two uranium energy peaks close to the main thorium peak. He asked if it were possible to discriminate against those uranium peaks in order to get the 'real' thorium intensity.

The author replied that the 2·2- and 2·4-MeV uranium series gamma emissions made a small but significant contribution to the count rate in the thorium channel. That should be calculated by reference to the uranium channel and subtracted.

In reply to R. D. Pratten's question as to whether the drill core spectrometer could accommodate large variations in grade over short distances within the length of core being scanned, the author said that the instrument was capable of analysing cores containing several thousand ppm uranium, but, if the count rate should become too great, the detecting crystal could be replaced with one of smaller size.

With regard to the problems involved in analysing broken core with the spectrometer, queried by G. R. Ryan, the author said that broken core could only be analysed by continuous scanning. Experiments with core containing homogeneously distributed uranium and thorium had shown that up to 30% of the core could be missing before there was a noticeable effect on the scanning records.

I. R. Wilson asked for further comments on the various U/Th/K ratios as an index of magmatic differentiation and for identification of some of the unfamiliar rock types mentioned in the paper.

B. L. Nielsen, in reply, said that the ratio between uranium, thorium and potassium could not generally be used to evaluate the state of magmatic differentiation. In the Ilímaussaq intrusion, south Greenland, they had found certain characteristics and trends within each main rock type; those, however, could not be extrapolated to magmatic rocks on a broader scale.

The unusual names of the rocks of the intrusion, e.g. naujaite and lujavrite, referred to undersaturated agpaitic nepheline syenites. Lujavrite, the main radioactive rock type, was a laminated melanocratic rock composed of the minerals albite, microcline, arfvedsonite, aegirine, analcite and nepheline. In addition, there were many accessory and secondary minerals characterized by such rare elements as Th, U, Nb, Be, Zr, Cl and F.

Dr. T. F. Schwarzer said that she would like to make two comments. First, regarding the ability to detect magmatic differentiation by gamma spectrometry, she and her colleagues had flown a pluton in Puerto Rico that was differentiated from perimeter to core, ranging from more basic rocks on the outside to more acid rocks in the interior. The airborne gamma-ray survey revealed regular variations of thorium, uranium and potassium from rim to core of the body, the absolute abundances of those elements increasing toward the centre. The ratios, however, showed little variation. Her second point was concerned with the rather pessimistic attitude which prevailed regarding measurement of overburden as against outcrop. It had been her experience that in many areas where *in situ* soils existed, measurements made on the soil were very similar to those on underlying bedrock, allowing the identification of underlying lithology and the location of contacts. That conclusion might present a more hopeful picture to people who had to work in areas with little or no outcrop.

Dr. F. Q. Barnes said that in his experience radioactivity at the edges of batholiths was related to the concentration of secondary uranium salts, and that feature was influenced primarily by structure, both major and minor (e.g. contacts and cooling joints), rather than by major changes in the clarkes in relation to the edge of batholiths.

He asked the author the type of down-the-hole logging equipment he used. Judging by the tube in the standardization cylinder, the sonde or probe would have to be of small diameter. The author said that the probe was part of an American portable borehole logging equipment with a $\frac{1}{2}$ in × $1\frac{1}{2}$ in scintillation crystal.

In reply to questions by G. R. Ryan on the effect of sample size and density on the analysis of ground and pulverized samples and the number of detectors employed, the author said that with 7 cm × 4 cm plastic containers there was no grain size effect, because the samples were too small to attenuate the high-energy radiations being measured. To be comparable with the standard samples it was most important that the containers be filled completely, but that requirement would not matter so much if a well-type detector were to be used. The mechanical arrangement of the automatic sample changer allowed only one detector to be incorporated in the equipment, but they had another non-automatic assayer in which the sample was placed between two crystals.

Dr. S. H. U. Bowie posed a question concerning radon effects. In the United Kingdom they had tried to measure the effect of so-called radon loss due to grinding, by use of glass containers, but had never been able to detect it. The loss did not seem to be very significant and was almost certainly due to the mineralogy of the particular type of material with which one was dealing. That was because the amount of radon that was lost was related essentially to the surface area of the particular type of radioactive mineral in question. In the western U.S.A. the grain size was very small and perhaps the consequent very considerable surface area of liberation produced effects very much greater than they would be at Ilímaussaq, where the radioactive minerals were refractory and enclosed within the ore gangue.

P. H. Dodd said that the grain size was important in placing the locus of the decay of radium within half a micron of the surface so that the radon could escape following recoil and diffusion.* Emanation also depended on the crystal lattice and to a lesser

*Tanner A. B. Radon migration in the ground: a review. In *The natural radiation environment* Adams J. A. S. and Lowder W. M. eds. (Chicago: University of Chicago Press, 1964), 161–90. Zimens K. E. Oberflächenbestimmungen und Diffusionsmessungen mittels radioaktiver Edelgase. *Z. phys. Chem.*, **192**, 1943, 1–55.

extent on the moisture content of the radioactive mineral.* Barretto had produced curves relating emanation to grain size and petrography.† He found that grain size was important below 74 mesh. Emanations of 1–25% were measured from rocks crushed to the same grain size class and values of 20–70% were noted in soils. Work by R. D. Evans (personal communication, 1971) on the radium content of teeth and bone had necessitated sealing the samples in copper containers before gamma counting. Samples for gamma counting at the AEC laboratory in Grand Junction were sealed in 'tin cans' to prevent loss of radon daughters, which had ranged from 4 to 54% for rocks, soils and ores, and averaged about 14% for most sandstone-type 'black ores'; thoron losses up to 7% were noted.‡

*Shashkin V. L. and Prutkina M. I. Mechanism underlying the emanation of radioactive ores and minerals. *Atomn. Energ.*, **29**, no. 1 1970, 41–2: *Soviet atom. Energy*, **29**, 1970, 724–6.

†Barretto P. M. C. Radon-222 emanation from rocks, soils, and lunar dust. M.S. thesis, Rice University, Houston, Texas, 1971.

‡Bush W. E., Higgins L. J. and Shay R. S. Determination of equivalent uranium (eU_3O_8) by 'gamma only' analysis. In *Handbook of analytical procedures. U.S. A.E.C.* RAMO-3008, 1970; *Nucl. Sci. Abstr.*, **24**, 1972, 20687.

Scott J. H. and Dodd P. H. Gamma-only assaying for disequilibrium corrections. *U.S. A.E.C.* RME-135, 1960, 21 p.

Airborne gamma-ray survey techniques*

A. G. Darnley M.A., Ph.D., M.I.M.M.

Geological Survey of Canada, Ottawa, Ontario, Canada

550.837.84:553.495.2

Synopsis

Airborne instrumentation was first extensively used in the search for uranium mineralization in 1949. During the next 22 years instrumentation developed from Geiger-counter installations, through small total-count radiation detectors to gamma-ray spectrometers with very large detector volumes. There are a variety of spectrometer systems available for commercial use, and in order to use them in the most effective way it is necessary to consider the problems involved. Districts containing uranium mineralization generally fall within or on the margin of regions containing above-average abundances of all the radioelements. These radioactive regions can readily be found by airborne surveys which measure only total radioactivity. Sometimes, the mineralization controls within an area may be so well known that it is unnecessary to use spectrometer systems. In general, it is advantageous to use the discriminating ability of a spectrometer to identify anomalies containing significant amounts of uranium. This selection of anomalies is essential wherever the costs of ground work are high.

The most sensitive spectrometers available are capable of measuring the mean ground level abundances of potassium, uranium and thorium, which can be expressed in conventional units of concentration. From an exploration viewpoint the most important parameters which can be measured are the relative concentrations of uranium to thorium and uranium to potassium. These are diagnostic for identification of zones of anomalous uranium concentration. Good counting statistics are essential in order to obtain meaningful ratio measurements. These are obtained by maximizing detector volume and minimizing ground speed of the aircraft. Minimizing flight elevation improves sensitivity for small area targets, but it entails closer line spacing for a given percentage coverage of an area. As a general principle, airborne gamma-ray surveys should be aimed primarily at finding target zones with linear dimensions of the order of hundreds or thousands of feet, where the mean enrichment of uranium at surface may be only one or two parts per million higher than the surroundings. It can be shown that high-sensitivity equipment, such as that operated by the Geological Survey of Canada, is most cost-effective for this purpose.

The availability of fully corrected data, with a comprehensive display of all the parameters in profile and map form, is essential for maximum value to be extracted from a survey with high-sensitivity equipment.

There have been two periods of rapid growth in the development of airborne uranium survey equipment: the first was between 1948 and 1950, and the second between 1966 and 1969.

The earliest reported proposal for the use of aircraft for uranium prospecting was made in a memorandum dated 4 November, 1944, from Dr. B. Pontecorvo to Dr. J. D. Cockroft, both being engaged at that time on wartime research for the atomic bomb.[14] Although a small Geiger counter is reported to have been flown in a Norseman aircraft in Saskatchewan, Canada, in the summer of 1948, and trials of a Geiger-counter array with an anti-coincidence circuit took place in the United Kingdom about the same time, the key moves towards success in airborne uranium exploration did not commence until October, 1948. At a meeting between Canadian and United Kingdom representatives held in Ottawa, it was decided to proceed with the development of ion-chambers, Geiger and scintillation counters. Scintillation detector and ion-chamber development was the responsibility of the Chalk River Laboratories in Canada. Lack of sensitivity to ground radiation led to the early termination of ion-chamber experiments, although the equipment was used to prepare a radiation intensity colour map of a small portion of the Gatineau Hills, near Ottawa, Ontario, and outcrops of radioactive pegmatite were located. It is interesting to note[14] that 'at the outset of the tests the importance of finding a method of distinguishing, whilst in the air, between localized outcrops of high grade ore and large areas of slightly abnormal activity was emphasized'.

During the summer of 1949 a seven-tube Geiger set, which trials had shown to have three times the effectiveness of the previously used ion-chambers, was flown over approximately 20 000 line miles of survey in the Northwest Territories of Canada, at a mean height above ground of 350 ft and with line spacings ranging from 1 to 8 miles. About 100 anomalies were recorded as a result of this work, but only a few were visited on the ground, and these proved to be areas of moderately radioactive granite. When the experiment was flown over known radioactive mineral occurrences in the Goldfields (now Uranium City) area of Saskatchewan, all except one were readily detected. By the end of October, 1949, a larger Geiger-counter installation with 22 tubes was ready for flight testing. Its performance, however, was not appreciably better than that of the seven-tube Geiger set.

In February, 1950, the Chalk River Laboratories produced an anthracene scintillation crystal assembly, a 2 in × 2 in crystal being used. It was tested, and gave results about equal to those obtained with the Geiger counters. In April, 1950, a thallium-activated sodium iodide crystal, 2 in × 2 in in size, became available, and was found to be about three times as effective as anthracene. In June, 1950, a 4 in × 2 in sodium iodide crystal with improved optical contact between the crystal and photomultiplier was tested and found to be about eight times as effective as the smaller crystal previously tried. This equipment was installed in an aircraft which operated in the summer of 1950 in the Northwest Territories, flying 17 300 line miles of survey, 55 anomalies being found. Ground parties made detailed investigations of more than 28 of these and in all cases abnormal radioactivity was found; but, generally, it was widespread and of low concentration. This assessment work seems to have discouraged those responsible for the continuation of this type of airborne survey. It was not recognized at the time that much of the area over which the trials had been taking place was not typical of the Shield, and was abnormal with respect to the numerous occurrences of granite of high radioactivity. This fact has only recently been fully appreciated as a consequence of airborne surveys commenced by the Geological Survey of Canada in 1969.[10]

The early burst of development in airborne radiometric survey methods reached a technical standard which was not substantially exceeded as far as routine surveys were concerned for the following fifteen years. The first airborne sodium iodide crystal had a volume of 12·5 in^3. Although the volume employed on surveys increased over the next decade to a maximum of 150 in^3, used by the United States Geological Survey in their Aerial Radiological Measuring Surveys Program, most work was done from small single-engined aircraft with detectors comparable in size to the first Chalk River products, coupled to the simplest possible rate meter circuits. It is notable that the uranium exploration boom of the mid-1950s produced no major discovery attributable to an airborne radiometric survey. The apparent lack of success of the early radiometric surveys contrasted unfavourably with the successes achieved by early aeromagnetic surveys in the search for iron ore deposits. This undoubtedly contributed to the almost complete cessation of airborne radiometric surveys when the uranium boom ended in 1959. These circumstances provided little inducement to further research into the method, but it should be noted that those organizations where relevant research was kept alive were in the most advantageous position when outside interest in the method reawakened in 1966.

Although there were proposals from the outset to use the energy discrimination ability of scintillation detectors to distinguish between radiation from potassium, uranium and thorium, the type and performance of electronic components then available discouraged attempts to employ this capability for routine survey purposes. Events which led to a demand for a much higher standard of portable instrumentation were the tests of nuclear devices in the atmosphere in the period around 1960. The need to know, both for health and military technology purposes, about the fission products created by these explosions forced the development of transportable instrumentation capable of energy discrimination. It was fortunate that during the same period solid state components were becoming available, and this made it practicable to construct instrumentation in a form which was sufficiently compact and rugged for airborne use, and at the same time had low electrical power requirements. The expertise involved in these developments became relatively widespread by the mid-1960s, and this contributed to the second phase of development of airborne radiometric survey techniques in response to renewed interest in uranium exploration.

Basic considerations

An airborne survey is not an exploration method which should be carried out or considered in isolation: it is part of a spectrum of exploration techniques. In most circumstances it should form the first phase of work in a new area, but details should be decided in the light of what is to follow. Planning should commence with consideration of what is already known about an area, covering every aspect from known mineral occurrences and geology to ground accessibility and climate.

Before some of the technicalities of airborne gamma-ray surveys are reviewed, it is necessary to outline the premises upon which they should be based. The justification for commissioning any type of airborne survey is that for a given degree of effectiveness it can cover ground at a lower cost per unit area than can a surface survey. If this first condition cannot be met, then there is no justification for it. The main advantage of an airborne survey is that it is possible to sample any predetermined proportion of a region by appropriate choice of line spacing and survey height, and it is possible to ensure that a consistent standard of measurement is applied. Ground level measurements in comparable detail are practicable only in very accessible areas, and inevitably require a large expenditure in man-hours. It is then more difficult to have a consistent measurement standard, or to use automatic data reduction. Because of the many assumptions that must be made it is not easy to provide estimates of the cost differential between ground and airborne surveys with comparable information value. From Geological Survey of Canada experience within the Canadian Shield ground measurements on a 200-ft grid, with a portable gamma-ray spectrometer, cost not less than $2000 per square mile; airborne measurements with a high-sensitivity gamma-ray spectrometer flown on a 0·25-mile spacing at 400 ft produce contour maps and profiles at a total cost of $80 per square mile. Comparisons indicate that the airborne system misses no significant features,[9] but the ground work provides better spatial resolution and the opportunity to collect samples.

EFFECTIVENESS

The effectiveness of any mineral exploration method should be judged by its ability to recognize targets of economic interest and to eliminate anomalies caused by non-economic features. It must possess sufficient sensitivity to identify poorly exposed targets without losing its selectivity. This is particularly important for airborne methods employed over regions where ground access is difficult—and, therefore, costly. An airborne survey which measures only the total radioactivity may be very sensitive, but it cannot be selective. A gamma-ray spectrometer provides selectivity because of its ability to discriminate gamma energies and thereby measure the mean ground level abundances of potassium, uranium and thorium. Although total radioactivity can be measured at between 20 and 30% of the cost of a high-sensitivity spectrometer survey, because the results lack selectivity they provide a far larger number of anomalies for investigation on the ground. In these circumstances either the costs of ground investigation become excessive or poorly exposed mineralization is almost certain to be overlooked.

PREFERENTIAL CONCENTRATION OF URANIUM

It is important to recognize the reasons why energy discrimination, i.e. the use of a gamma-ray spectrometer, provides the advantage of selectivity. Ore deposits represent anomalous concentrations of particular elements in the earth's crust and their existence is the result of some comparatively rare process causing preferential concentration of a particular element. It is characteristic of almost all economic uranium deposits that uranium is preferentially concentrated relative to thorium, and this preferential concentration may exceed normal relative abundances by several orders of magnitude. In all uranium deposits uranium is concentrated relative to potassium. The potassium content of rocks normally ranges between 0·1 and 10%, whereas uranium commonly ranges between 0·4 and 4 ppm. Before a uranium concentration can be of economic interest, it must exceed 4 ppm by a factor of 250. Experimental surveys over several types of Canadian uranium deposits show that the uranium to thorium ratio is generally more specific for potentially economic uranium deposits than is the uranium to potassium ratio, although the latter provides a wider target halo. Therefore, it is very important that any airborne gamma-ray survey provide the means of obtaining statistically significant ratio measurements. These will only be forthcoming if counts per unit distance travelled by the aircraft are maximized. Counting statistics have been discussed further by Darnley and co-workers.[7]

ABUNDANCE RATIO MEASUREMENTS

It is important to stress that the practical value of ratio measurements as a key to uranium exploration lies in the fact that, in general, the abundances of the three radioelements K, U and Th vary sympathetically with one another over a wide range of rock types.[8] It is only on rare occasions that some uncommon process results in the enrichment of one of these elements unaccompanied by enrichment in the others. Almost every such case merits investigation (not necessarily for its uranium potential) because of its rarity.

One of the advantages of monitoring abundance ratios is that this provides a means of filtering count rate anomalies, besides recognizing unusual radioelement distribution where high count rates may not occur. This is important since the measured count rate is influenced by the area of exposed rock, and the solid angle it subtends relative to the aircraft, as well as by absolute abundances in the rock. It should be noted that an increase in absolute abundance, of uranium, for example, is probably only a distant indicator of potential mineralization unless it is accompanied by changes in relative abundance.

REGIONAL ABUNDANCE OF URANIUM

Experience in Canada and elsewhere indicates that districts containing uranium deposits are generally characterized over tens or hundreds of square miles by above-average radioactivity relative to their surroundings. Uranium deposits are not known to occur within districts of below-average radioactivity.[10] Therefore, because of their relatively low cost, total radioactivity measurements can serve a useful purpose by eliminating ground from further examination by more expensive methods. This can be done by use of widely spaced flight lines—for example, 5 miles. Within districts of above-average radioactivity identification of the zones where uranium exhibits the greatest relative concentration requires a discriminating system. Other zones within the district may show enrichment in other radioelements, as occurs in the Bancroft area of Ontario.[9]

COMPOSITION OF SURFACE MATERIALS

Gamma radiation in the range 1·46–2·62 MeV, which is of concern in most measurements of rock radioactivity, is quite strongly attenuated by rock and overburden. Approximately 90% of the total gamma radiation from a rock of density 2·7 g/cm^3 comes from the upper 6–9 in; from dry overburden, with a density of 1·5 g/cm^3, 90% of the radiation is received from 12–18 in.[15] Thus, radioactivity measurements only respond to radioelements contained in a relatively thin surface layer. It is therefore of vital importance prior to the contemplation of a survey in a region to establish that the surface materials are reasonably representative of bedrock composition. This means that in an area where overburden is predominant, i.e. greater than 60% of the area, it must be either residual or locally derived in order that surface measuring radiometric methods can be useful. It also means that both outcropping rock and overburden must not have suffered removal of significant constituents by weathering. Uranium is most susceptible to this type of alteration, especially under hot, humid conditions. It should be pointed out, however, that since radiometric detection of uranium hinges on the measurement of ^{214}Bi, which follows on radium and radon in the decay series, the recent weathering and removal of uranium itself is of no consequence as long as there is residual radium. Similarly, the lateral migration of radium over distances of up to 200 or 300 ft is of no consequence to airborne measurements, since this is the order of magnitude of resolution. In fact, it is advantageous, by increasing the area of the target to be detected. The author is aware of no reliable evidence to indicate that under Canadian climatic conditions the migration of uranium and/or its decay products has presented any practical problems as far as airborne methods are concerned.

GEOCHEMICAL HALOES

It is the existence of primary and secondary dispersion haloes in the vicinity of any type of mineralization that provides the justification for all geochemical exploration techniques. Uranium is no exception to this, and the primary halo is one of the reasons for the existence of targets of regional extent. The success of airborne methods in locating zones which are hosts to mineral occurrences, and, occasionally, pinpointing the actual occurrences, stems from the existence of local dispersion haloes covering thousands of square yards. The average equivalent uranium content in these zones may be only one or two parts per million greater than in the surrounding area, but with a sensitive system it is sufficient to form an airborne exploration target. An airborne method must give first attention to finding these zones rather than aim to locate actual point sources of uranium mineralization which may or may not be exposed. In the early days of airborne radiometric surveys undue attention was focused on the problem of locating 'point sources'. This stemmed from insufficient appreciation of geochemical considerations, combined with the inability to distinguish between the radioelements. The difficulty of locating 'point sources' should not be regarded as a serious handicap to an airborne method because point sources do not occur in isolation.

Method

THE GAMMA SPECTRUM

The gamma-radiation spectrum of the complex uranium and thorium decay series contains numerous photo peaks over the range from a few keV to 2·8 MeV. Potassium (^{40}K) possesses a single photo peak at 1·46 MeV. A recent review of the physics of gamma rays has been given by Adams and Gasparini.[2]

For total radiation measurements all photons are counted above some energy threshold set by the noise level of the photomultiplier assembly. It is important that this threshold does not drift, as the number of photons at the lower end of the spectrum increases very rapidly with decreasing energy. The key photo peaks for separating the uranium and thorium contributions are 1·76 MeV (^{214}Bi) for uranium and 2·62 MeV (^{208}Tl) for thorium. As was mentioned above, there are many other photo peaks in each series, but they are either of too low energy (and thus strongly attenuated in air) or they cannot be readily resolved by NaI(Tl) scintillation detectors either from one another or the Compton scatter background. Solid state detectors, such as Ge(Li), could provide the necessary resolution to enable some of these peaks to be used, but the count rate efficiency of such detectors is low and they suffer from the operational disadvantage of requiring continuous cooling by liquid nitrogen.

The persistence of the key photo peaks at a height in excess of 600 ft above the ground is illustrated in Fig. 1.

INSTRUMENTATION

There have been three different instrumental methods for obtaining spectral information from airborne (and ground) survey measurements: (1) to obtain the count rate from the full spectrum in a series of discrete steps or channels (multi-channel spectrometer); (2) to obtain the count rates only from portions of the spectrum (windows) in the vicinity of photo peaks of interest (window spectrometer); and (3) to sum the counts from the spectrum in a series of overlapping steps by varying the energy threshold at which counting commences: for example, a threshold of 0·5 MeV would be used to measure the sum of K+U+Th, a threshold of 1·6 MeV would be used to measure U+Th and a threshold of 2·4 MeV to measure Th alone (threshold spectrometer).

A multi-channel spectrometer defines the full spectrum, resolution depending on the number of channels used. Because the full spectrum is being recorded,

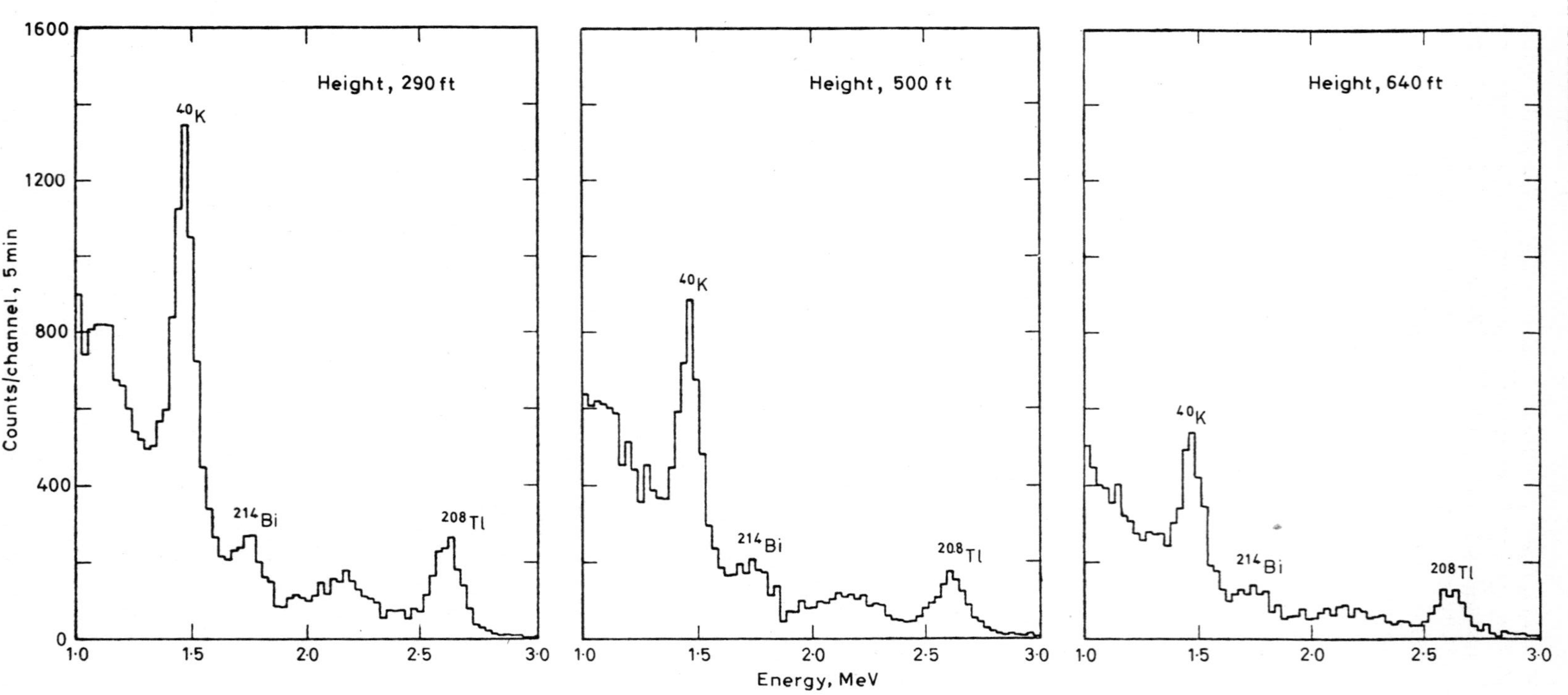

Fig. 1 Radiation spectrum at three different heights

it can be reconstituted and corrected to allow for any small drift in calibration which may have taken place. The general form of the natural spectrum remains unchanged and only the spectral details, i.e. the relative heights of the peaks, change in the course of a survey. To record the full spectrum, however, magnetic tape must be used and, because of the number of channels that must be recorded, the equipment is relatively complex. There may also be design limitations with regard to permissible sampling rates due to the time required to read out data from the spectrometer to recorder.

Window spectrometers are the type most commonly used for airborne surveys. The use of four windows minimizes equipment complexity and at the same time avoids the collection of excessive data. It is not possible to reconstitute the spectrum, however, in order to monitor calibration, and the system must therefore either have some form of automatic stabilization or be very carefully designed to minimize electronic instabilities.

Variable-threshold spectrometers are electronically somewhat simpler than the types discussed above, and are therefore cheaper. Because a wider portion of the spectrum is being utilized for the collection of counts, a variable-threshold spectrometer exhibits a high count rate for a given detector size. Because the characteristic spectral information relative to, for example, uranium constitutes only a small part of the spectrum, the critical signal to noise ratio is not being significantly improved. Electronic stability for the variable-threshold spectrometer is as important as for other types of spectrometer.

In order to minimize instability caused by drift in the high voltage supply and temperature-dependent variations in the gain of the photomultiplier tubes, some spectrometer designs make use of automatic gain adjustment by use of an isotope of known energy as monitor. The two isotopes most commonly employed are ^{241}Am, which is an alpha emitter, or ^{137}Cs, a gamma emitter. Recently, ^{57}Co has been advocated to avoid some of the disadvantages inherent in the other two. The gamma energy equivalent of the alpha emission of americium is observed by different detectors at slightly different energies, and so is not satisfactory for multi-crystal assemblies. Its gamma equivalent energy is in the range 3–4 MeV, which is above the range of natural terrestrial radiation. The energy of ^{137}Cs (0·662 MeV) is below the photo peaks used for energy discrimination, but it encroaches upon a portion of the spectrum normally wanted for total radiation measurements. Depending on the resolution of the detectors, its use requires the total-count threshold to be set at not less than 0·8 MeV. Therefore stabilization with ^{137}Cs entails a loss in total radiation sensitivity. Newly introduced for stabilization purposes, ^{57}Co avoids this problem. A recent review by Denham[11] of the present status of instrumentation designed for airborne radiometric surveys considers in some detail the use of ^{137}Cs and ^{241}Am as reference sources.

The principal cause of spectral instability is temperature change in the photomultiplier–preamplifier assembly, and for this reason it has become common practice to maintain the whole detector system at a constant temperature by means of heating jacket and thermostat. Large changes in ambient temperature, which may be experienced in airborne survey operations, can affect other components of the spectrometer system, and with window- or threshold-type spectrometers this requires them to be monitored and adjusted either manually or automatically.

Table 1 lists airborne scintillation counters and spectrometers currently available for purchase. Differential (window) spectrometers are more common than variable threshold (integral) spectrometers. Four 6×4 NaI(Tl) crystals provide the maximum detector volume normally offered, and although some

Table 1 Airborne scintillation counters and spectrometers available for purchase. After Hood[17]

Manufacturer (country)	Model designation	Number, size and type of detector, in diameter × length (volume in in^3)	Type of spectrometer (D, differential; I, integral; Ch, channel)	Time constant, sec, or counting times	Recorder system (Ch, channel)	Power requirements	Spectrum stabilization	Compton correction facility
Abem (Sweden)	ABA 1501	One 5 × 6 or 10 × 5 NaI(Tl) crystal	D 4 Ch	Channel 1: 1, 2·5, 5, 10 sec channels 2, 3, 4: 2·5, 5, 10, 20 sec	5 Ch Analog	12–24V, dc; 50 W	—	—
Commercial Products Div., AECL (Canada)	CPD 322	Up to twelve 9 × 4 NaI(Tl) crystal Temp. regulated	D 128 Ch 4 integration regions	0·1–90 Dual timer	6 Ch Analog and Magnetic tape 7-track IBM	28 V, dc, or 115 V, ac; 800 W with all options	—	X
	CPD 287	One or more 9 × 4 NaI(Tl) crystal Temp. regulated	D 128 Ch 4 integration regions	1 to 10 sec count time. Automatic for U/Th ratio	6 Ch Analog or Digital	28 V, dc, or 115 V, dc; ac; 40 W plus recorder power	—	X
	CPD 192D	One 5 × 4 NaI(Tl) crystal for helicopter	D One integrating region	0·3 to 30 sec count time	Analog or Digital	28 V, dc, or internal rechargeable batteries	—	—

Exploranium (Canada)	DGRS-1001	Two to four 6×4 NaI(Tl) crystal	D or I selectable from front panel 4 Ch	1, 2, 5 and 10 sec each Ch	Analog signal 0–3 V 4 Ch	24 V, dc, 35 W; 12 V, dc, optional	^{137}Cs or ^{57}Co	X
	DiGRS-2000	Two to four 6×4 NaI(Tl) crystal	D or I selectable from front panel 4 Ch	Count times of 0·1, 0·2, 0·4, 0·5 and 1 sec	Digital 3BCD characters/Ch Analog signal 0–3 V per Ch	12 or 24 V, dc; 100 W	^{137}Cs or ^{57}Co	X
	DiGRS-3000	Two to four 6×4 NaI(Tl) crystal	D or I selectable from front panel 4 Ch	Count times from 0·1 to 9·9 sec in 0·1 sec increments	Digital 4 BCD characters/Ch Analog signal 0–3 V per Ch	24 V, dc; 100 W; 12 V dc optional	^{137}Cs or ^{57}Co	X
McPhar Geophysics (Canada)	AV-4	Up to three 6×4 or 8×4 NaI(Tl) crystal	I 4 Ch		4 Ch Analog	23–30 V, dc, 12 W	^{241}Am	—
Nuclear Enterprises (United Kingdom)	NE 8424	Four 6×4 NaI(Tl) crystal	D 4 Ch	1, 2·5 and 5 sec	Analog	28 V, dc	—	—
Scintrex (Canada)	GIS-2	One to four 5×4 NaI(Tl) crystal	I Single Ch	2, 8 and 16 sec	Single Ch	18 V, dc 3 V, dc	—	—
	GDSA-4	One to four 5×4 NaI(Tl) crystal	D 4 Ch	1, 3 and 5 sec	6 Ch Analog	28 V, dc; 7 W	^{241}Am	X
	GISA-4	One to four 5×4 NaI(Tl) crystal	I 4 Ch	1, 3 and 5 sec	6 Ch Analog	28 V, dc; 7 W	^{241}Am	X

digital recording systems are available, analogue systems are the most numerous at the present time (1971).

Until recently, it has been the usual practice to record the radiation levels measured by scintillation counter or gamma-ray spectrometer as counts per second by means of a rate meter. This method, though relatively simple, is a major source of inaccuracy because of the difficulty in achieving a high degree of linearity in the rate meter and also because of the delayed response inherently associated with the time constant normal to this technique. For this reason, at even quite low survey speeds (80–100 miles/h) a radioactive target is past before true peak intensities are recorded. Also, the fact that output from this type of counting system is in analogue form is a disadvantage if subsequent data processing or automatic compilation is required. The alternative to a rate meter measuring technique is to accumulate counts for precisely predetermined periods of time. Provided that the aircraft's ground speed is known, each sample then relates to a known strip of ground. As the data from the analyser are accumulated pulse by pulse, the sum of the radiation provides a precise indication of the radioactivity of the terrain covered by the preset sampling period. The sum is in digital form and is thus suitable for recording on magnetic tape, or numerical display, and conversion into analogue form for visual strip chart recording.

CORRECTION OF DATA

It has been demonstrated in other publications[8,9] that airborne spectrometric measurements can be directly related to the mean ground level abundances of the radioelements. To do this, however, several corrections must be applied to the count rate as measured by the spectrometer. The first of these is subtraction of background radiation. Background radiation is made up of a cosmic contribution, a contribution from the structure of the aircraft and its equipment and, the most difficult to determine, a part arising from the presence of radon in the atmosphere. ^{214}Bi responsible for the 1·76-MeV peak used to measure uranium abundance is a decay product of radon. The background contribution commonly forms between 60 and 80% of the ^{214}Bi count rate measured at 400 ft above ground level. The variability is due to changes in atmospheric composition stemming from environmental factors: porosity and composition of ground, recent rainfall, wind strength, pressure gradient, etc. The correct determination of the atmospheric contribution is of great importance. A background correction must be applied to airborne data as a preliminary to obtaining any quantitative data relevant to ground level concentration.

The second step which must be taken is to normalize measurements to a standard distance above ground. This is usually the nominal terrain clearance of the aircraft. As was mentioned previously, considerable attention has been given in the past to the effect the geometry of the source has upon the ideal attenuation formula which should be applied. It has been found from empirical observations over the Canadian Shield that for such topography a uniform exponential correction can be applied without causing any serious errors: greater errors will arise if no correction is applied. In general, the consequences of applying a terrain clearance correction are to smooth data and reduce highs and lows caused by variations in ground clearance. The half-height, that is to say the distance over which the count rate is reduced by half, is approximately 300 ft for radiation in the 1·46-MeV window radiation and 400 ft for radiation in the 1·76-MeV and 2·62-MeV radiation windows. This is primarily due to a build-up of Compton scatter from the 2·62-MeV photo peak included in the 1·76 window compensating for the attenuation by absorption. Over the height range 150–800 ft the simple exponential function $N = N_0 e^{-\bar{\mu}H}$ is a close approximation to the observed attenuation. Values for $\bar{\mu}$ are given in Table 2. For first-order approximations the attenuation can be taken as 20% per 100 ft. It is important that it is only meaningful to apply the height correction with a system which provides an adequate count rate. If one standard deviation for the mean count rate is about 20%, there is little purpose in attempting to correct for clearance deviations of less than 100 ft.

Table 2 Experimental overall attenuation coefficients

	Radiation, MeV	$\bar{\mu}$, ft^{-1}
^{40}K	1·46	$2{\cdot}3 \,.\, 10^{-3}$
^{214}Bi	1·76	$1{\cdot}7 \,.\, 10^{-3}$
^{208}Tl	2·62	$1{\cdot}7 \,.\, 10^{-3}$
Σ	0·4 – 2·82	$2{\cdot}0 \,.\, 10^{-3}$

The third correction which must be applied is to compensate for Compton scattering within the detector crystals which contributes to both uranium and potassium count rates. Scattered thorium radiation contributes counts to both the low-energy uranium and potassium channels, whereas scattered uranium contributes to the potassium channel only. The precise value for Compton scattering corrections depends on the window width employed and size, number and spacing of radiation detectors. Typical values are shown in Table 3, together with the formulae used for correction.

Table 3 Formulae and sample values for Compton scattering corrections. From Grasty and Darnley[13]

$$N_{\text{Th corr.}} = N_{\text{Th}} - \text{Bgd}_{\text{Th}}$$

$$N_{\text{U corr.}} = N_{\text{U}} - \text{Bgd}_{\text{U}} - \alpha\, N_{\text{Th corr.}}$$

$$N_{\text{K corr.}} = N_{\text{K}} - \text{Bgd}_{\text{K}} - \beta\, N_{\text{Th corr.}} - \gamma\, N_{\text{U corr.}}$$

where N_{Th}, N_{U} and N_{K} represent the observed count rates in the relevant windows, Bgd is the background count rate, the subscript indicating the window. α = counts in the U window per count in the Th window; β = counts in the K window per count in the Th window; γ = counts in the K window per count in the U window.

Sample values

	Helicopter 3 (5 × 5) in NaI	Skyvan 12 (9 × 4) in NaI
α	0·43	0·35
β	0·62	0·33
γ	0·91	0·56

CONVERSION TO ELEMENT CONCENTRATION

With background, terrain clearance and Compton scattering corrections applied, counts can then be converted directly to element abundances. The conversion factor is dependent on the dimensions and volume of the detector system, window width, sampling time and flying height. Although airborne radiometric survey measurements have not often been expressed in units other than counts per unit time, there is a strong argument for requiring them to be expressed in terms of mean ground level abundance of the respective elements.

A measurement reported in terms of count rate is peculiar to a particular survey. Count rates from two different surveys cannot normally be compared in any meaningful way, and so maps based on a reproducible unit cannot be compiled. Although there are assumptions and problems involved in converting count rates to element abundance, the advantages of such a presentation outweigh the disadvantages, and the act of making the correlation encourages careful consideration of the problems involved. If and when regional airborne radiometric maps are to be made in the manner now practised for aeromagnetic surveys, such standardization will be essential.

In Canada the correlation between corrected airborne count rates and mean ground level element abundance has been approached by the Geological Survey of Canada[9] in two ways: (*a*) calibration flights at known heights over a carefully selected strip of ground which has reasonably uniform known abundances of the radioelements, determined by systematic ground measurements; and (*b*) systematic ground measurements over different sites, representing different levels of radioactivity, within a survey area.

Table 4 Element concentration–airborne count rate correlation: GSC Skyvan system with 3055-in^3 detector volume at 400 ft

Window widths			
	K	(^{40}K)	1·36–1·56
	U	(^{214}Bi)	1·66–1·86
	Th	(^{208}Tl)	2·42–2·82
	1% K	≡	80 counts/sec
	1 ppm *e*U	≡	10 counts/sec
	1 ppm *e*Th	≡	5 counts/sec

Table 4 is based on results obtained by the GSC Skyvan spectrometer system and shows the relationship between element concentration at ground level and count rate at a height of 400 ft above ground for the specified window widths. It emphasizes the comparatively low count rates obtained even with a large detector volume (3055 in^3) and underlines the necessity for accumulating counts over a period greater than 1 sec (2·5 sec has been adopted as standard for this system).

Comparison of systems

An airborne survey system comprises all the equipment necessary to perform the survey, including the aircraft, height, position and radiometric instrumentation. In planning a radiometric survey, a number of parameters can be varied—namely detector volume, flying height, air speed and line spacing; it is assumed that spectral window widths have optimum setting which should remain fixed whatever the other operational parameters of the survey, whereas rate meter time constants or sample integration times are optimized after other parameters have been decided. It commonly occurs that a choice of equipment is available for a survey, and some performance criterion, apart from cost, is desirable for assessment of the comparative effectiveness of different systems.

Counts per unit distance travelled by the aircraft are a measure of overall effectiveness. Factors which influence this value are the detector volume, mean height above ground and mean ground speed. Systems can be compared with respect to their effectiveness for both large and small sources. The best index of comparison is provided by the square root of the normalized count rate for each system, since it is the relative magnitude of their standard deviations which determines the sensitivity of each system to changes in element concentration. The square root of the normalized count rate is called the 'figure of merit of the system'.

Table 5 Spectrometer systems: figures of merit

System	Detector*		Terrain clearance, ft	Speed, miles/h	Merit	
	Size	Volume			Large area	Small area
GSC mapping	12 (9 × 4)	3055	400	120	10·0	10·0
GSC reconnaissance	6 (9 × 4)	1527	500	140	6·0	5·2
GSC helicopter	3 (5 × 5)	295	250	30	7·1	10·0
Commercial						
A	4 (6 × 4)	453	300	90	4·9	5·0
B	1 (8 × 4)	201	125	80	4·0	10·0
C	1 (6 × 5)	141	150	60	3·75	6·7
D	3 (6 × 4)	340	150	130	4·0	7·1

Figures of merit calculated from the following formulae:

Large area $$\left[\frac{V^x}{3055}\cdot\frac{120}{S_x}\cdot\frac{100}{e^{-\bar{\mu}(400-Hx)}}\right]^{\frac{1}{2}}$$

Small area $$\left[\frac{V^x}{3055}\cdot\frac{120}{S_x}\cdot\frac{(400)^2}{(H_x)^2}\cdot 100\right]^{\frac{1}{2}}$$

Vx is volume of detector crystals, *Sx* is mean ground speed, *Hx* is mean terrain clearance and $\bar{\mu} = 1{\cdot}8 \times 10^{-3}$.

* Detector dimensions in inches (diameter × thickness).

It can be demonstrated that given a minimum crystal dimension of 4 in, the count rate from a NaI(Tl) detector assembly is approximately proportional to its volume. Table 5 gives the relative count rates per unit distance for a number of survey systems. Various commercial systems are designated by the letters *A–D*. In this example all are compared against the Geological Survey of Canada's Skyvan-mounted spectrometer system (rated as 10). For large area targets the large detector volume of the Skyvan system provides a clear advantage: for small targets the low terrain clearance employed by some other systems compensates for small detector volume. It has been pointed out earlier, however, that a large area of weak enrichment is a more realistic target for airborne exploration than individual 'hot spots', so sensitivity to weak large area targets is the most desirable feature of any system intended for use on a regional scale. If necessary, a small installation in a low-flying helicopter can be used during a later phase of exploration to find the 'hot spots'. One factor to be considered in connexion with very low-level surveys (150 ft), apart from their hazard, is that they are strongly influenced by the inhomogeneity of surface details, that is to say outcrop distribution, surface geometry and the presence of small patches of water. Ratio measurements provide the best means of minimizing these factors, which will have a substantial effect on absolute count rate. At 200-ft flying elevation approximately 65% of 2·62-MeV radiation originates within a radius of 300 ft of the aircraft. At 440-ft elevation the same percentage originates within a radius of 500 ft of the aircraft. A diagrammatic representation of the radiation yield at different heights as a function of source influence has been given by Cook and co-workers.[5] In general, the lower the terrain clearance, the closer the line spacing must be to achieve comparable search effectiveness.

Available data suggest that, currently, there is a factor of between 2 and 4 difference in cost between commercially conducted gamma-spectrometry surveys with systems such as *A–D* (Table 5) and high-sensitivity equipment comparable to the GSC Skyvan system. This cost difference must be taken into account in assessing cost-effectiveness.

Results

TOTAL GAMMA-RADIATION SURVEYS

In the past total radiation surveys have been recorded almost entirely on analogue strip charts, and for compilation purposes it has not generally been thought necessary to make any corrections (nor has it been practicable to do so). For compilation into contoured maps, however, especially where these may relate to large areas and the data may be collected over a period of days, weeks or even months, it is necessary that there should be corrections for the inevitable variations in natural atmospheric background radioactivity. Variation in cosmic background and man-made radioactivity may also occur. Corrections for terrain clearance variations are also highly desirable. Recently, electromechanical equipment for digitizing analogue records has become available, and some surveys are now being processed in this way in order that corrections can be applied by digital computer prior to contouring. Direct digital recording avoids the errors inherent in manual digitizing.

SPECTROMETRIC SURVEYS

To obtain maximum value from spectrometry data it is necessary that the information should be fully corrected and used to prepare both corrected profiles along the flight lines, which are the primary source of information for further detailed exploration, and also isorad contour maps, which indicate the regional pattern of radioelement distribution and provide the most convenient means of correlating with other forms of earth science data.

The first commercially available spectrometers contained no internal means of making corrections and relied upon analogue recording charts. Consequently, only grossly simplified data presentations were possible, and contouring was not attempted or was limited to thorium. One of the most common forms of data presentation has been to use a symbol superimposed on a flight line plan to indicate an anomaly which exceeds by some given factor an arbitrary background value (Fig. 2 is an illustration of this type of presentation; this indicates uranium 'anomalies'). This type of inexpensive presentation is unsatisfactory from several points of view. First, it is somewhat subjective; secondly, since the particular instrumentation being used had no provision for Compton scattering correction, there is no indication of the extent to which the uranium anomalies are a consequence of high associated thorium values. Thirdly, the concept of background as used in this situation relates to the average overland radiation level. It thus ignores the information which the overland radiation base level can provide about the general geochemical environment. Fourthly, since no terrain clearance correction is applied, the user of these data must assume that the clearance was within acceptable limits and that the anomaly was not caused by a topographic high—an assumption which may be incorrect.

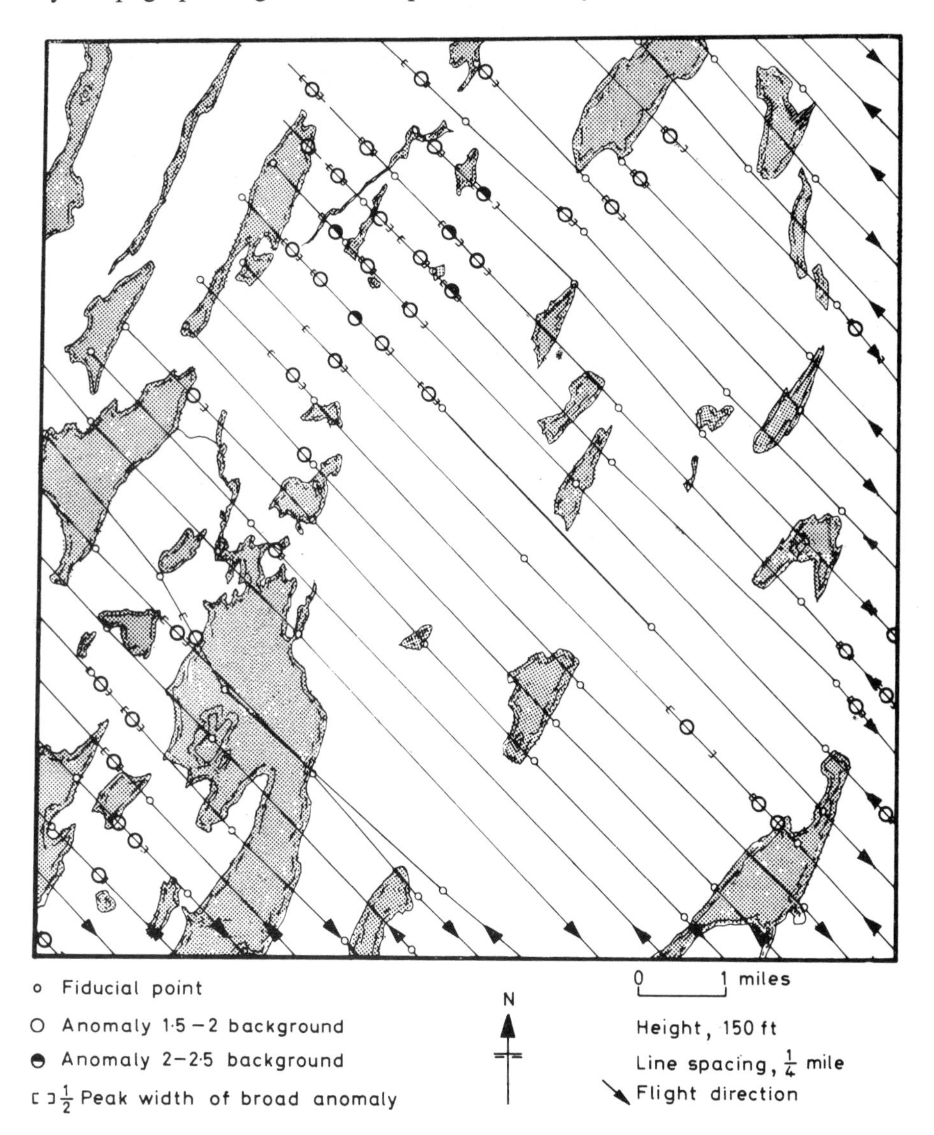

Fig. 2 Presentation of results—simplest form: uranium anomalies shown by symbol

Slightly more information is shown by the example given in Fig. 3. The uranium count rate in the anomaly peak is shown alongside the uranium background (used in the same sense as the first example) and the thorium value in the anomaly is also shown against its background value. The report which accompanied this presentation did not state whether a Compton scattering correction has been applied to the uranium value; the relative uranium and thorium background values suggest that it has not.

A further elaboration is to contour thorium content and add this to the display of anomalies. This takes account of the fact that the thorium count is usually more reproducible and statistically more significant than the uranium count and it can therefore be used as an aid in the interpretation of the geology.

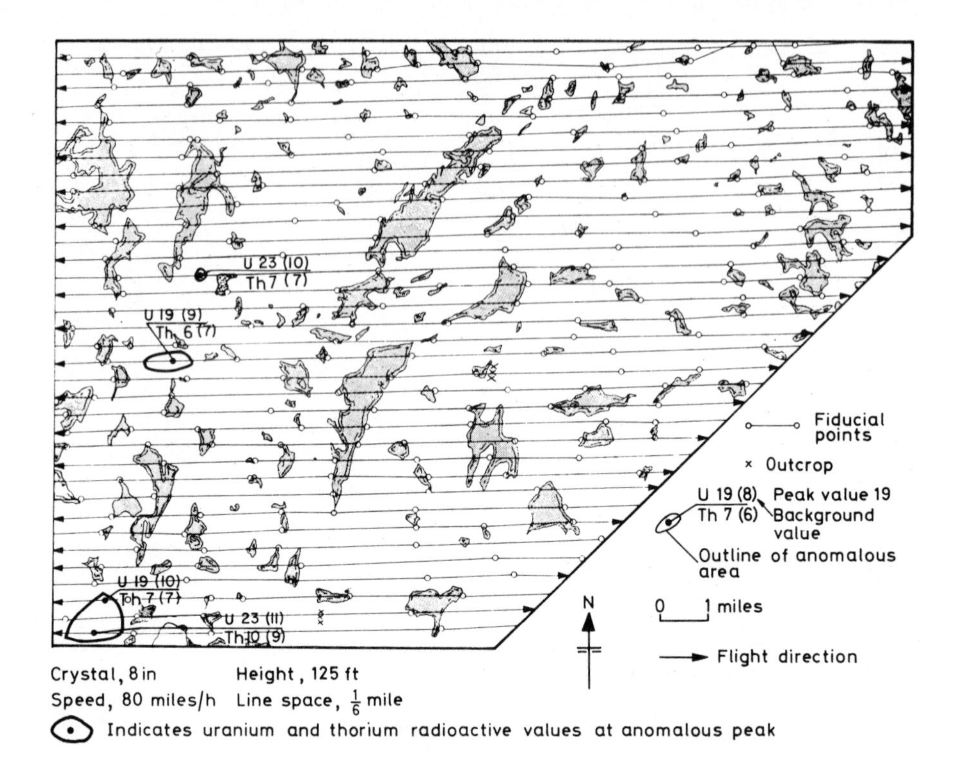

Fig. 3 Presentation of results: uranium anomalies with annotation of uranium and thorium count rates inside and outside anomaly area

A portion of a map with this type of presentation is shown in Fig. 4. It is probable that a Compton scattering correction has been applied to the uranium value, but it is not explicitly stated in the information which accompanied the map.

Fig. 5 shows another type of presentation, known as offset profiles. The strip chart data have been transferred from the original flight record on to a map and plotted alongside the flight lines. Ideally, this allows the user to see the relationship in radiometric pattern from line to line and link up features which he considers similar. In this particular example, which is taken from a contractor's finished presentation, no allowance has been made for the lag in plotted positions due to the time constant employed in the survey. Thus, anomalies on adjoining lines are laterally displaced relative to one another because adjoining lines were flown in opposite directions. This is clearly unsatisfactory, as is the failure to record all scale values.

With the exception of this last example, in which the contractor provided separate offset profile maps for total count, potassium, uranium and thorium, the commonest form of presentation has consisted of only one map compilation in order to minimize costs. It must be emphasized that an elaborate presentation of data is not warranted if either the counting statistics of a spectrometer system or its stability are inadequate; but, assuming that these requirements

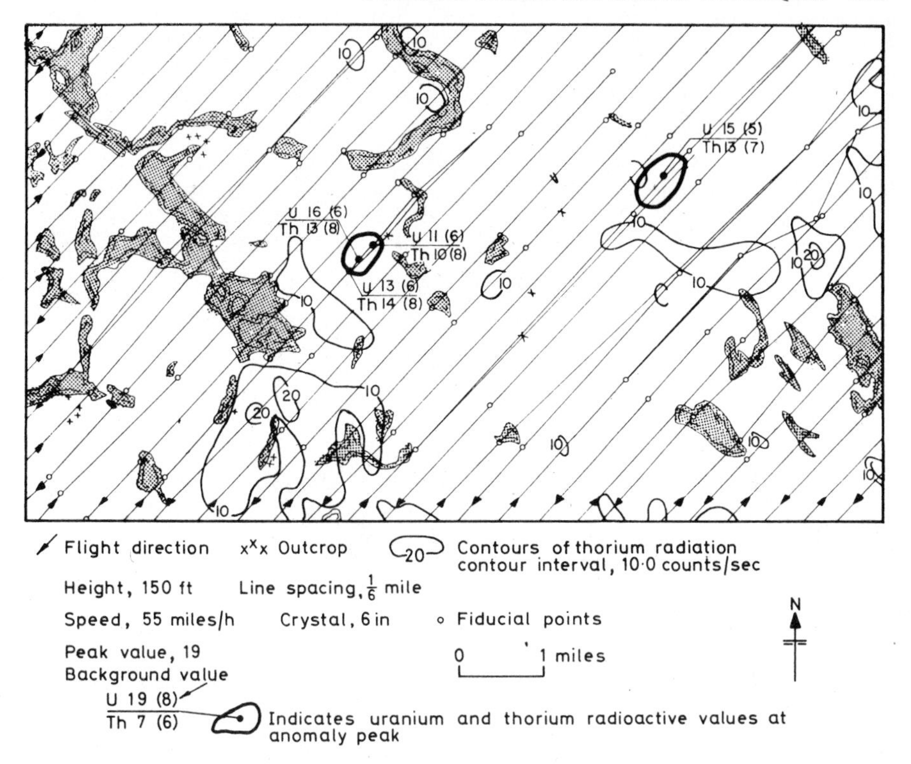

Fig. 4 Presentation of results: uranium anomalies with annotation of uranium and thorium count rates, and thorium contouring

are met, automatic data processing is the only practical solution to the problem of complete and informative presentation of results. The examples illustrated are typical of many commercial presentations of spectrometer data, and these show that, in general, satisfactory presentation has not yet been achieved. Information necessary to judge the quality of data collection is not usually available, and only a portion of data is being presented.

Inadequacies of both instrumentation and data presentation are only partly economic and technical in origin. They are also a reflection of lack of knowledge by users about the capabilities and limitations of airborne radiometric techniques. As in many other fields, consumer education is a prerequisite to obtaining value for money expended.

Appendix 1 contains a sample specification for a high-sensitivity gamma-ray spectrometer survey suitable for detailed exploration and mapping of an area. Appendix 2 lists some of the reasons why unsatisfactory results have been obtained in the past.

HIGH-SENSITIVITY SPECTROMETRIC SURVEYS

The illustrations which follow demonstrate the advantages of comprehensive data presentation based on automatic data processing and indicate the effectiveness of the corrections applied. They show why gamma-ray spectrometry can be an effective uranium exploration tool. They are made up principally of profiles obtained with the Geological Survey of Canada's high-sensitivity gamma-ray spectrometer system. The data were collected in the 1969 and 1970 seasons and all the examples shown are from data which have been released

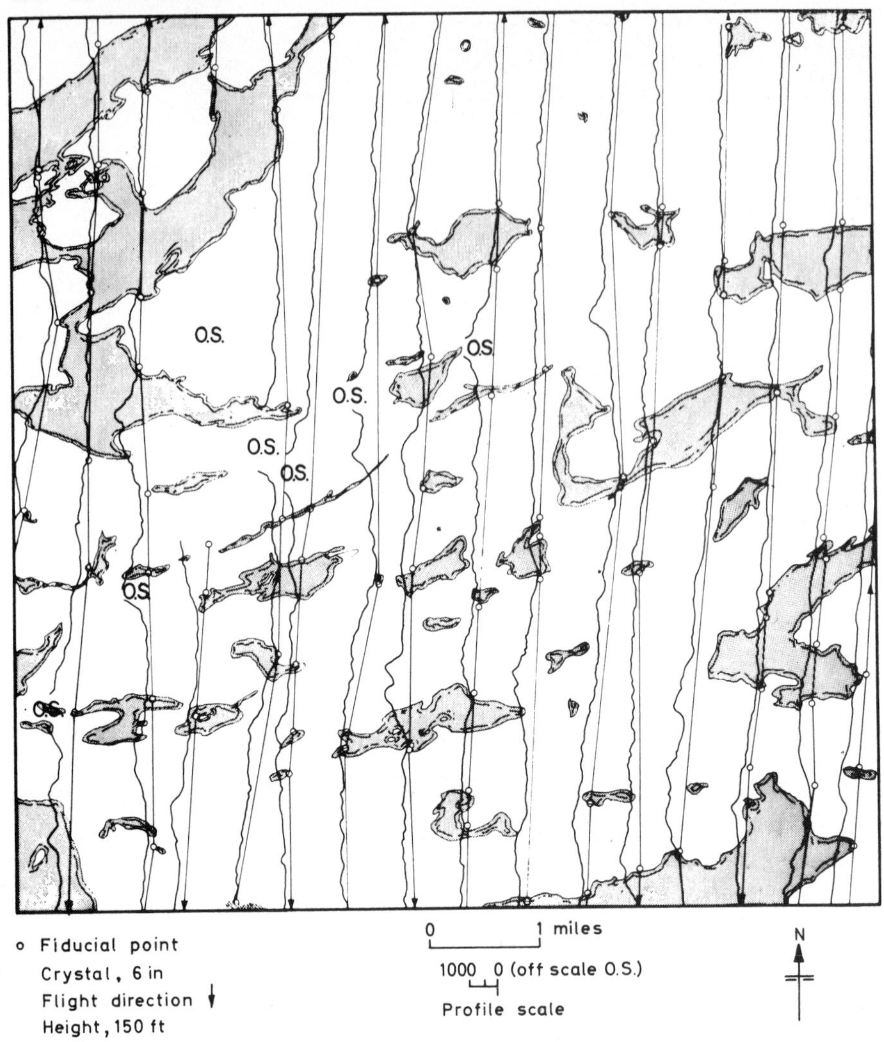

Fig. 5 Presentation of results: offset profiles

on open file. They have all been computer-plotted and follow a standard format, the distance in statute miles from some arbitrary base line being plotted from left to right. Fiducial numbers are also shown for the Athabasca and Elliot Lake surveys. The integral (total radiation), potassium, uranium and thorium corrected counts per unit counting time are plotted, as are the uranium to thorium, uranium to potassium and thorium to potassium ratios, and the terrain clearance. The sample counting time was: integral, 0·5 sec; other channels, 2·5 sec for both Bancroft and Elliot Lake surveys; integral, 1 sec; other channels 5 sec for the Athabasca survey. A longer sampling time was employed for the Athabasca survey because only half the full Skyvan detector volume was being carried, and to maintain constant counting statistics relative to other Skyvan operations the sampling time was doubled.

Fig. 6 is a test line across the St. Andrews East, Quebec, carbonatite about 35 miles west of Montreal. This niobium-bearing body is enriched in thorium, with no accompanying enrichment in either uranium or potassium (R. Lambert, personal communication). The profile across it demonstrates the effectiveness of both the Compton scattering corrections and the ratio measurements.

Fig. 7 demonstrates the effectiveness of the height correction in smoothing count rate variations caused by topographic effects, while responding to features

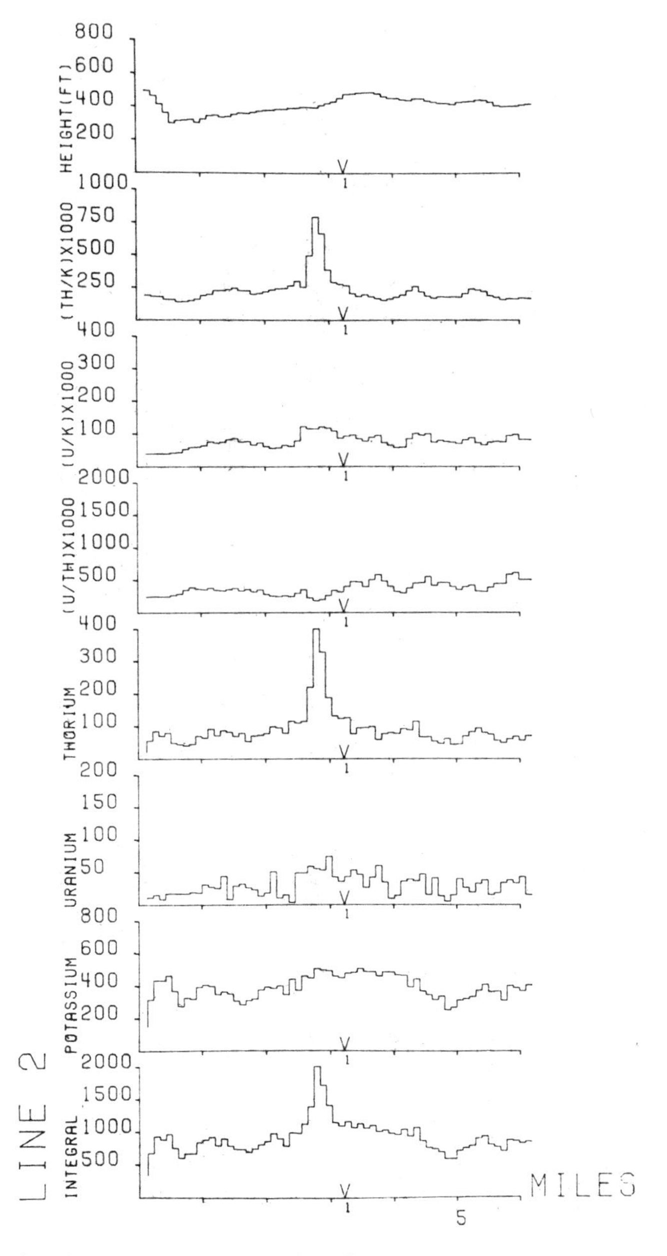

Fig. 6 St. Andrew East carbonatite, Quebec: prominent thorium anomaly

which might be missed if height corrections were not applied. This example is line 49 from the Geological Survey of Canada's Elliot Lake survey, where, at the northern end (right-hand side), there is relief of the order of 800 ft with the ancient peneplained surface dissected by irregular steep sided valleys. On this particular line the terrain clearance varies between 250 and 750 ft. The height

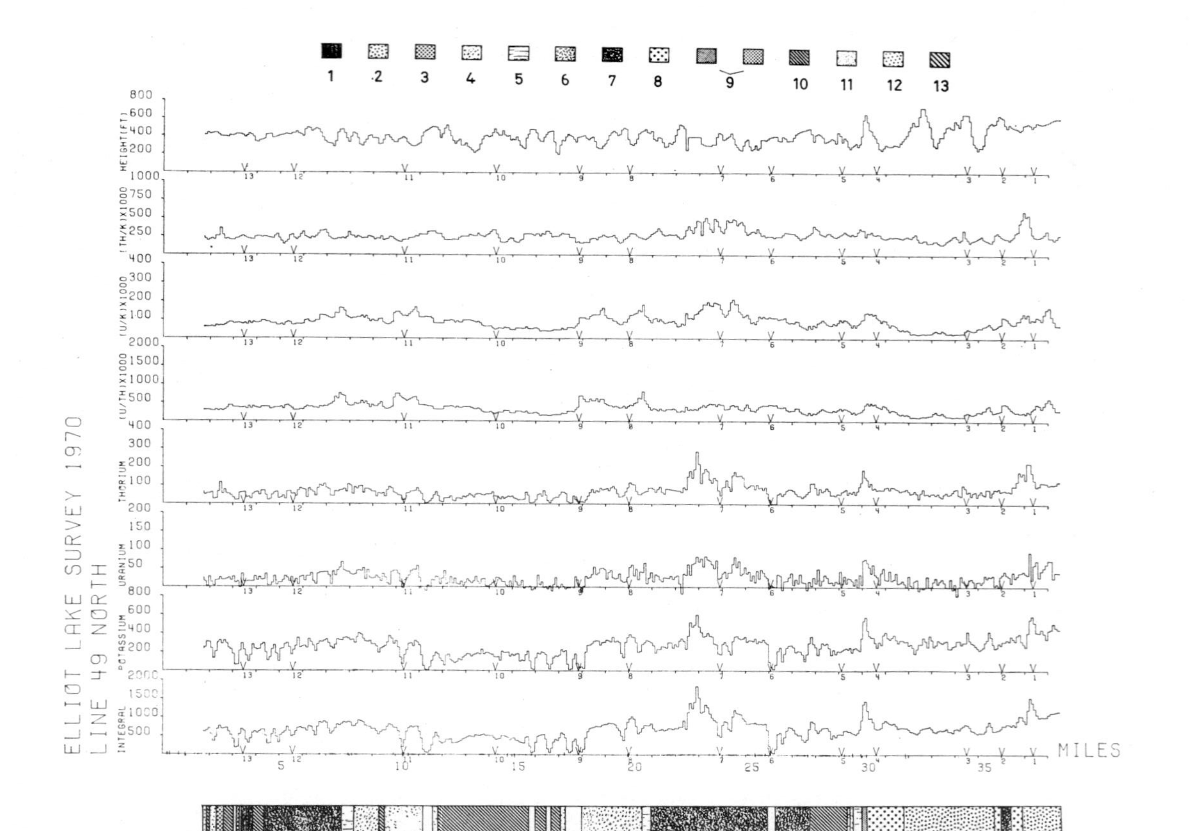

Fig. 7 Elliot Lake survey line 49: demonstration of effectiveness of height correction procedure. Fid. 9 is by Matinenda Is. in Matinenda Lake. Fid. 13 is Blind River. 1, Metavolcanics; 2, acid intrusive; 3, Elliot Lake Group; 4, Hough Lake Group; 5, Bruce & Espanola; 6, Serpent; 7, Gowganda; 8, Lorrain; 9, Upper Cobalt Group; 10, L. post-Huronian; 11, schists, quartzite; 12, Cutler granite; 13, diabase

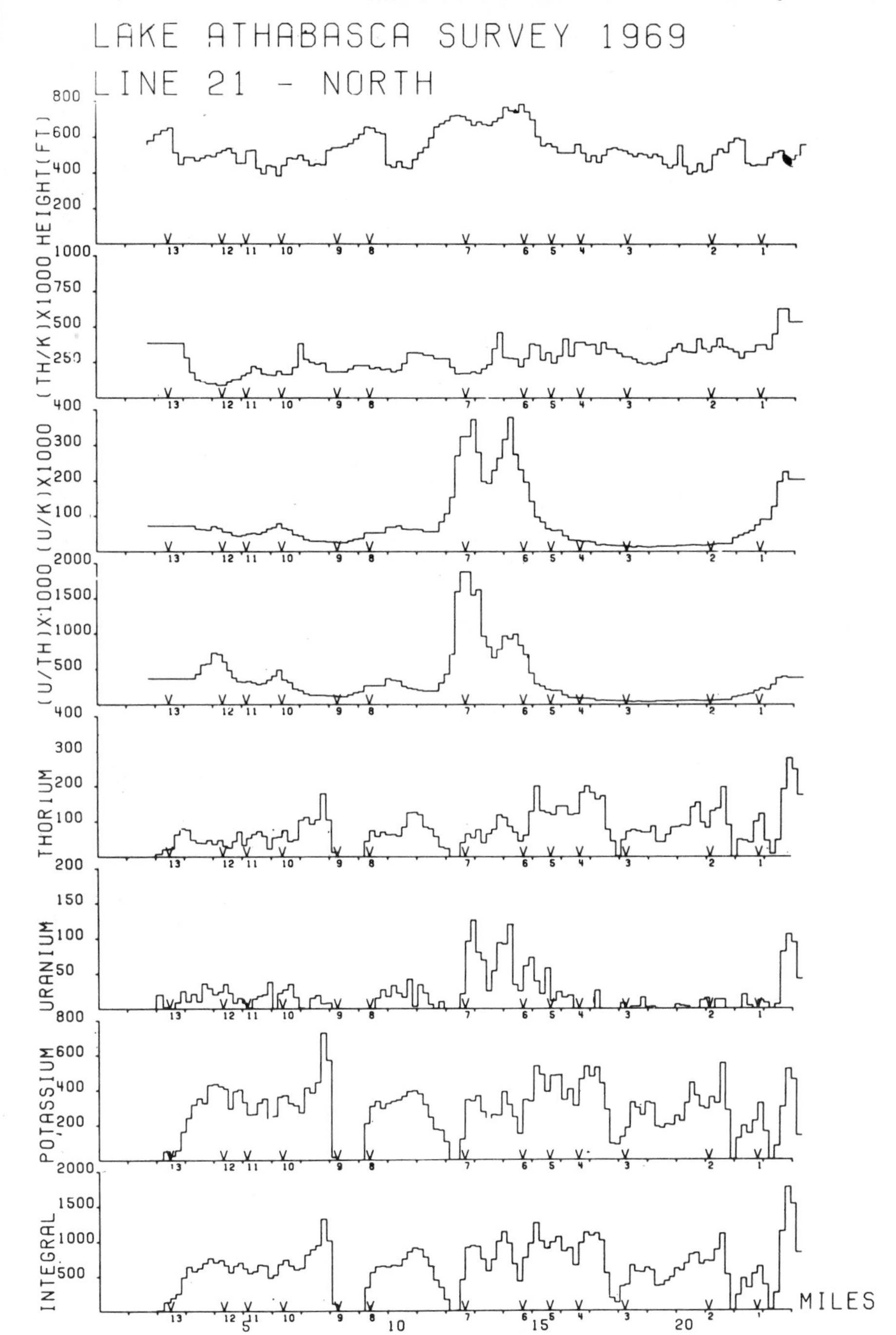

Fig. 8 **Athabasca line 21: strong uranium anomaly lost in integral plot, showing advantage of spectrometry**

profile is a rather subdued mirror image of the topographic relief, since the aircraft is flown so as to follow the contours. It should be noted that the height profile between fiducials 1 and 5 shows three prominent ridges and three steep-

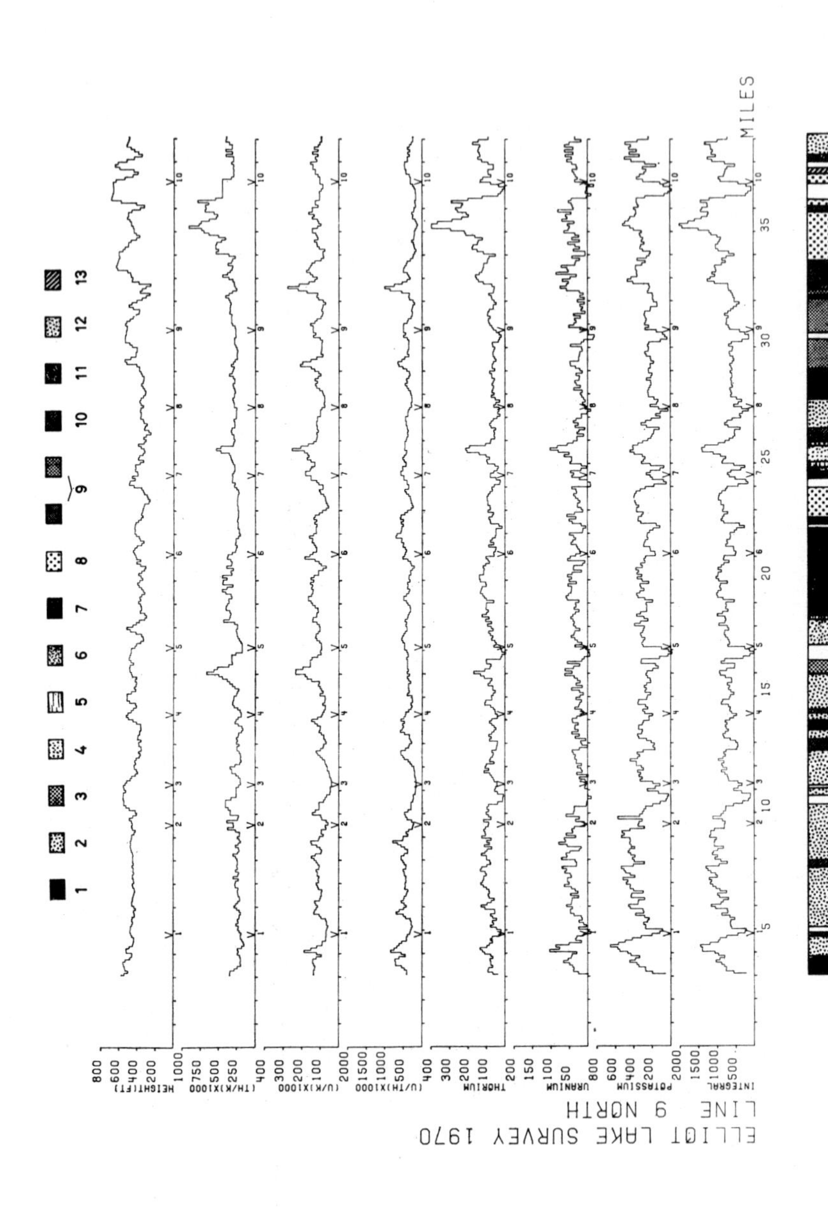

Fig. 9 Elliot Lake survey line 9. Fid. 5 is on Elliot Lake, north side. Fid. 1 is west of Pronto mine. For explanation of key see Fig. 7

sided valleys, with one shallower valley. None of the ridges gives rise to any spurious ‘anomalies’ and none of the valleys gives rise to any spurious ‘lows’, but one valley, by fiducial 4, does give a peak in all primary channels (integral, potassium, uranium and thorium). Since the ‘half-height’ for the uranium and thorium radiation measured is about 400 ft, the variations in terrain clearance would be sufficient to create more than twofold variations in count rate over the ridges and valleys if the correction were not applied.

Despite the large detector volume of the GSC system, it should be noted that after corrections have been made, ‘average’ count rates per 2·5 sec are approximately as follows: K, 300; U, 25; Th, 50; integral, 750 per 0·5 sec.

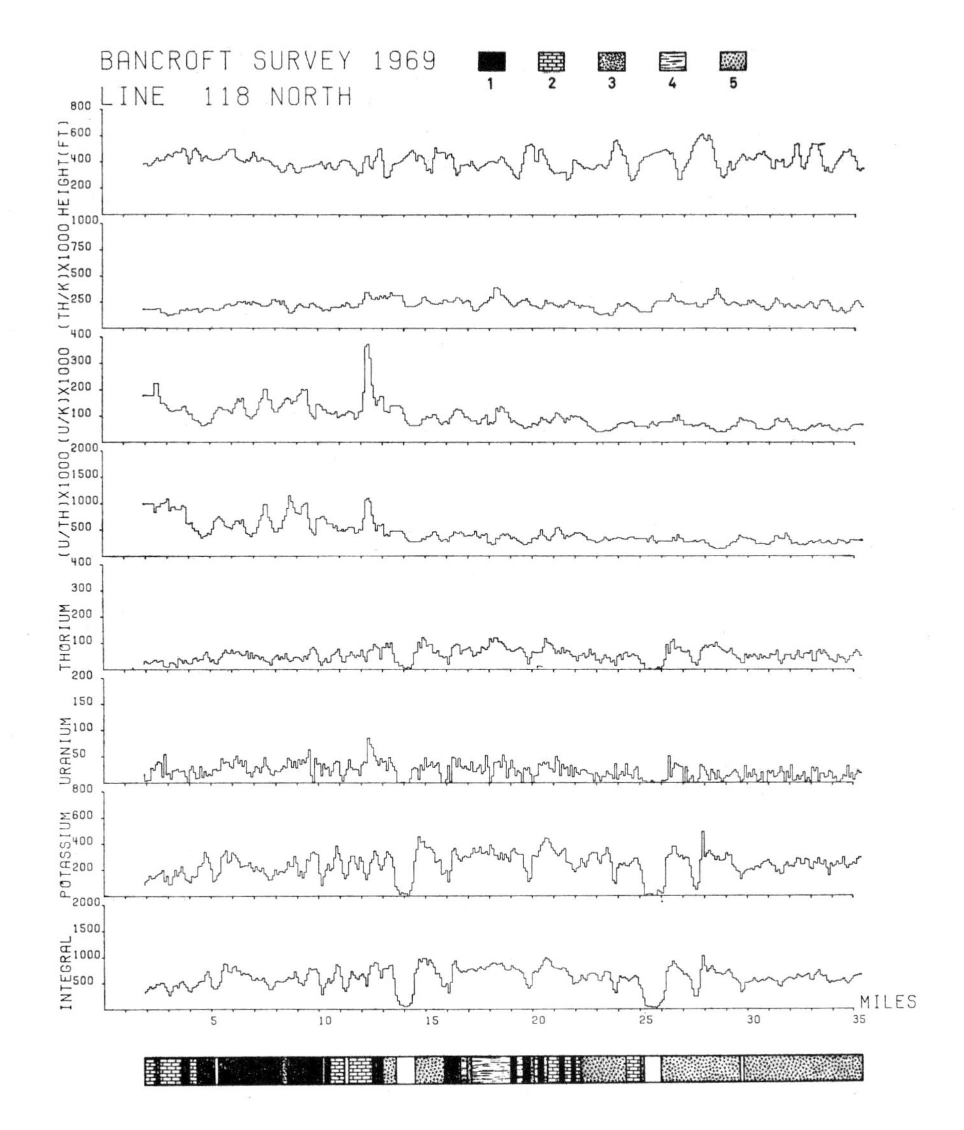

Fig. 10 Bancroft line survey 118. Paudash Lake (North Bay) is at 14-mile mark. Baptiste Lake (northwest arm) is by 25-mile mark. 1, Quartzites and conglomerates; 2, limestones; 3, diorite, gabbro; 4, syenite gneiss; 5, granite, granite gneiss

This underlines the fact that uranium is the most difficult to determine.

Fig. 8 is a line from the Uranium City area of Saskatchewan which demonstrates how a very significant uranium anomaly can be lost in a total gamma survey because it coincides with low thorium and potassium values. The existence of the uranium anomaly would not be suspected if only the integral

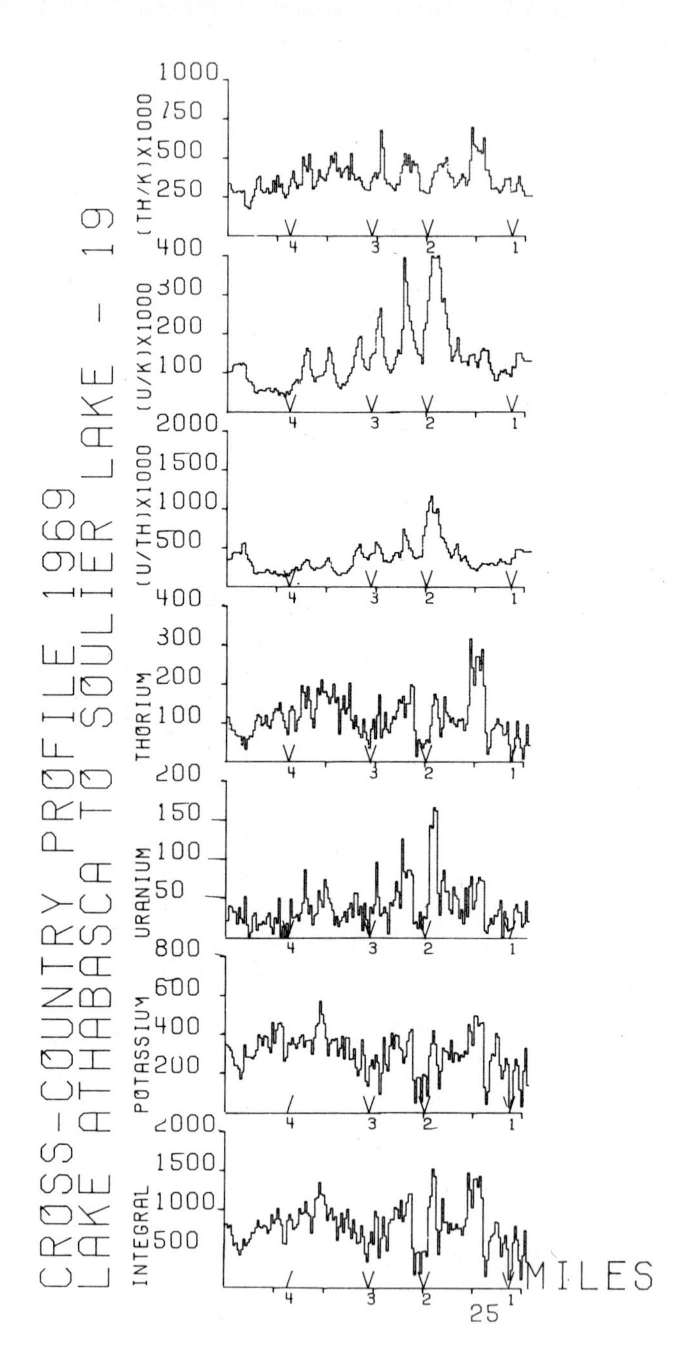

Fig. 11 Athabasca to Soulier Lake: cross country. (Note sequence of equispaced uranium enrichment zones picked out by U: K display)

plot were available, and if the integral plot were being used to guide exploration, attention would probably be focused, quite unnecessarily, on anomalies in the vicinity of fiducials 1 and 9.

Fig. 9, from the Elliot Lake area, provides another illustration of how the discriminating ability of a spectrometer survey permits the breakdown of anomalies into groups according to their geochemical nature. Thus, it shows,

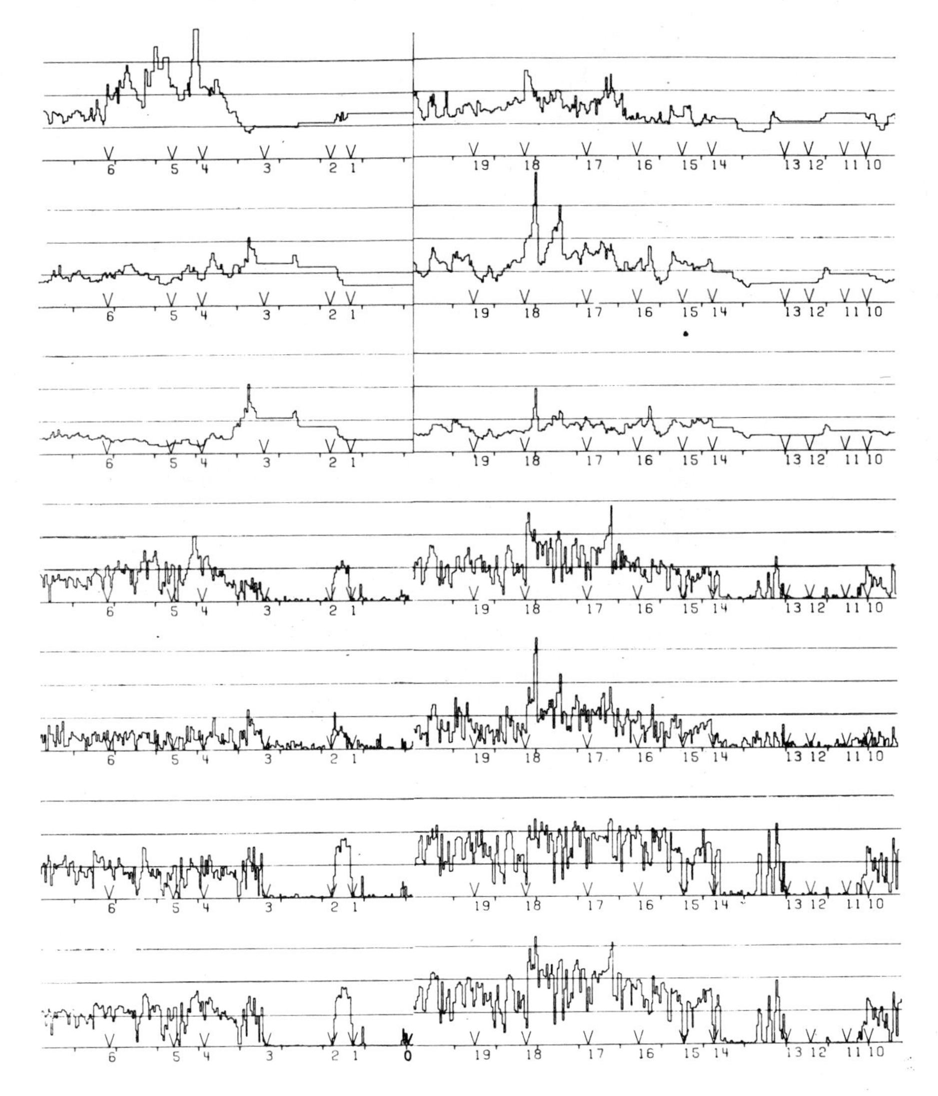

Fig. 12 Part of cross-country profile over northeast Saskatchewan. Wollaston Lake (between fiducials 3 and 0) and Reindeer Lake (fiducials 14 to 10). Displayed in same form as previous figures (Note prominent U:Th anomaly (third profile from top) between fiducials 3 and 4, which is 4 miles north of Gulf Minerals Rabbit Lake property). Base line is marked at 5-mile intervals

by the 35-mile mark, a thorium high over part of the Lorrain Formation; by the 25-mile mark there is a combined uranium and thorium peak over the Hough Lake Group sediments; and by the 5-mile mark there are combined potassium and uranium peaks over pre-Huronian granite. Note that there is a

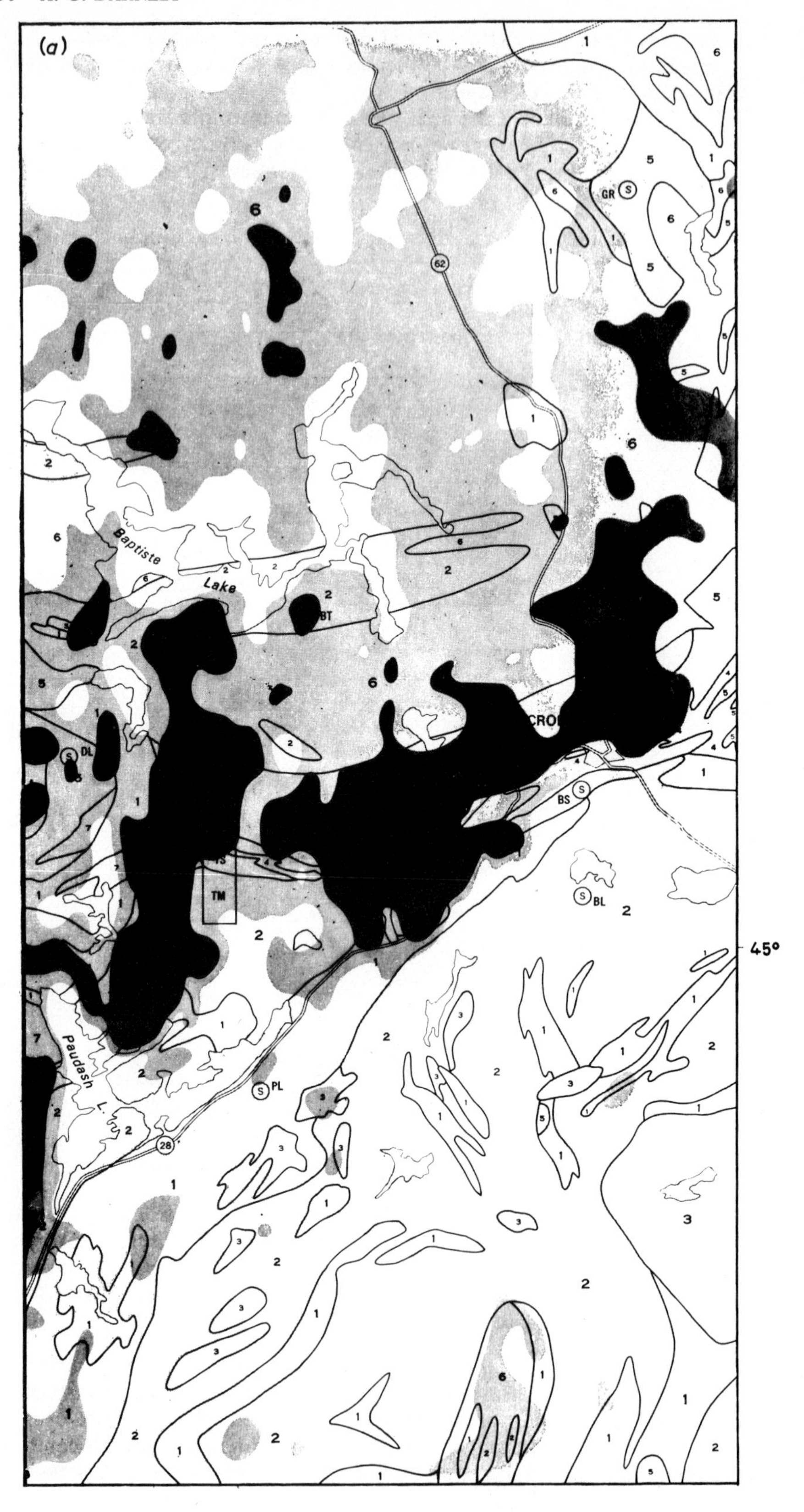
(a)
Baptiste
Lake
Paudash
L.
62
28
GR
DL
BT
BS
BL
PL
TS
TM
CRO
45°

(b)
Baptiste
Lake
BANCROFT
Paudash
L.
TG
TM
GR
BT
DL
AL
BS
BL
PL
62
28
78°

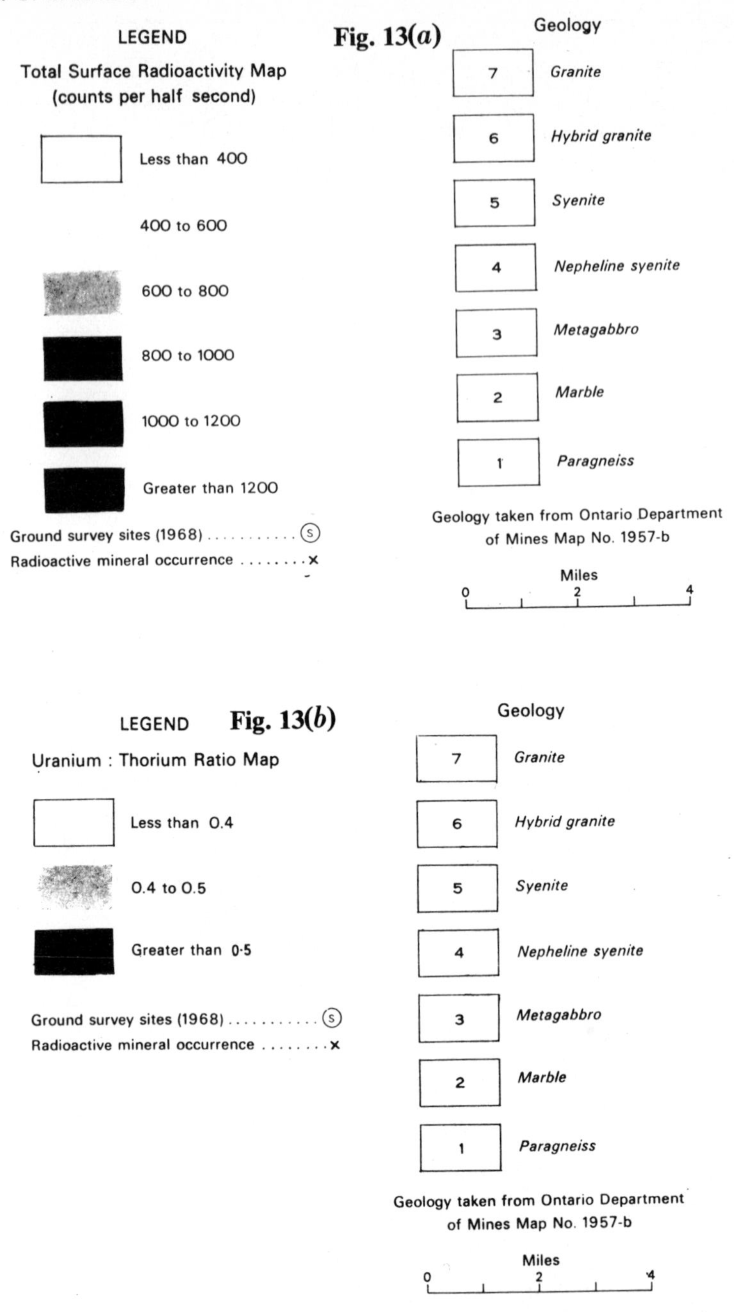

Fig. 13 Bancroft area: comparison of integral and U:Th maps showing how use of ratio reduces area to be searched. From Darnley and Grasty[9]

distinct U:Th ratio anomaly by the 32-mile mark spatially related to what is mapped as Lower post-Huronian. This is about 0·5 miles west of Pronto mine. The U:K ratio picks out the same feature, and an horizon in the Cobalt Group close to 29 miles, the Hough Lake group at 25 miles previously mentioned and the Elliot Lake group by fiducial 5. This latter point is more clearly distinguished by the Th:K ratio.

Fig. 10 is an illustration from the Bancroft area of a small uranium occurrence, at 12 miles, which is readily distinguished by a spectrometric survey, but which would not be found by a total radiation survey. Note also on this profile how the arenaceous metasediments at the south end (left-hand side) contain zones with anomalous U:Th ratios, although these would hardly be recognized from examination of the primary data. The marble at the extreme south end also shows a high U:Th ratio, which is quite common over this lithology.

Fig. 11 is of interest because of the succession of uranium to potassium ratio anomalies at approximately two-mile intervals (note that scale is different from previous profiles), commencing at fiducial 2 and extending 14 miles to fiducial 4 with diminishing intensity. They seem to correlate with topographic lineaments in the area. The ratio plot makes these features readily recognizable; they would not be identified if only the total radiation results were available. Information such as this is clearly of potential value in any detailed study of mineralization controls.

Fig. 12 is taken from a compilation of flight lines obtained on a cross-country reconnaissance from Ottawa to Yellowknife,[10] and provides another example of the way in which a small anomaly, in this case by fiducial 3, could be easily overlooked without the pointer provided by the ratio plots, especially U:Th. This profile, which is approximately 105 miles long, is aligned northwest–southeast across Wollaston and Reindeer Lakes in northeast Saskatchewan. It crosses the west side of Wollaston Lake about 4 miles to the north of the Gulf Minerals, Ltd., Rabbit Lake deposit, and the anomaly between fiducials 3 and 4 is approximately on the strike of the Rabbit Lake discovery. It should be noted that close by fiducial 18 there is a comparable anomaly for which no ground control information is available. Fig. 12 also illustrates a point made earlier concerning the way in which zones enriched in uranium occur within or on the margins of radioactive 'highs' of regional dimensions. In this cross-section the 'high' has a width of about 40 miles between Wollaston and Reindeer Lakes. The central anomalous block between fiducials 18 and 16 appears to coincide with the Compulsion River fold belt. The regional high can be traced in a northeast–southwest direction for a distance of more than 300 miles across Saskatchewan.

Profiles provide the most precise means of correlating data, but contour maps permit more rapid examination by giving the overall distribution of a particular parameter.

Fig. 13 shows the total surface radioactivity map of the Bancroft area of Ontario, being compared with the uranium to thorium ratio map of the same area. This illustrates the extent to which ratio maps can reduce the size of target area for detailed ground investigation.

Fig. 14 is a comparison of the total radioactivity map of the Eldorado area of northern Saskatchewan with the uranium to potassium ratio contour map of the same area. The maps in Fig. 13 were redrafted for display purposes from the original computer printouts, but in this case the maps are reproduced in their original form. The closely spaced contours in the left centre of the total radioactivity map are over the Eldorado mine property, and other properties in the vicinity of Uranium City. Neglecting the man-made anomalies in the vicinity of the mine properties, the total radioactivity distribution shows no clear pattern over the area as a whole, but the ratio contour map focuses attention on less than 25% of the area. As with the Bancroft area, the areas with anomalously high ratios enclose all known uranium mineral occurrences. It should be noted that data for the Eldorado area survey, which was flown at 2-km line spacing, were collected in eight flying hours.

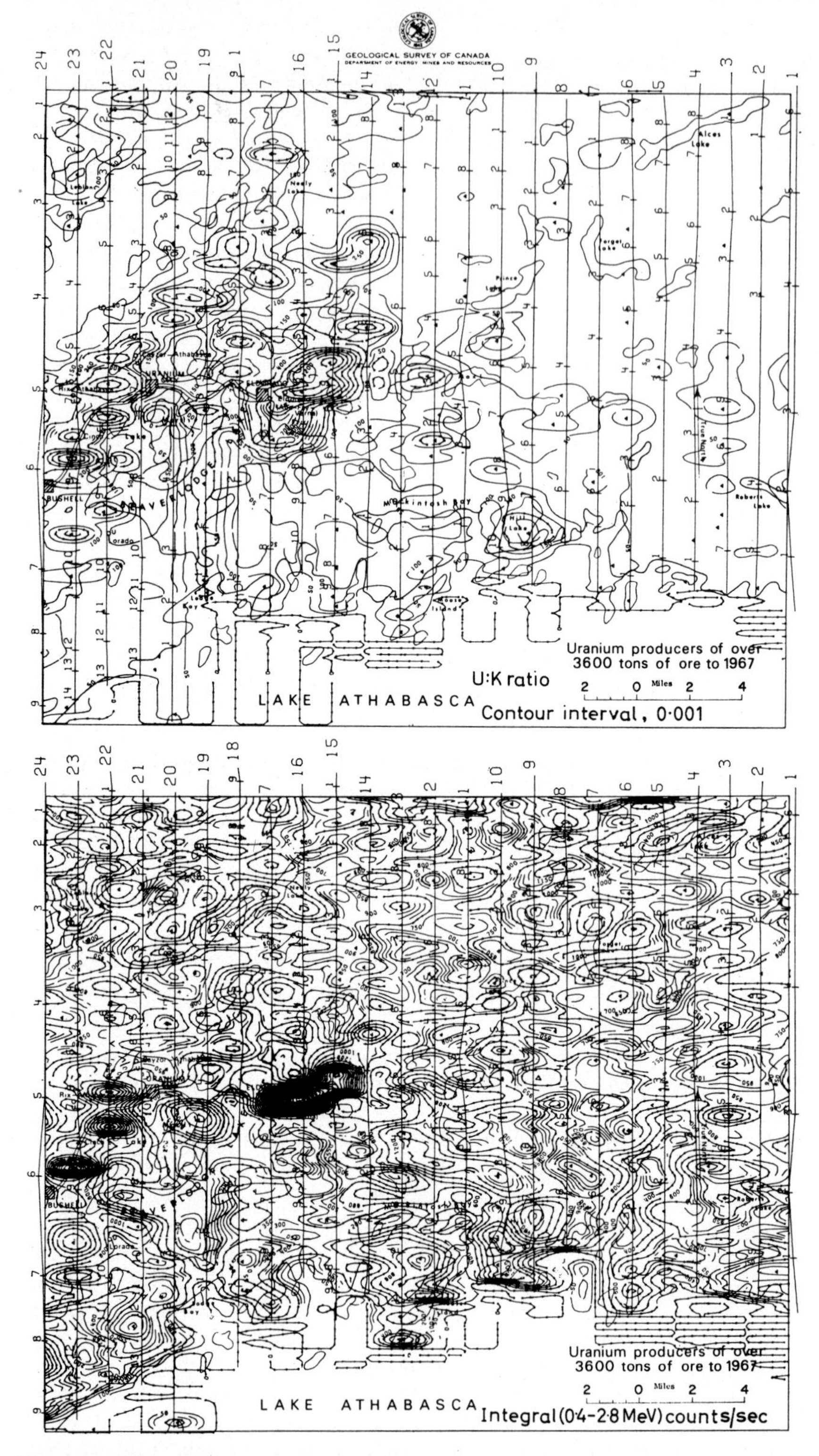

Fig. 14 Eldorado area: comparison of integral and U:K maps showing how use of ratio reduces area to be searched

Conclusion

Airborne gamma-radiation survey techniques provide an efficient means of exploring for new uranium deposits. The following considerations should be kept in mind.

(1) Total radiation measuring systems are suitable for (*a*) preliminary reconnaissance surveys to recognize broad areas requiring more detailed examination by airborne systems capable of energy discrimination and (*b*) target identification within areas where the mineralization controls are well known and the possibility of confusion with non-mineralization targets can be eliminated.

(2) Spectrometer systems should be used (*a*) wherever ground access is difficult and there is the need to classify and select different types of radioactive anomaly prior to ground examination and (*b*) wherever significant showings may be very small or very diffuse at the surface and are recognizable only because of a small increase in the relative concentration of uranium.

(3) Comparable results may be obtained from different systems by use of different combinations of detector volume, height, speed and line spacing. Costs, however, may vary substantially.

(4) Any type of airborne radiometric system, and spectrometer systems in particular, must provide a sufficiently high count rate after all corrections have been applied to be statistically meaningful at the required sensitivity. The maximum number of counts per unit distance must be achieved. This is obtained by maximizing detector volume and minimizing air speed.

(5) Data should be of a sufficiently high standard to merit comprehensive presentation by automatic compilation methods.

(6) In order to standardize results between surveys, results should be expressed in terms of mean radioelement abundance at ground level.

(7) The most important parameters in uranium exploration by airborne gamma spectrometry are the U:Th and U:K ratios. They facilitate selection of uranium anomalies.

Acknowledgment

This review of airborne gamma-radiation survey techniques has been made possible by the participation of many individuals at various stages in the development of the Geological Survey of Canada's airborne, ground and laboratory gamma-spectrometry projects. Credit is due to all who have been involved. Especial acknowledgment is due to Dr. R. L. Grasty for the development and application of the computer routines for automatic data presentation.

References

1. Adams J. A. S. and Lowder W. M. eds. *The natural radiation environment* (Chicago: The University of Chicago Press, 1964), 1069 p.
2. Adams J. A. S. and Gasparini P. *Gamma-ray spectrometry of rocks* (Amsterdam: Elsevier, 1970), 295 p.
3. Charbonneau B. W. and Darnley A. G. A test strip for calibration of airborne gamma-ray spectrometers. *Pap. geol. Surv. Can.* 70–1, pt B, 1970, 27–32.
4. Charbonneau B. W. and Darnley A. G. Radioactive precipitation and its significance to high-sensitivity gamma-ray spectrometer surveys. *Pap. geol. Surv. Can.* 70–1, pt B, 1970, 32–6.
5. Cook B., Duval J. and Adams J. A. S. Progress in the calibration of airborne

gamma spectrometers for geochemical exploration. In *Geochemical exploration* (Montreal: CIM, 1971), 480–4. (*Proc. 3rd Int. Geochem. Explor. Symp., Toronto 1970*) (*CIM spec. Vol.* 11)
6. Darnley A. G. and Fleet M. Evaluation of airborne gamma-ray spectrometry in the Bancroft and Elliot Lake areas of Ontario, Canada. In *Proc. 5th Symp. remote Sens. Environm.* (Ann Arbor: University of Michigan, 1968), 833–53.
7. Darnley A. G., Bristow Q. and Donhoffer D. K. Airborne gamma-ray spectrometer experiments over the Canadian Shield. In *Nuclear techniques and mineral resources* (Vienna: IAEA, 1969), 163–86.
8. Darnley A. G. Airborne gamma-ray spectrometry. *Trans. Can. Inst. Min. Metall.*, **73**, 1970, 20–9.
9. Darnley A. G. and Grasty R. L. Mapping from the air by gamma-ray spectrometry. In *Geochemical exploration* (Montreal: CIM, 1971), 485–500. (*Proc. 3rd Int. Geochem. Explor. Symp., Toronto 1970*) (*CIM spec. Vol.* 11)
10. Darnley A. G. Grasty R. L. and Charbonneau B. W. A radiometric profile across part of the Canadian Shield. *Pap. geol. Surv. Can.* 70–46, 1971, 42 p.
11. Denham G. M. *Instrumentation for airborne radiometric surveying* (Toronto: Exploranium Corporation of Canada, 1970), 3 p.
12. Frondel C. Systematic mineralogy of uranium and thorium. *Bull. U.S. geol. Surv.* 1064, 1958, 400 p.
13. Grasty R. L. and Darnley A. G. Calibration of gamma-ray spectrometers for ground and airborne use. *Pap. geol. Surv. Can.* 71–17, 1971, 23 p.
14. Godby E. A. *et al.* Aerial prospecting for radioactive materials. *AEP (Can.)* CRR–495; MR–17, 1952, 93 p.; *Nucl. Sci. Abstr.*, **6**, 1962, 6596.
15. Gregory A. F. and Horwood J. L. A laboratory study of the gamma-ray spectra at the surface of rocks. *Res. Rep. Mines Brch Can.* R85, 1961, 52 p.
16. Heinrich E. W. *Mineralogy and geology of radioactive raw materials* (New York: McGraw-Hill, 1958), 614 p.
17. Hood P. J. Mineral exploration: trends and developments in 1970. *Can. Min. J.*, **92,** Feb. 1971, 185–214.

Appendix 1

SAMPLE SPECIFICATION FOR A HIGH-SENSITIVITY GAMMA-RAY SPECTROMETER SURVEY*

Operational parameters

Mean flying height	400 ft ± 25 ft
Local height deviations	+400 ft, −200 ft
Air speed	To be held ± 5 mph in range 60–120 miles/h; detector size and counting interval to be determined according to speed of aircraft to be used. The Contractor may select the type of aircraft to be used, but this will determine certain details of the equipment, especially detector volume

Detector system

Crystal size	Minimum diameter of individual crystals, 5 in Minimum thickness of individual crystals, 4 in
Detector volume	The number of crystals employed must provide a total of approximately 1500 in³ of NaI(Tl) at 60 miles/h or approximately 3000 in³ of NaI(Tl) at 120 miles/h or *pro rata* at intermediate speeds
Detector stabilization	Maintain at constant temperature and, if single-channel analysers are used in spectrometer system, a radioisotope calibration–source spectrum stabilizer is necessary. This calibration source must not interfere with the specified window widths

Spectrometer system

Spectrum analysis (*a*) Preferred system: multi-channel analyser with optional detector stabilization with windows corresponding to 1·36 → 1·56 MeV (^{40}K);

*Based on the Geological Survey of Canada–Atomic Energy of Canada, Ltd., Skyvan system.

1·66 → 1·86 MeV (^{214}Bi); 2·42 → 2·82 MeV (^{208}Tl); 0·4 → 2·82 MeV (Σ) with oscilloscope display of spectrum and window positions for calibration monitoring purposes. (*b*) Alternative system: stabilized detector assembly with single-channel analysers corresponding to above window widths.

The system is to be calibrated by setting up the spectrometer to provide three adjoining windows (each 50 keV wide) to bracket any one of several photo peaks of specified energy. For prime calibration ^{88}Y and ^{137}Cs are necessary; for routine calibration ^{40}K (from the natural environment) should be used.

The stability of the system to be demonstrated by (*a*) drift in count rate over 8 continuous hours test period to be less than $\pm 2{\cdot}5\%$ of mean count rate without manual adjustment and (*b*) mean count rates from a natural radiation test source at a fixed distance to be reproducible within $\pm 2{\cdot}5\%$ throughout survey operating time using a standard calibration procedure.

The Compton scattering corrections and sensitivity of the system in terms of count rate per unit concentration for each radioelement are to be established on the GSC test pads at Ottawa and by a test flight over a calibration strip.

Sampling time Counts to be accumulated over sampling times in the range 0·5–5·0 sec; to be preselected according to operational requirements.

Terrain clearance A suitable means of continuously monitoring terrain clearance, e.g. radar altimeter, must be used in conjunction with the spectrometer, and mean terrain clearance during each sampling period must be determined and recorded with the data from the spectrometer. The performance of the altimeter must be demonstrated to be satisfactory for the purpose. Unserviceability of this instrument will render all results unacceptable.

Line spacing Detailed survey: 0·25 mile (or 0·5 km). Reconnaissance survey: 2·5 or 5·0 km; to be determined by scientist in charge.

Track recovery Automatic track recovery method of proven reliability under all normal survey conditions and/or TV/video tape recording and/or photographic strip film camera. Fiducial points must be entered on the magnetic tape record at the start and finish of every line; at least every four miles if an approved automatic track recovery system is employed, and at least every mile if reliance is placed solely on a track recovery camera.

Recording Digital recording of raw uncorrected data on magnetic tape is required. Contractor either to be responsible for provision of his own data recording, processing and presentation programs or he may use GSC gamma-spectrometry data reduction programs. In either case data must be in a standard format as recommended by GSC. In-flight analogue presentation of data is also necessary in order to monitor correct functioning of all parts of the system and make preliminary assessment of results whilst survey is in progress. There must be provision for operational information such as line number, date, time of day, type of measurement and fiducial points to be recorded on the tape. There must be provision for in-flight or immediate post-flight verification of magnetic tape digital recording. An airborne or ground read-out/display unit is mandatory.

Data correction

Raw data must be corrected as follows (1) For background (cosmic and atmospheric) radioactivity: this must be determined at the beginning and end of flight by flying an approved test line at least three miles long. (2) For deviation from mean terrain clearance by use of the recommended formula. (3) For Compton scattering contributions to the observed ^{40}K and ^{214}Bi count rates. Correction factors must be confirmed by making observations on the GSC calibration pads, Uplands Airport, Ottawa.

Data presentation

Data must be presented in the following forms.

(1) Digital printout in tabular array of raw data (computer listing).

(2) Profile for each line flown to show terrain clearance, integral, potassium, uranium and thorium count rates and the ratios of the U/Th, Th/K and U/K count rates plotted against distance along the line. Fiducials to be shown superimposed on the distance axis.

(3) Plots of flight lines superimposed on base map of the area to show exact location of fiducial points.

(4) Contour maps to be prepared for each of the parameters shown in profile form at a scale to be confirmed. Contour interval to be specified following production of profile.
(5) The programs to be used, provided by the Contractor, must be of equivalent quality to programs developed by GSC. The right is reserved to require the Contractor to demonstrate the accuracy of his computer programs by reducing satisfactorily test data provided by the Geological Survey of Canada.
(Note: This draft specification is for general information only. Details of the digital tape format and data reduction programs can be discussed on request.)

Appendix 2

COMMON CAUSES OF UNSATISFACTORY AIRBORNE RADIOMETRIC SURVEYS

(1) Installation close to radium-luminized equipment causing excessive background values.
(2) Shielding of detectors by aircraft fuel tanks causing reduced sensitivity when tanks are full.
(3) Intermittent electronic noise causing spurious anomalies.
(4) Inadvertent or erroneous setting of high voltage supply, window widths or energy thresholds.
(5) Improperly adjusted base lines on analogue records.
(6) Binary bits missing or added in digital recording.
(7) Unreliable radar altimeter.
(8) Failure to monitor atmospheric background and fly test line.

DISCUSSION

M. G. Huggett asked (1) how well the flight tracks were recovered by the Geological Survey of Canada and to what accuracy anomalies were pinpointed; (2) if the author could give more details about the test line for making background corrections; and (3) the volume which was considered to be insensitive—he referred to the author's discussion of height correction.

The author said that with regard to track recovery, they estimated that they could recover their position to within one-tenth of a mile in Shield areas with the television system. Track recovery over the Shield was relatively easy, the large number of lakes, with their characteristic shapes, making it quite straightforward. With experience, the operator watching the monitor screen could get good track recovery, but when difficulties were expected they turned on the video tape recorder. The reason for not running the video recorder continuously was due to the problem of changing tapes in the air. When cassette-type video tapes became available, they would run the recorder continuously; but, at present, they planned to use one tape per flight. They were also employing Doppler, and the along-track measurements of that were very satisfactory; but they had yet to achieve good cross-track measurements. Automated track recovery methods were probably one of the major improvements required for airborne survey work in the future.

With regard to obtaining corrections for background from a test line, they had always been in the habit of measuring the atmospheric background over a lake at the beginning and end of each flight. In addition, they had found it desirable to run a test line which also went over land for a distance of several miles in order to take account of the variations in ground moisture, for instance, and the effect that that might have on uranium count rates. Prolonged periods of heavy rain depressed the uranium channel and it was as well to have some measure of that effect. The test line thus went over water and land; and its path was chosen not only to be easy to follow but orientated so that the structure was at right angles to the line of flight. In those circumstances deviations of a few hundred feet on either side of the line would not be of great importance.

Turning to the question of 'insensitive' systems, Table 5 (p. 187) compared the various methods of operating. If one were concerned with a very small area, then one could be effective by flying very low and very slowly with a comparatively small detector, but flying a big area in that manner was financially disastrous. There was a need to include costing in the overall assessment of efficiency. The profiles obtained showed that the average uranium and thorium channel counts were about 50 per 2½ sec. A common uranium count was only 25; and potassium 400 counts per 2½ sec. Thus, even with the volume which they were using, which seemed large to most people, by the time all the corrections had been applied only relatively few counts were available. For their purposes a detector system of one-tenth the size would be a waste of effort.

Dr. G. Lawrence asked how the background correction for ^{214}Bi was applied. The author said that, on average, between 60 and 80% of the observed counts in the uranium channel were due to background. They flew over an area of water at the beginning and end of the flight; and, although for two-thirds of the time in any area the values obtained were sufficiently consistent, the remaining third of the time they were not. To overcome that they updated the background count every time the aircraft flew over a sheet of water more than ¼ mile in diameter and there were sufficient lakes of that size within the Shield to obtain a true background all the time. That was possible only in Shield conditions; where no water was available, consideration had to be given to in-flight methods of direct radon measurement.

Dr. Lawrence said that in parts of Africa and Europe it had been found that radon built up overnight and could settle over lakes beneath inversion layers. Indeed, some of those lakes gave distinct broad anomalies when overflown in the early morning, but those were dissipated by mid-morning.

Dr. T. Schwarzer, with regard to calibrating an airborne system to convert count rate data to equivalent abundances, said that at Rice University they had also done that and found it to be very useful for identifying lithological types and building up geological maps from the air. They had also found that the use of radioelement ratios tended to remove the lithological distinctions. That point, together with the observation that uranium anomalies stood out very well when expressed as the U/Th ratio, was very important. Because of that consideration, she wondered if it were necessary to calibrate count rate data quantitatively and convert them to concentration units.

The author said that for localized exploration purposes the calibration need not be too precise; but, where regional compilation and comparison of results were necessary, standard units were required.

Dr. H. W. Little suggested that the comparison of costs of ground gamma-spectrometer survey to airborne was not altogether fair. In the particular survey mentioned the equipment used was rather primitive (it weighed about 40 lb) and the area surveyed was unusually heavily vegetated. With more modern, lightweight equipment the cost would be a fraction of the figure quoted. It had been noted, however, that ground surveys should be total count, followed, if desired, by testing of interesting anomalies by gamma spectrometer.

The author agreed that the equipment used was heavy, but they did not measure a significantly greater number of stations per day with lighter gear at Fort Smith in 1970. In the Bancroft area the team could travel by car to within a mile of any measuring station. Ground and air spectrometry were not equivalents as the method employed obviously depended upon the size of the area. All his evidence indicated that the cost of ground work was very high—at least $Can.1500 per square mile—wherever a systematic pattern of measurements required line-cutting.

Dr. W. Domzalski expressed admiration for the tremendous amount of excellent work and research evident in the paper. At the same time he felt that it was necessary to focus more attention on the problems which bedevilled industry and on the factors which really controlled an average 'commercial' airborne radiometric survey. To start with, the most which could be expected at present from a commercial installation was up to four crystals of 5 in × 5 in or 6 in × 4 in. It was true that it might be possible to have additional crystals installed, but that would be a 'special' system, adding prohibitively to the cost of the survey and considerably to the weight of equipment. Regarding money, a multi-crystal detector (say, twelve crystals) might cost anything between £20 000 and £25 000, on top of spectrometer, recorder, etc., costs. Extra expenses would be involved in automatic stabilization systems and modifications in general. That was not, however, the only limiting consideration. One detector and housing weighed about 70 lb, and a multi-detector system could not be accommodated in a small aircraft. When selecting the aircraft it would be necessary to consider the following weight items: spectrometer system; magnetometer (almost always flown simultaneously in commercial surveys); ancillary instrumentation (positioning cameras, Doppler, radio-altimeter); long-range tanks in areas where airfields were not readily available outside main centres (Africa, Asia, Middle East); and pilots, navigator, operator and, frequently, a security officer!

It was obvious that that was not a task for a small aircraft which was to fly at 80 miles/h at less than 250 ft. Thus, practical considerations and availability of equipment clearly imposed a definite limit on the type of equipment. In consequence, the parameters which could be varied on a commercial survey were line spacing and time constants of the equipment: even altitude and speed were confined to narrow practical limits. Accordingly, more attention should be given to the problems of line spacing, instrument time constants and controls of radiometric surveys. Even the line spacing was more frequently the function of funds available for the survey than of purely technical considerations. One was usually torn between a host of considerations, involving time deadline for carrying out the work, availability of aircraft and equipment, disposability of funds, weather factors, operational aspects, necessity for optimizing the survey specifications and idiosyncrasies of half a dozen people involved in the situation.

The suggestion that one flew a scintillation counter first and, later, a spectrometer used selectively, would not be accepted gracefully by financial managers. One therefore went for a spectrometer, realizing that conditions during at least part of the survey would not be amenable to reliable spectrometer results. It was largely a question of compromise. In order to minimize statistical fluctuations it was necessary to increase the time constant, with the result that point sources might not register adequately. Both the 'minimum time constant' (time constant for full response) and the 'optimum time constant' increased with altitude of flying.

When choosing line spacing it was necessary to consider the possible size of target and the flying altitude. It was true that an average reconnaissance survey had greater value from the point of view of delineating areas of relatively higher and lower radioactivity than specific sources. It was normally to be expected, however, that survey planning would allow for the likelihood of detection of such specific sources.

To appreciate the problems involved, the following example might be useful. If a response above the source (in the 'zenith') was 100%, then at a lateral distance away from the source (line displacement) equal to the altitude, the response fell to about 30%. That was true strictly for a 'point' source and for a range of altitudes between 300 and 500 ft. In other words, at a 'nominal' survey altitude of 300–500 ft, and with line spacing of about 1000 ft, it should be possible to record not less than 30% of full response for all 'off-line' sources. With a spacing of 1 mile, there would be roughly a one in five chance of all 'off-line' sources giving more than 30% of peak response.

With regard to the presentation of results in terms of mean ground level element abundance, considerable difficulties might arise due to the fact that the gamma count in the air depended so much on the atmospheric conditions. That was one of the reasons why a control line was usually flown at survey height at the beginning and end of each sortie. With all those uncertainties, however, the 'element-abundance' calibration might be somewhat problematical.

Dr. S. H. U. Bowie, with regard to Table 5 (p. 187) said that the figures of merit did not seem to have taken into account the cost, which was clearly a very important factor. That element should include both the initial and operating costs of the equipment.

The author agreed that the figures of merit did not take cost into account. That consideration should follow on after the merit had been obtained from the technical factors. The major cost component was the flying cost, i.e. the number of line miles required: with a small system that number could easily be double that needed to cover the same area with a larger system. The initial cost of the equipment, when spread over a large survey, was quite a small percentage of the flying cost. When 20 000–50 000 line miles per year were being flown over several years (some commercial companies did twice that amount), flying costs were predominant. If, on the other hand, only a few hundred square miles were to be flown, then equipment price was important and a small system flown low and slowly would be more suitable.

Radon methods of prospecting in Canada*

Willy Dyck B.E., M.Sc., M.C.I.C., M.A.E.G.
Geological Survey of Canada, Ottawa, Ontario, Canada

550.839: 546.295: 553.495.2

Synopsis
Within two years of the publication of the first results describing the application of radon to prospecting for uranium in Canada, radon detectors have evolved from the clumsy laboratory instrument to portable field models weighing less than 10 lb. Most units have a detection limit of 0·2–0·5 pCi of radon and are versatile enough to permit, with minor modifications, the determination of radon in soil emanations as well as radon in surface waters. Ionization chambers of several hundred litres' capacity are being tested for use in aircraft to monitor the radon content of air masses.

Lack of topsoil in much of the Canadian Shield limits the application of the radon method. Under favourable conditions soil radon has been found to outline uranium mineralization more clearly than the scintillometer. Radon surveys of surface waters from streams and lakes outline uranium-rich regions. Because of the abundance of small lakes and streams in the Canadian Shield, water radon surveys have been proved to be quite helpful in uranium prospecting in Canada on a regional as well as on a detailed scale.

The radon content in natural water bodies varies inversely with the distance between sample and sediment and temperature, and, to a lesser degree, with flow and turbulence of the water.

Comparison with the uranium method shows that neither is infallible and, when possible, both should be used. Uranium in lakes gives a better regional picture; radon pinpoints uranium sources more clearly. The radon method also has the advantage of being a true field method in that the result can be obtained at the site.

Radon has occupied man ever since its discovery by F. E. Dorn in 1900.[22] Some may recall the radon advertisement that health resorts used to attract tourists; it is reported that one of the most radioactive springs in England was at one time reserved for the Queen and her court.

Although people were hardly prospecting for uranium around the turn of the century, a rather intriguing method of collection and detection of radon in soils and stream waters was described in 1904.[1] Radon methods of prospecting for uranium, particularly those involving soil radon, have been employed for more than three decades in eastern Europe.[2,21,29] In western Europe and North America the radon method came into use somewhat later.[4,12,13,20,23,32,35]

It was known to the author from radiocarbon dating work that certain soils contained much radon, which interfered in the dating of soils and, therefore, had to be removed from the processed samples. Shortly after the author joined the Geochemistry Section of the Geological Survey of Canada (G.S.C.), the search for uranium was intensified in order to meet the uranium demand, which was thought to be likely to increase as a result of a speed-up of the reactor construction programmes. On the basis of observations about radon in soils A. Y. Smith encouraged the author to set up an instrument and try it out in Bancroft in the summer of 1967. Both the soil and surface water tests were so encouraging[9,10] that a better instrument and a larger sampling programme were warranted.

A careful literature survey soon revealed that many people had previously thought of using radon for uranium prospecting; however, there seemed to be no reports of the method having been tried in North America, although Rogers and Tanner have reported in some detail on the behaviour of radon in natural environments.[25,26,33] The regional surveys carried out in the Bancroft and Elliot Lake areas of Ontario[31] and the detailed studies of the Elliot Lake, Ontario, and Gatineau Hills, Quebec,[8] confirmed the preliminary findings that radon could be used to detect uranium mineralization. A final test of the method was carried out in the Beaverlodge area of Saskatchewan[6] to compare the radon results with the established uranium methods. In addition, careful attention was paid to environmental factors to determine their influence on radon in the surface environment.

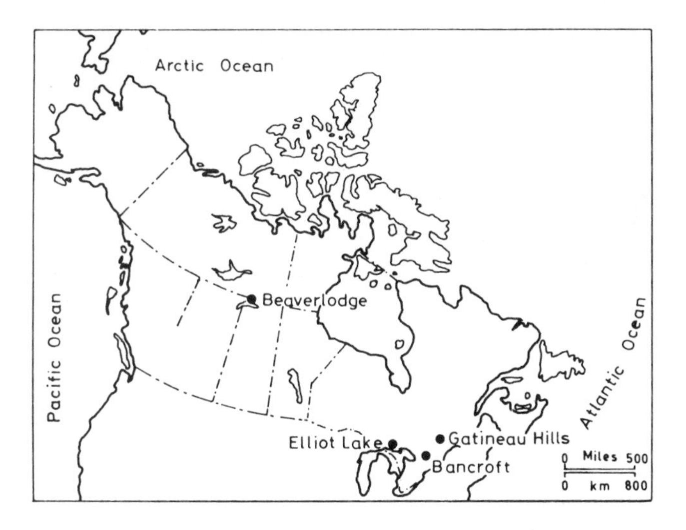

Fig. 1 Location of major known uranium deposits in Canada

The present paper is essentially a summary, in chronological order, of the development of instrumentation, sampling methods, highlights of soil tests and regional and detailed water surveys carried out by the author in the three principal uranium-producing areas in Canada, i.e. Elliot Lake, Ontario, Bancroft, Ontario, and Beaverlodge, Saskatchewan (Fig. 1). Results of a number of detailed studies carried out in the Gatineau Hills National Park, Quebec, are also included. The area lies just outside the city limits of Canada's National Capital and contains several uranium showings. Both factors add to the convenience of testing the radon method. The showings are just a little too small to make them an economic venture.

Development of radon detectors

The first radon detector of the Geochemistry Section, G.S.C., was, in essence, an adaptation of a system devised by Higgins *et al.*,[14] and consisted of an old tube-type scaler–timer–amplifier with a 2-in photomultiplier. The cells were made by coating the inside walls of a 150-ml Erlenmeyer flask with a thin layer of silicone grease, leaving a clear window in the bottom and dusting the greasy wall with copper-activated zinc sulphide powder. Scotch tape was used to make a light-tight covering around cell and photomultiplier tube.

Parts of this radon counter are shown in Fig. 2. The apparatus had severe shortcomings. For example, the tape and grease would slide slowly in very hot weather, and the zinc sulphide took around 15 min to de-excite after exposure to light. It did, however, function well enough to demonstrate the usefulness of the radon method.

Fig. 2 First radon counter of Geological Survey of Canada

The second-generation detectors (Fig. 3) consisted of portable scaler–timer–amplifier assemblies with silver-activated zinc sulphide cells made from 2-in diameter copper tubing. For soil tests open-end sleeves were used; for water tests the sleeves were sealed with a glass disc at one end and a valve at the other to permit evacuation and, hence, increase the sensitivity of the system. At the

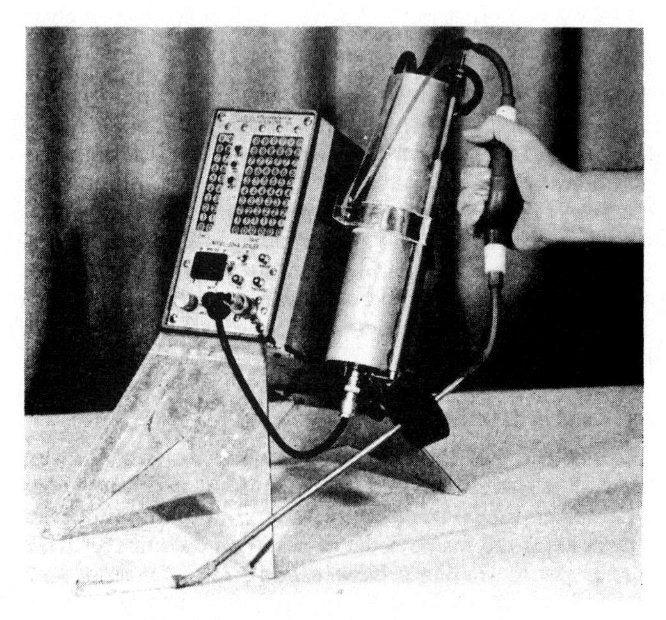

Fig. 3 Second radon counter of Geological Survey of Canada

time of writing, the author knows of four makes of portable radon counters produced in Canada. Three of these work on the activated zinc sulphide principle and one employs the ionization chamber principle. The zinc sulphide devices have a sensitivity of about 0·2–0·5 pCi, and are sensitive and versatile enough for analysis of surface waters as well as soil gases. The ionization chamber unit is suitable for soil gas work, but does not have the sensitivity required for radon levels encountered in surface lake waters. A modern portable radon counter developed by A.E.C.L. is shown in (Fig. 4).

Fig. 4 Modern portable radon counter

Fig. 5 Giant (250-l) ionization chamber

More recently, the author, in parallel with Krcmar,[17] is experimenting with giant ionization chambers similar to those used by Israël and co-workers[15,16] to see if a continuous method of detecting and recording of radon in air masses is feasible. It is thought that it may be possible to trace air masses containing higher radon concentrations to the region from whence they emanate. A 250-l ionization chamber currently under test in the G.S.C. laboratory is shown in Fig. 5. The unit consists of an intake port, a blower, an absolute filter, the ionization chamber proper and exhaust port. The outer shell is kept at ground potential and an inner shell, 1 in inside the outer one, at 250 V. The central electrode is connected to a vibrating reed electrometer. Although it has been shown[15] that stationary models can detect natural concentrations of radon in air near the ground, the unit shown here can only measure about 0·2 pCi/l when put in a moving vehicle, such as a helicopter or truck. It appears that mechanical vibrations introduce much larger signals than the radon content of air. The above sensitivity is roughly within a factor of ten of the sensitivity required to detect natural variations of radon in the atmosphere.

It should be remembered that elaborate equipment does not always ensure spectacular results. The alpha counter Lord Rutherford used consisted of a watch and light-sensitive eyeball—but with what results! More than fifty years ago Satterly and Elworthy,[30] by the use of a gold leaf electrometer and glass carboys as gas pumps, measured the radon content of the Carlsbad Springs near Ottawa with the same accuracy as the author did only a few years ago with transistors, photomultipliers and noiseless vacuum pumps.

Methods of collecting radon

COLLECTING RADON FROM SOIL EMANATIONS

The first G.S.C. radon profile was obtained by analysing soil gas samples which had been trapped in one-quart oil cans. The open ends of these cans were pushed firmly into the soil below the organic matter and the sides and tops were covered with leaves or organic matter. After several hours, or longer, the air from the can was transferred to an evacuated 1-l flask through a hose inserted into a ½-in hole punched into the top end of the can. The flask was then taken to the laboratory, where a portion of the gas was put into a radon cell and counted. This method gave good results, but it was too slow for routine work. With the portable instrument it was possible to determine radon at the site. Most subsequent results were obtained by making a 1- to 2-ft deep hole in the ground with a 6-ft crowbar, which had an enlarged, smoothly tapered, three-edged tip. As soon as the bar was withdrawn from the soil, a ¼-in tube with a fritted plug at the intake end was inserted into the hole and the air in the hole was pumped through the zinc sulphide chamber (see Fig. 6).

The withdrawal of a sample in this manner from the soil, followed by counting, permits an evaluation of the relative amounts of radon and thoron in the sample. By counting the sample for several thoron half-lives, a reasonably good estimate of the dominating activity in the soil can be obtained. By use of 1-ft deep holes spaced 100 ft apart, relatively small pegmatite dykes were outlined in overburden 2–3 ft thick. Tests deeper than 2 ft were difficult to make either because of a lack of soil or the abundance of boulders in the glacial till. Soil augurs were found to be rather useless in overburden containing boulders. Tests with a ¾-in diameter hexagonal hollow drill rod and a 6-lb sledge were carried out to a depth of 6 ft. The rod was prepared as follows: a pointed steel tip was pressed into the bottom end and a handle was welded to the top end. The soil gas sample was pumped with a small piston pump through a number of fine holes near the bottom end of the rod, through the rod itself, and thence into the counter via an interconnecting nipple at the top of the rod. This method gave satisfactory soil gas samples, but strains in the weld of the handle and nipple produced by the sledgehammer broke these pieces off in a short while. It is quite apparent that unless a quick and easy method of making holes is found,

Fig. 6 Detection of radon in soils

the soil gas method may prove unfavourable. It would seem that a $\frac{3}{4}$-in hollow hexagonal steel rod and hammer, and loosely fitting attachments for withdrawal of a sample and the rod, should be a good way to sample difficult overburden.

COLLECTING RADON FROM NATURAL WATERS

When radon work was started in the G.S.C. laboratory it soon became apparent that it was not easy to contain radon. Even in dilute concentrations in water, radon would disappear from plastic bottles at an alarming rate. For this reason all water samples for radon analysis were collected in flint glass bottles with plastic screw caps. These are very cheap and radon loss or bottle breakage has not been a problem. It may be of interest to prospectors in cold climates that radon loss from plastic bottles is negligible if the sample is frozen at the time of collection and thawed just prior to analysis in room-temperature water.

Taking a water sample in built-up areas is done most economically by motor car at bridges. In inaccessible regions sampling the bays of small lakes with a small helicopter is most economical. One man sampling in a Bell 47G2 can take between 20 and 30 samples per hour, depending on the sampling density

Fig. 7 Sampling lakes in detail in inaccessible areas

and the distance from home base. Detailed lake sampling can be done by strapping a depth sampler to the running board of the helicopter (see Fig. 7). It is important, however, that the sampler treat stream samples and lake samples separately. When lake samples are taken from shore at shallow depth the difference between them and stream samples is not so great. The reason for the difference stems from the fact that the radon content of surface water samples depends greatly on the distance between sample site and the sediment. The person sampling should therefore try to standardize this distance as much as possible or record the distance and correct for it mathematically. Since stream sampling is usually carried out from shore, this distance effect is less pronounced; however, for streams the temperature of the water varies considerably. This affects the solubility of radon, and, hence, the radon content of the water. A very cold sample invariably points to a spring source nearby. Thus, apart, from the greater solubility of radon in the colder water, the sealing effect that the water experiences underground also leads to an increased radon content compared with surface runoff.

Experience has shown that a density of one sample per one or two square miles outlines uraniferous areas in sufficient detail for detailed follow-up work. In fact, one lake sample per five or even ten square miles seems to indicate where the 'big' deposit is (see below). Detailed sampling of lake shores or streams shows that sampling at 200- to 300-ft intervals will reveal small individual uranium showings.

The actual analysis of the water samples involves the following steps: removal of radon from an aliquot of the sample by passing air through it, filling a ZnS cell with this air–radon mixture, and counting the alpha particle activity with a radon counter described above and in detail earlier.[7] A simple air-circulating system can also be used to analyse waters on site. This feature is of particular interest to the traditional prospector, who likes to work by himself.

Results and discussion

SOIL GAS TESTS

As was pointed out earlier, soil conditions in much of the Canadian Shield are far from ideal for radon emanation tests. Usually, swamps or bogs alternate with soil-poor hills or outcrops. Where overburden does occur it often consists of compacted clay or gravel interspersed with boulders. Soil radon tests have been made in four areas in Canada in which uranium is known to occur—Bancroft and Elliot Lake, Ontario, Gatineau Hills, Quebec, and Beaverlodge, Saskatchewan. For contrast or comparison a few tests were also carried out in the Sudbury, Ontario, area. All except the Beaverlodge tests have been described in detail earlier.[8,10] The Beaverlodge tests were made across the St. Louis Fault, just outside the town of Eldorado, in a section of uncontaminated sand and gravel covering the fault. These tests gave slightly higher radon values on each side of the position of the fault as located by Tremblay.[34] This increase, however, was not significant in view of the reproducibility of the tests, and time did not permit repetition of these measurements.

Two radon profiles are presented, one from the Gatineau Hills and one from the Elliot Lake area, to illustrate the relative merits of the gamma-ray and the radon probes. The profiles shown in Fig. 8 are from the Gatineau Hills area. The top profile was obtained in the spring (10 June, 1968) and the lower two in the autumn (13 September, 1968). It is believed that the smoother lower profile obtained in the spring is due to moisture in the ground. Although the radon and gamma profiles correspond, the radon peak gives a more positive indica-

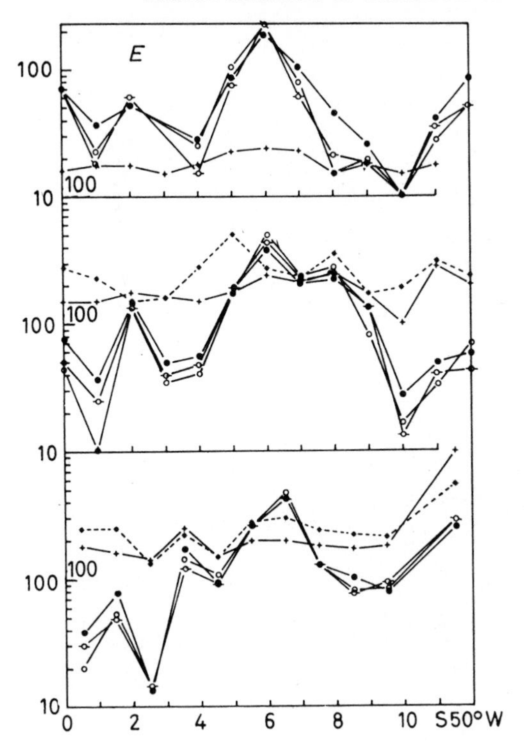

Fig. 8 Radioactive emanations from soils in Gatineau Hills. Plots show alpha activity from selected traverses in soil gas and gamma-ray activity (counts/min, outer vertical scale; counts/sec, inner vertical scale) as a function of distance along a traverse in hundreds of feet. Gamma-ray activity at soil gas sampling sites is shown by circles with four tails and outcrop or boulder gamma-ray activity by stars. Alpha activity in soil gases from 1-ft holes for three successive 1-min counting intervals is represented by solid, double-tailed and plain circles, respectively

tion of the pegmatite dyke than does the gamma peak. In Fig. 9 the activities over a subcrop of uranium ore just east of the No. 1 shaft of Rio Algom Quirke mine are shown. Again, the radon profile gives the larger contrast. Other radon profiles obtained at various locations have revealed persistent fluctuations, which could not be detected with the gamma probe. Without actual drilling to check for causes of such fluctuations, one can only speculate as to their origin.

SURFACE WATER TESTS

The encouraging results obtained with radon in surface waters and the abundance of small streams and lakes over large areas of the Canadian Shield prompted the G.S.C. to put most effort into radon surveys of drainage systems on a regional scale. Thus, the 1968 regional surface water radon and uranium surveys of the Bancroft and Elliot Lake areas[31] outlined the uranium zones equally well, radon giving a somewhat more focused picture of the uranium mineralization. To gain a better understanding of the factors affecting the distribution of radon in the natural environment Morse[20] undertook a more detailed study of the Bancroft region and the author took a field party to Beaverlodge to study the behaviour of radon in the natural surface environment and compare it to uranium for relative merits for prospecting. Except for one

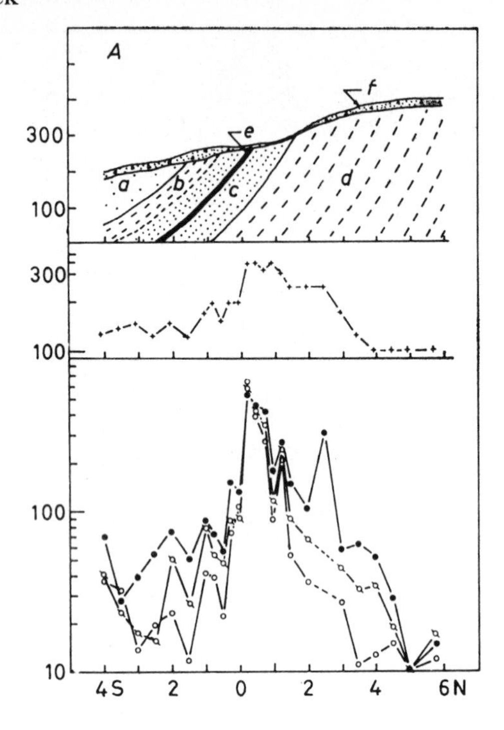

Fig. 9 Activities across subcrop of conglomerate ore at Rio Algom Quirke mine. Position zero is located 250 ft east of No. 1 shaft. Geological cross-section at top was taken from Robertson.[24] *a–f* denote Mississaği quartzite, Pecors and Ramsay Lake greywacke argillite, Matinenda quartzite, ore conglomerate, basement–andesite–diorite, etc., main conglomerate ore and gravel, respectively. Vertical scale for cross-section denotes relative elevation (ft). For explanation of scales and symbols see Fig. 8

illustration of the use of radon in detailed work in lakes, discussion is confined to the results of the Beaverlodge study.

The shore radon profiles in Kingsmere Lake, Gatineau Hills (Fig. 10), exhibit two peaks: both correspond with a radioactive pegmatite dyke near the shore. The dykes are upstream from the lake, judging from the slope of the topography. Pegmatite dykes downstream did not give radon peaks in the lake. A similar situation with similar results was observed in Fortune Lake, a few miles away from Kingsmere Lake. From these and other tests one can conclude that whenever surface or ground water comes in contact with uranium mineralization the waters become loaded with radon and, thus, will give a clue as to the whereabouts of the mineralization. This overriding conclusion must always be kept in mind, even though environmental variables will sometimes distort the true picture.

To show the effect of certain environmental variables, such as temperature, depth of water, alkalinity, pH, size of water body, elevation, etc., on the radon content of a sample the author has reworked, condensed and added data to the results of the 500 square mile survey of the Beaverlodge area. Since the presentation of the raw data at the Third International Geochemical Exploration Symposium,[6] radium, iron and manganese analyses of the stream sediments have been obtained, 125 stream water and sediment samples and 125 lake water and sediment samples collected from a 30 square mile area within the regional survey area have been examined and computer techniques have been applied to determine correlations between variables and multiple-regression analysis to determine the dependence of trace-element content on environmental factors and minor-element content of samples. Although considerable time and effort went into the manipulation of computer programs to obtain background-corrected data, the contoured data were little different from the raw contoured data obtained initially. This is somewhat surprising, but also

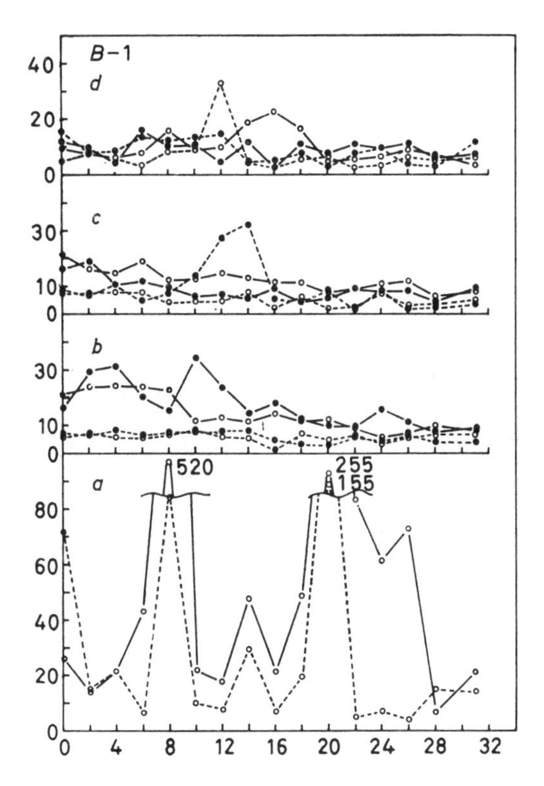

Fig. 10 Radon profiles along north shore of Kingsmere Lake, Gatineau Hills ^{222}Rn concentration of lake water samples (pCi/l, vertical scale) plotted as a function of distance along lake shore in hundreds of feet. Plots *a*, *b*, *c* and *d* give results of analyses of samples located 0, 20, 50 and 100 ft offshore, respectively. Solid circles denote samples from bottom and open circles samples from surface of lake. Solid lines give results of samples collected 15–26 July; dashed lines 24–26 September. Average depth (ft) and range of depth of lake at traverses *b*, *c* and *d* are: July, 4, 1–7; 10, 2–15; 15, 3–20; September, 4, 1–8; 9, 2–15; 15, 1–22

reassuring. It suggests that environmental factors tend to cancel one another to some extent in their influence on the trace-element content of geochemical samples—at least if we consider average effects. To be sure, there are noticeable changes in the distribution of the elements, as is shown later, but they are exceptional.

The 500 square mile region surveyed in the Beaverlodge area is shown in Fig. 11. The rectangle in the centre outlines 30 square miles which were re-sampled in more detail. The geology shown in this and the succeeding maps was taken from Christie,[5] Blake[3] and Tremblay.[34] The topography is rugged

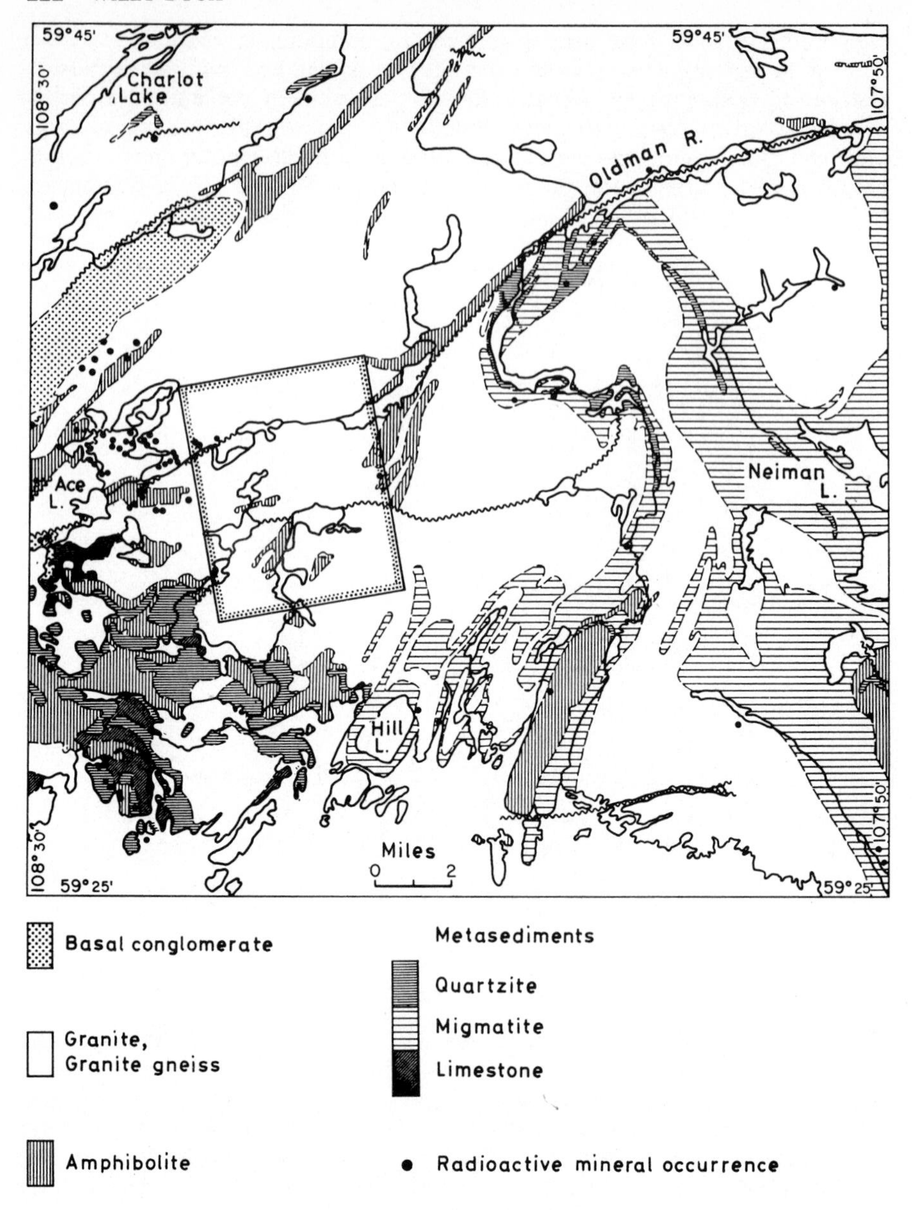

Fig. 11 Map outlining regions surveyed in Beaverlodge, Saskatchewan

and slopes downward to the southwest. Between one-quarter and one-half of the bedrock is exposed, the more numerous outcrops being in the southern half of the region. Granite and granite–gneiss make up most of the surface showings. In the southwestern corner quartzite and limestone dominate. This is also where the uranium has been concentrated in economic quantities. The deposits of importance are of the epigenetic type with pitchblende as the principal uranium mineral. Vegetation is moderate, but persistent, making helicopter landings on land difficult. Spruce is predominant. Birch, alder and willow are usually present along streams beds. The streams are generally small and short, serving mainly as interconnecting links between innumerable small lakes. Fig. 12 gives an aerial view of the area. The marked change of the tone in the centre could be due to mining operations; or could it be due to radioactivity? Stream waters and sediments were collected at an average density of one sample per square mile, and lake waters at a density of about 1·3 samples per square mile, in an area stretching for approximately 10 miles to the south, 12 miles to the north and 22 miles to the east of the mine. The 30 square miles near the centre of the area were re-sampled at a density of four samples per square mile for stream and lake water and sediment. The temperature of water, width,

Fig. 12 Typical landscape in Beaverlodge area, Saskatchewan

depth, level, flow and turbulence were recorded at the site; radon, alkalinity and pH were determined in the field laboratory; and uranium, radium, iron, manganese and volume of 250 mg of −60-mesh sediment in the home laboratory. Highlights of the results are summarized in Tables 1–3.

Because the chi square test for normality shows that radon, radium and uranium in the sediment tend toward a lognormal distribution, all data given here were derived from log-transformed data. Uranium in waters is neither normally nor lognormally distributed. Correlations, means, chi square, histograms, etc., were obtained by use of a modified version of GESTAT, a computer program developed by Garrett.[11] The problem of contamination is very difficult to resolve. To meet the criticism of contaminated samples, all sites where there was any trace of man's activity, such as trenching, blast pits or road fill, were coded so that the computer could reject and treat such samples separately. The numbers of samples rejected thus, and the respective means of their trace-element content, are listed in Table 1 for comparison with the uncontaminated samples. This does not mean that only or all contaminated samples were rejected, but at least an attempt was made. That contamination is real is quite evident. Whether it is man-made or natural is another matter. To avoid the charge that the data may be loaded, none of the 'contaminated' samples was used in any of the plots given here. If it is assumed that these samples had higher concentrations of trace elements because of natural contamination, we could rejoice even more when we say, 'Look how well geochemistry works!' Although the difference in the means of 'contaminated' and 'uncontaminated' samples is significant, the large difference in the radon means of stream and lake samples is striking—a factor of 9 in the case of the regional survey and 15 in the case of the detailed survey separates these means. This explains why a sampler should pay close attention to the location of the sample site, and, in particular, the distance between sample and sediment. In the case of uranium the sample location or medium (lakes versus streams) is not as critical, even though the differences are noticeable.

The means from the 30 square mile area sampled in more detail are appreciably higher than those from the region as a whole. The author would like to think that this means that the area is closer on the average to the 'big' deposit. But, again, contamination cannot be ruled out entirely. In this case wind-blown dust from the open-pit mining at Zora Lake, about two miles east of the study area, may have settled in the area contributing to these higher levels of uranium and radon.

Rather interesting results and conclusions were derived from correlation and multiple-regression analysis of the data from the various sampling media and sampling areas. Tables 2 and 3 are an attempt to summarize numerically the effect of environmental factors and minor elements on the trace-element content of a geochemical sample. The numbers in Table 2 are a measure of the degree of influence of the individual independent variables on the trace-element content

Table 1 Geometric means, threshold and anomalous values of uncontaminated samples, and geometric means of contaminated samples

	Uncontaminated samples				Contaminated samples	
Variable and sample medium	Number	Back-ground (mean)	Threshold (mean $+\sigma$)	Anomalous (mean $+2\sigma$)	Number	Mean
Rn, pCi/l						
–rss	540	11·2	84·0	630·0	50	38·5
–dss	124	32·4	312·0	3000·0	2	4620·0
–rls	681	1·2	5·2	22·9	132	2·7
–dls	124	2·0	8·7	40·1	1	1311·0
$U(H_2O)$, ppb						
–rss	542	0·5	1·2	2·8	50	0·9
–dss	124	1·0	1·9	3·7	2	2·1
–rls	681	0·4	0·7	1·4	132	1·4
–dls	124	1·0	2·3	5·0	1	1·0
Ra, pCi/g						
–rss	483	0·7	1·9	4·9	41	1·0
–dss	124	1·2	3·7	11·2	2	3·3
–dls	124	1·0	1·9	3·7	1	2·5
U(Sed.), ppm						
–rss	483	5·1	23·6	109·0	42	7·0
–dss	124	22·7	88·6	345.0	2	65·3
–dls	124	16·8	65·4	254·0	1	205·6

–rss, regional stream survey; –rls, regional lake survey; –dss, detailed stream survey; –dls, detailed lake survey; σ, standard deviation.

Table 2 Percentage of variance of dependent variable accounted for by individual independent variables

Independent variable	Per cent variance accounted for by independent variable														
	Dependent variable Rn				$U(H_2O)$				Ra			U(Sed.)			
	–rss	–dss	–rls	–dls	–rss	–dss	–rls	–dls	–rss	–dss	–dls	–rss	–dss	–dls	
Width or area	−14	− 8	− 1	− 0	− 2	+ 2	− 0	+ 1	− 2	− 2	− 0	− 1	− 1	− 7	
Depth	− 6	− 2		− 6	− 4	+ 1		+ 0	− 1	− 2	+ 0	− 0	− 0	+ 5	
Flow	− 3	− 0			− 1	− 2			+ 0	+ 2		− 1	+ 5		
Temperature	−14	−29	− 2	+ 1	− 0	− 0	+ 1	+ 3	− 0	− 1	+ 0	− 0	− 4	+ 3	
Elevation	+ 1	− 3	+ 1	−10	− 2	− 5	+ 0	−13	− 0	− 4	− 3	+ 3	− 6	− 3	
Alkalinity	+10	+11	+ 1	+11	+14	+ 3	+ 9	+ 9	+ 3	+ 3	+ 4	+ 7	+ 9	+16	
Mn	+ 1	+ 2		+ 1	− 0	− 1		− 0	+17	+35	+ 9	+ 2	+ 0	+12	
Fe	− 1	+ 2		+ 0	− 0	− 1		− 0	+13	+13	+ 3	− 0	− 3	+ 1	
Volume	+ 1	+ 0		+ 2	− 1	+ 0		− 2	+ 3	− 0	+ 6	+13	+28	+30	
pH	− 0	−14	+ 0	+ 1	+ 1	+ 3	+ 0	+ 4	+ 0	+ 0	+ 0	+ 1	+ 0	+ 2	

For explanation of letters see footnote to Table 1.

Table 3 Summary of regression analysis of log-transformed data (correlation coefficients in brackets; $r_{\cdot 99}$ = ·12 for regional survey = ·23 for detailed survey)

Dependent variable	Percentage variance accounted for with all independent variables	Percentage variance accounted for with five independent variables	Five independent variables with most apparent significance as determined by multiple-regression analysis (correlation coefficients in brackets)				
Rn							
–rss	33	32	Temp. (−·36),	Alka. (+·25),	Width (−·37),	Elev. (+·10),	Flow (−·03)
–dss	53	53	Temp. (−·52),	pH (−·37),	Alka. (+·30),	Elev. (−·18),	Mn (+·15)
–rls	6	6	Alka. (+·12),	Temp. (−·13),	Elev. (+·11),	Area (−·11),	pH (+·05)
–dls	26	24	Alka. (+·33),	Elev. (−·31),	Depth (−·24),	Vol. (+·12),	pH (+·12)
$U(H_2O)$							
–rss	19	18	Alka. (+·37),	Depth (−·23),	pH (+·16),	Elev. (−·16),	Flow (−·05)
–dss	11	10	Elev. (−·20),	Flow (−·14),	Fe (−·12),	Alka. (+·18),	pH (+·18)
–rls	13	13	Alka. (+·31),	Temp. (+·10),	Elev. (+·02),	Area (−·06),	pH (+·17)
–dls	20	18	Elev. (−·35),	Alka. (+·29),	Vol. (−·13),	Depth (+·05),	Temp. (+·18)
Ra							
–rss	23	22	Mn (+·37),	Fe (+·33),	Alka. (+·17),	Vol. (+·15),	pH (+·02)
–dss	47	45	Mn (+·56),	Fe (+·36),	Alka. (+·17),	Elev. (−·30),	Depth (−·13)
–dls	22	21	Mn (+·31),	Vol. (+·24),	Alka. (+·20),	Elev. (−·18),	pH (+·05)
U(Sed.)							
–rss	28	27	Vol. (+·39),	Alka. (+·26),	Elev. (+·18),	pH (+·15),	Width (−·09)
–dss	55	51	Vol. (+·52),	Elev. (−·24),	Alka. (+·30),	Temp. (−·19),	Width (−·12)
–dls	65	58	Vol. (+·55),	Alka. (+·40),	Mn (+·34),	Temp. (+·18),	Elev. (−·18)

For explanation of letters see footnote to Table 1.

under consideration, and the sign indicates whether the influence is contributing to (+) or inhibiting (−) the content, that is, whether the correlation is positive or negative. They are the percentages of the variance of the dependent variable which can be accounted for as a result of its dependence on the independent variable. The closer they are to 100%, the more perfectly is the variability of the element explained by its dependence on the independent variable. In Table 3 the combined effect of all measured, as well as of the five most apparently significant, independent variables is entered. Tables 2 and 3 indicate not only which of the measured independent variables affects most the distribution of uranium, radium and radon in the surface environment but also the extent to which they affect the distribution. It is seen that most associations are relatively weak (small correlation coefficients in brackets), but most are significant by use of the 95% confidence level. Also, the majority of the samples have element contents which are close to the detection limit of the method in the case of radon in lakes, radium in sediments, and uranium in lake and stream waters; much of the variability of the data is therefore due to analytical error. For example, the radon and uranium means in lakes from the regional survey are 1·2 pCi/l and 0·4 ppb, respectively. The detection limits for these elements were 1 pCi/l and 0·5 ppb. One can see that the uranium mean is an approximation, 0·3 ppm being used as the value for all those samples which read less than 0·5 ppb. Noticeable and persistent trends emerge, however, to permit certain reasonable conclusions to be drawn. Unfortunately, the depth of water at the sample sites from the regional lake survey was available from only one-quarter of the sites, so this important variable could not be used in the regression analysis. But its importance is quite evident from the size of the correlation coefficient and from the detailed lake survey. The importance of temperature, particularly in the case of stream waters and radon, is also apparent.

Alkalinity shows an unusually strong positive association with all four dependent variables. It is, of course, well known that the carbonate ion complexes the UO_2^{++} ion, so one could expect this in the case of uranium in the waters. But uranium in the sediments should, if anything, show the opposite effect and radon and radium no effect. The evidence leads one to conclude, as Macdonald[19] did, that the uranium mineralization is related to calcite. In fact, calcite is a sizable component in the gangue of the mined ore. This association can also explain the seemingly contradictory correlation of elevation and radon. As is evident, it is positive for the regional survey and negative for the detailed survey. It is believed that in the case of the regional survey the large area of granitic rocks to the north, away from the uranium mineralization, situated at a higher elevation, where temperatures were colder on the average and streams somewhat smaller, outweighed the smaller mineralized, calcite-bearing corner in the southwest of the study area. In the detailed study area the more alkaline streams and lakes, lying at lower elevations, contained the higher radon contents. In other words, it looks very much as though the lithology of the region is reflected by the elevation variable. (Note also the relatively strong positive correlation, or dependence of radium on iron and manganese, particularly manganese.) Uranium, on the other hand, is quite strongly bound by the organic matter in the sediment as represented by the variable volume (the volume of 250 mg of −60-mesh sediment). Because clastic samples were often not available, organic rich samples were collected and the volume of the sample used as a measure of the fraction of organic matter in them. Iron does not correlate and manganese correlates only weakly with organic matter. The relative strength with which trace elements are bound to organic matter in the Beaverlodge samples, as shown by correlation coefficients and percentage reductions in variance, is

$$Cu > Pb > U > Zn > Ni > Ra$$

If it is assumed that there are no other environmental variables, except mineralization, to be accounted for, the remaining variance, i.e. the variance not accounted for by the element associations, can be subdivided into analytical error variance, random variance and variance due to uranium mineralization.

The multiple-regression analysis program used to obtain the above data has been described by Krumbein *et al.*,[18] applied by Rose and his co-workers,[27,28] and was adapted and modified for the G.S.C. computer by R. G. Garrett and J. D. Hobbs. The calculations assume a linear dependence of the trace element on the chosen independent variables. From this dependence a predicted value for the dependent variable of interest is computed. This predicted value can be subtracted from the measured value of a sample and a large positive difference is taken as evidence of mineralization. In this work, ratios of measured over predicted values were used so that tests on the log-transformed data could be carried out to see if regression analysis had, in fact, reduced the dependence of the trace elements on the independent variables. Correlation coefficients of ratios were, with a few exceptions, below the 95% probability level in the case of dependent elements Rn, Ra, U(H_2O), U(Sed.) versus the independent variables. Correlations between the dependent variables remained essentially the same. Although this test gives confidence that the analysis was successful, there is one drawback to ratio calculations. It is theoretically possible to obtain a large ratio if the predicted value turns out to be much smaller than the measured value. No such false anomalies were found, however, in the twelve ratio contour plots obtained in this work.

To illustrate the effect this 'background' correction has on the Beaverlodge data a number of figures of the raw as well as of the ratio plots were prepared. In Fig. 13((*a*)–(*d*)) are shown four contour plots of data from the detailed stream survey. Plots *a* and *b* are contour levels of the raw or measured data and plots *c* and *d* of ratios. The predicted values for the ratio calculations were obtained by the multiple-regression analysis using all the independent variables listed in Table 2. The four plots in Fig. 14((*a*)–(*d*)) give the same information for the detailed lake survey. The foremost conclusion to be drawn from these eight plots is that the four elements from the uranium decay chain congregate in four areas and that the difference between raw and computed data, although noticeable, is not strikingly different. The largest mineralized zone is around Yahyah Lake. The three smaller showings are located southwest of Prince Lake, south of Shmoo Lake and at Gibbs Lake. Examining the plots in more detail, noticeable differences emerge. For example, the radon anomaly in the upper left corner of the stream data plots is enhanced in the ratio plot because of the higher than normal temperature of the water and an abnormally low iron and manganese content in the sediment at those sites. In other words, the computer has corrected for the inverse correlation between radon and temperature and the direct correlation between iron and manganese and radium (hence radon). The values at these sites were not included in the calculation of the contour levels for two reasons. First, the radon levels were so high that they literally blanked the whole area if left in the calculations for the contour levels. Secondly, the high levels could have been due to contamination; 10–15 year old trenches were visible upstream from these sites. The points were included in the plots, however, to see what the relative effect of these on the remaining sites would be. Two other radon showings in the raw data plot are not revealed in the ratio plots. The one in the upper right corner disappears because of a temperature correction (6 and 3°C) and the other near the middle because of an unusually high iron content (>3%). Similarly, a small radium anomaly in the raw stream data map disappears when regression analysis is applied, due to an

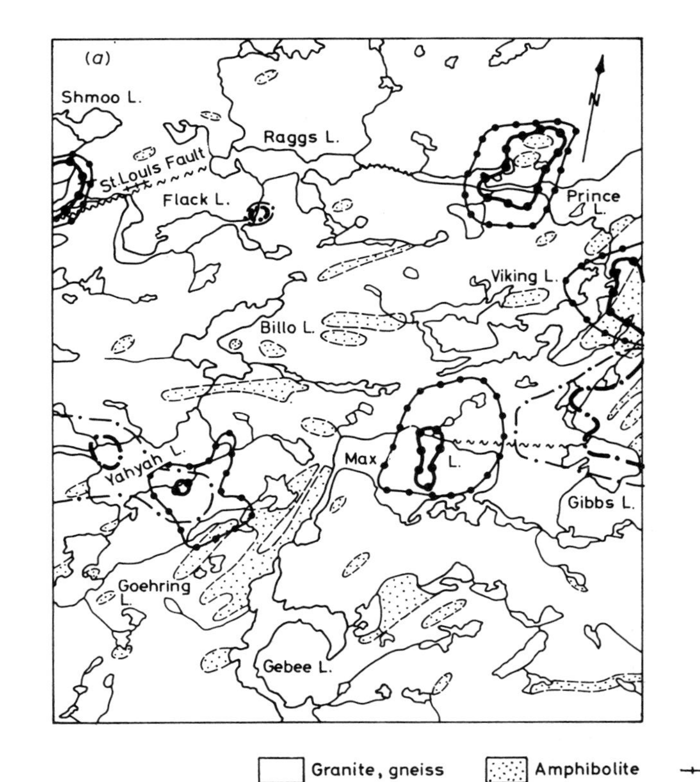

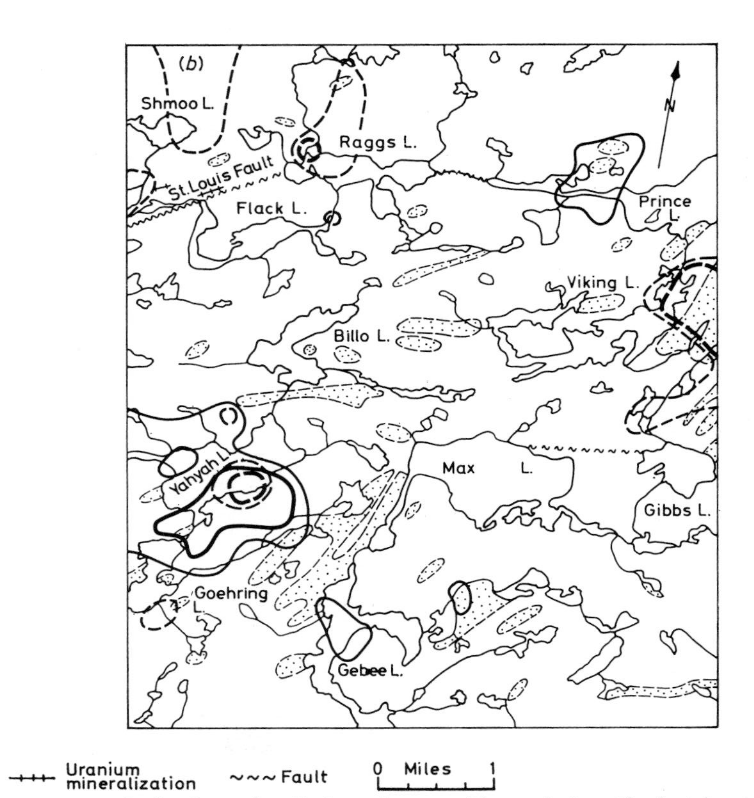

Granite, gneiss Amphibolite Uranium mineralization Fault 0 Miles 1

Fig. 13 (*a*) **Radon and radium content of stream water and sediment, respectively, from detailed study area, Beaverlodge, Saskatchewan. Rn, dotted solid line (pCi/l); Ra, dotted dashed line (pCi/g); thin line, $\bar{X}+\sigma$; thick line, $\bar{X}+1{\cdot}5\sigma$; (*b*) uranium content of stream water and sediment from detailed study area, Beaverlodge, Saskatchewan. U(H_2O), dashed line; U(Sed.), solid line; thin line, $\bar{X}+\sigma$; thick line $\bar{X}+1{\cdot}5\sigma$**

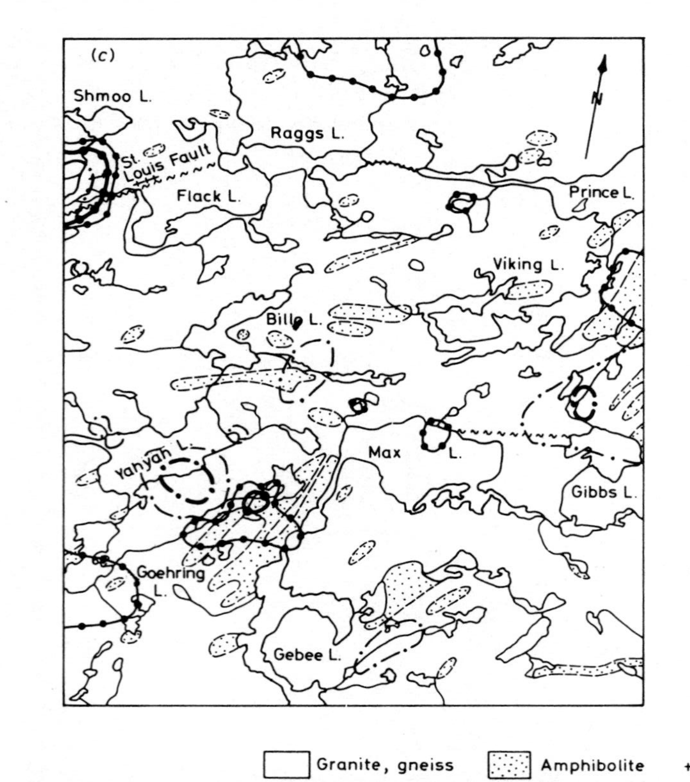

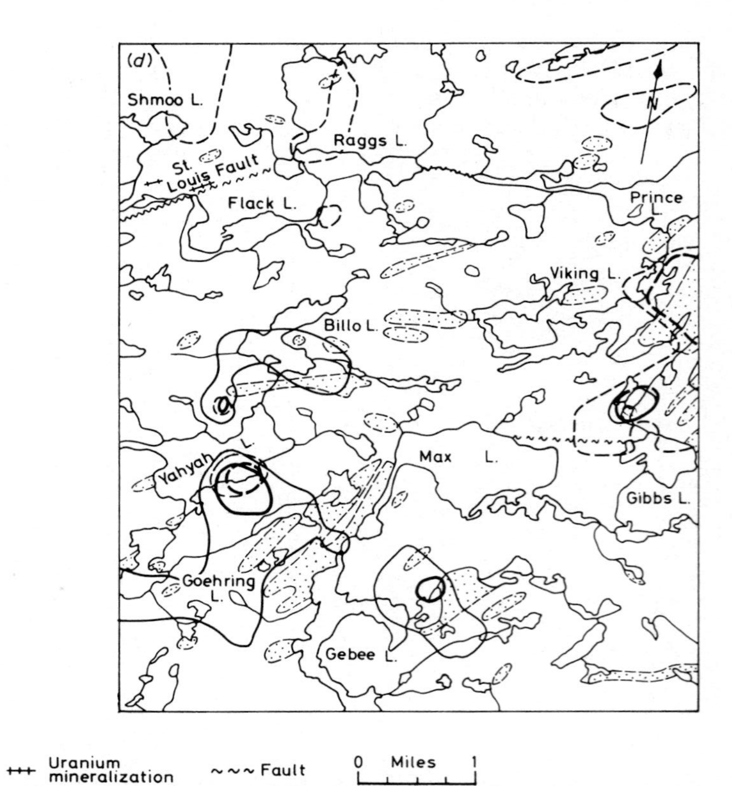

Fig. 13 (*c*) Radon and radium concentration ratios in stream water and sediment, respectively, from detailed study area, Beaverlodge, Saskatchewan. Rn, dotted solid line; Ra, dotted dashed line; thin line, $\bar{X}+\sigma$; thick line $\bar{X}+1{\cdot}5\sigma$; (*d*) uranium concentration ratios in stream water and sediment from detailed study area, Beaverlodge, Saskatchewan. U(H_2O), dashed line; U(Sed.), solid line; thin line, $\bar{X}+\sigma$; thick line, $\bar{X}+1{\cdot}5\sigma$

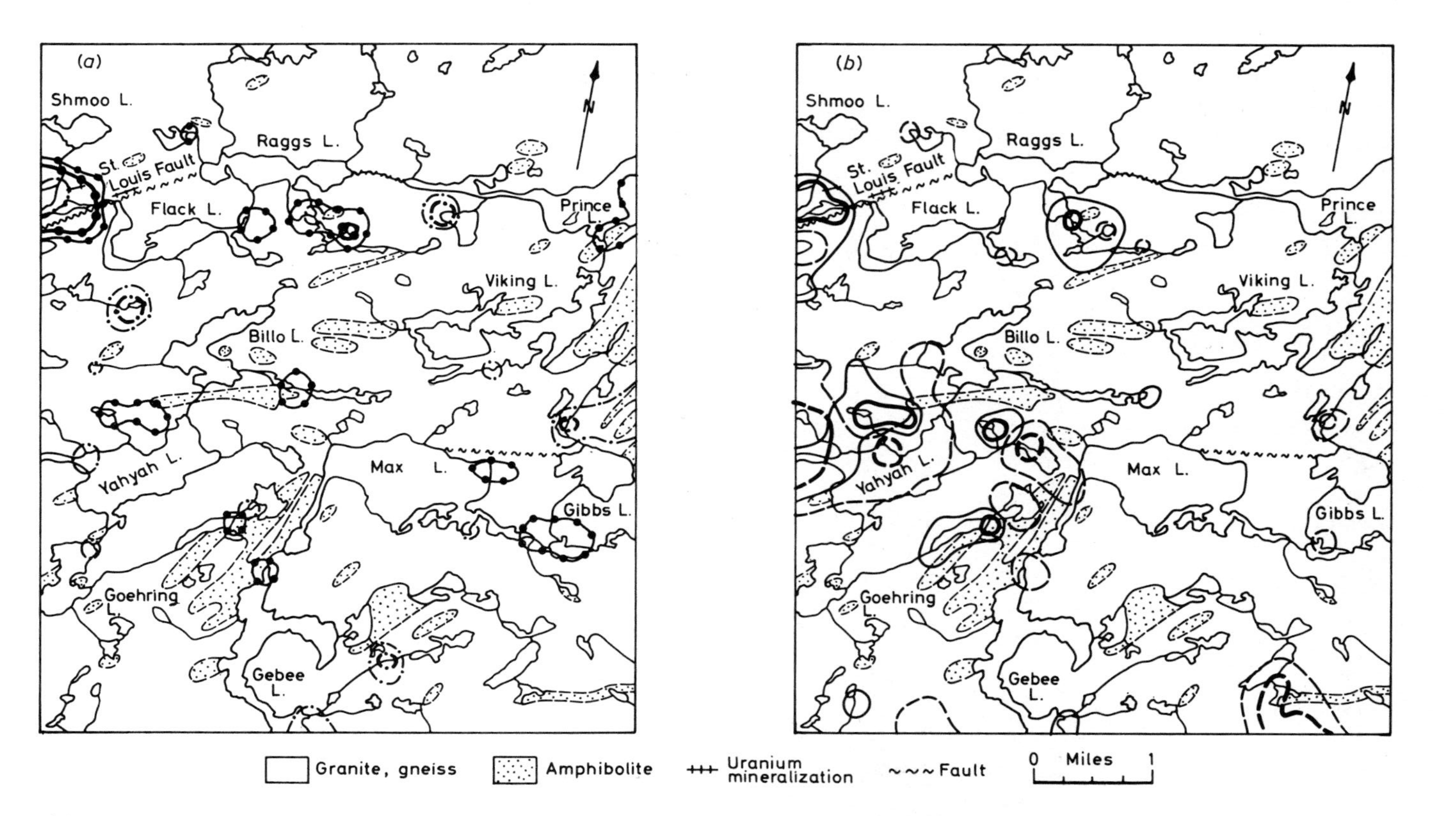

Fig. 14 (*a*) Radon and radium content of lake water and sediment, respectively, from detailed study area, Beaverlodge, Saskatchewan. Rn, dotted solid line (pCi/l); Ra, dotted dashed line (pCi/g); thin line, $\bar{X}+\sigma$; thick line, $\bar{X}+1{\cdot}5\sigma$; (*b*) uranium content of lake water and sediment from detailed study area, Beaverlodge, Saskatchewan. U(H_2O), dashed line; U(Sed.), solid line; thin line, $\bar{X}+\sigma$; thick line, $\bar{X}+1{\cdot}5\sigma$

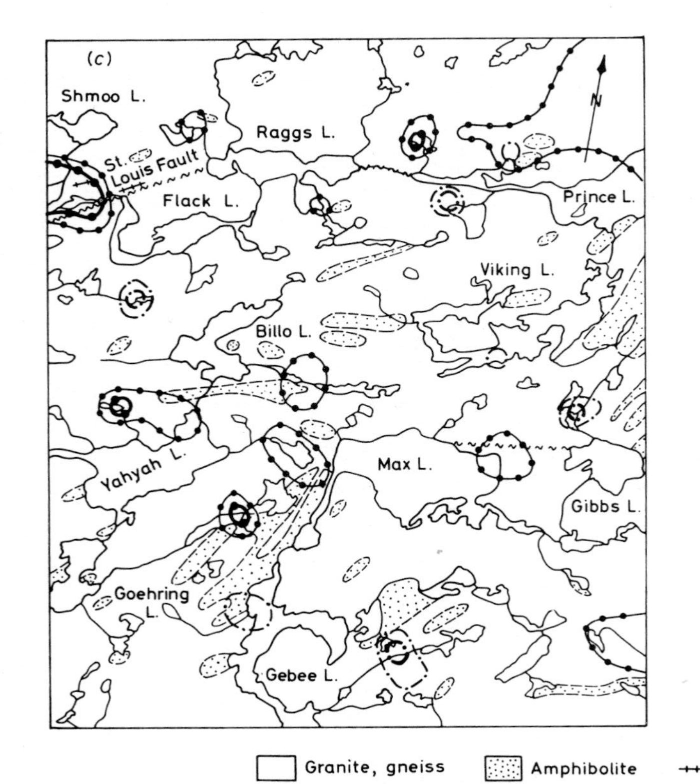

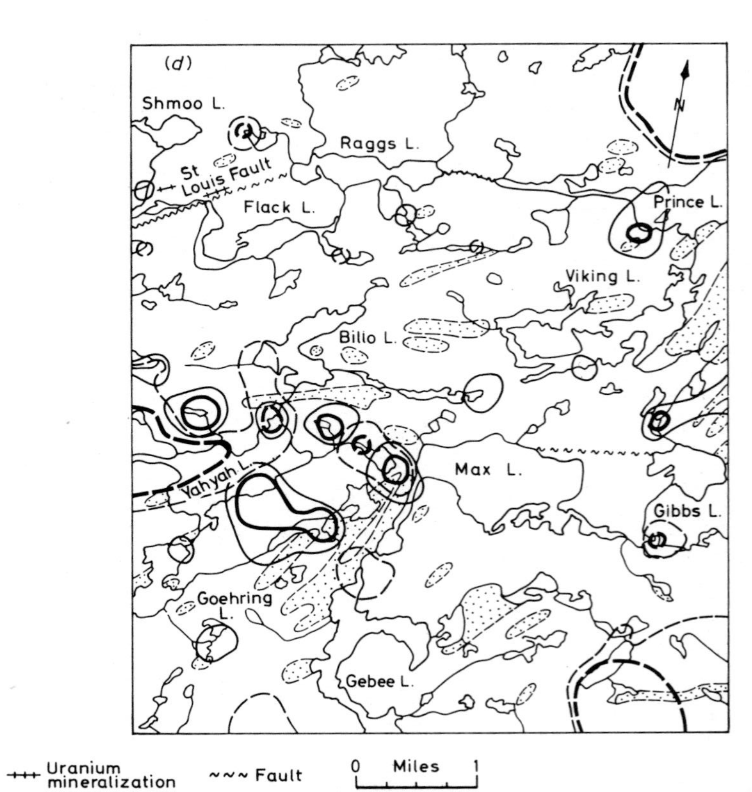

Fig. 14 (*c*) **Radon and radium concentration ratios in lake water and sediment, respectively, from detailed study area, Beaverlodge, Saskatchewan. Rn, dotted solid line; Ra, dotted dashed line; thin line, $\bar{X}+\sigma$; thick line, $\bar{X}+1{\cdot}5\sigma$; (*d*) uranium concentration ratios in lake water and sediment from detailed study area, Beaverlodge, Saskatchewan. U(H_2O), dashed line; U(Sed.), solid line; thin line, $\bar{X}+\sigma$; thick line, $\bar{X}+1{\cdot}5\sigma$**

unusually high manganese content in the sample (31 800 ppm). In the plots from the detailed lake survey the difference between raw and regressed data is less conspicuous. Probably this is because the independent variables vary somewhat less from lake to lake than they do from stream to stream. The above results show that multiple-regression data give more reliable results than do raw data; however, an experienced prospector soon learns by trial and error which factors influence most the distribution of the element he is seeking and allows for them in the evaluation of an anomaly. There can also be drawbacks, if the technique is applied indiscriminately: for example, the association of calcite gangue with

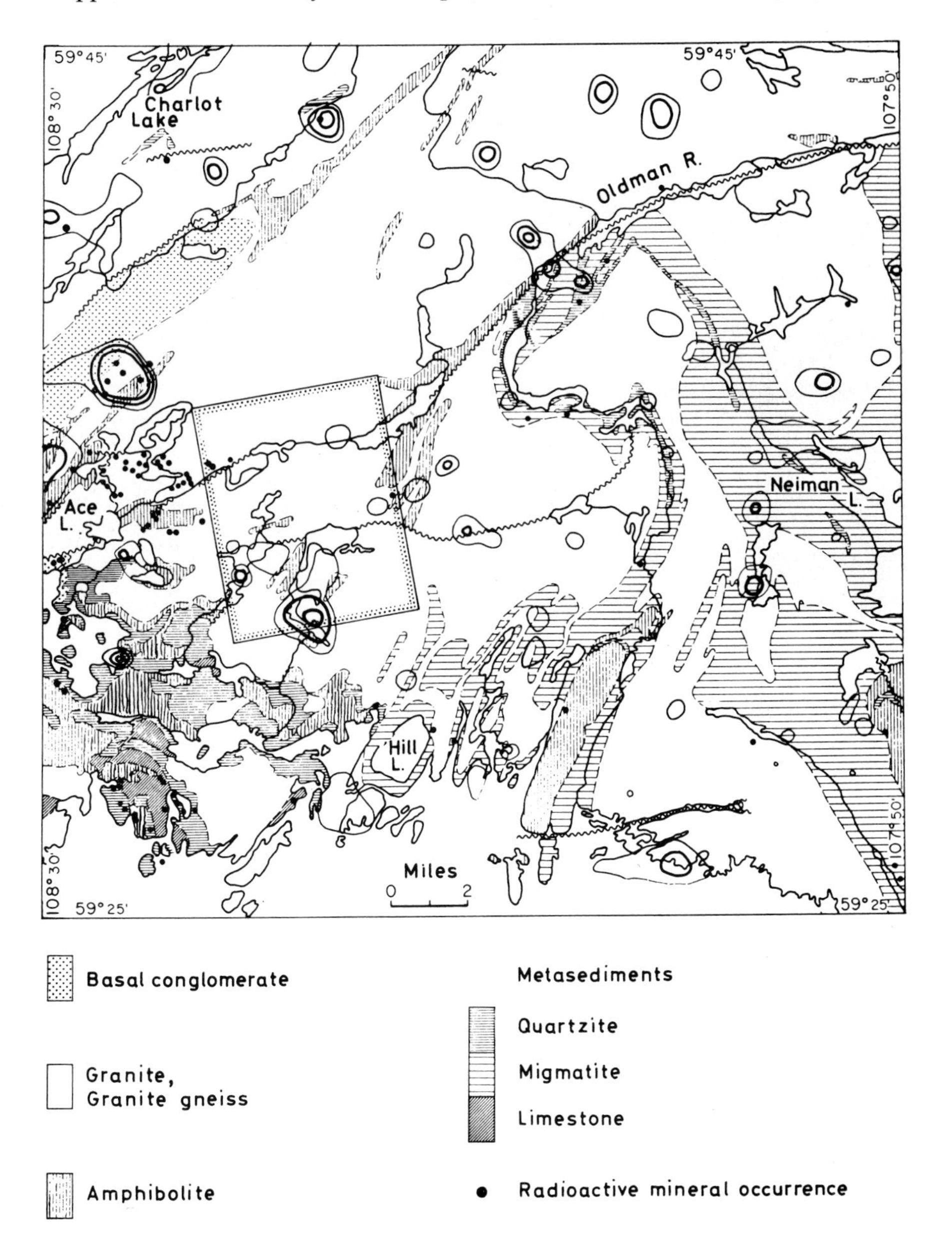

Fig. 15 Radon concentration ratios of stream waters, Beaverlodge area, Saskatchewan. Sample density, 1 per square mile; contours: lowest, 4·2; intermediate, 8·9; highest, 19·0

uranium mineralization is well known; that the carbonate ion complexes with the uranyl ion is also well known. The uranium–alkalinity correlation in waters already observed by Macdonald[19] results in reduced uranium levels in the ratio plots because the computer cannot sense this natural association of uranium and calcite in the rocks. The ratio plots of the uranium and radon concentrations in stream waters from the regional survey are shown in Figs. 15 and 16. Their similarity is unmistakable. The raw data plots of these elements published earlier[6] do not differ appreciably from the ratio plots.

In conclusion, reference should be made to the matter of sampling density.

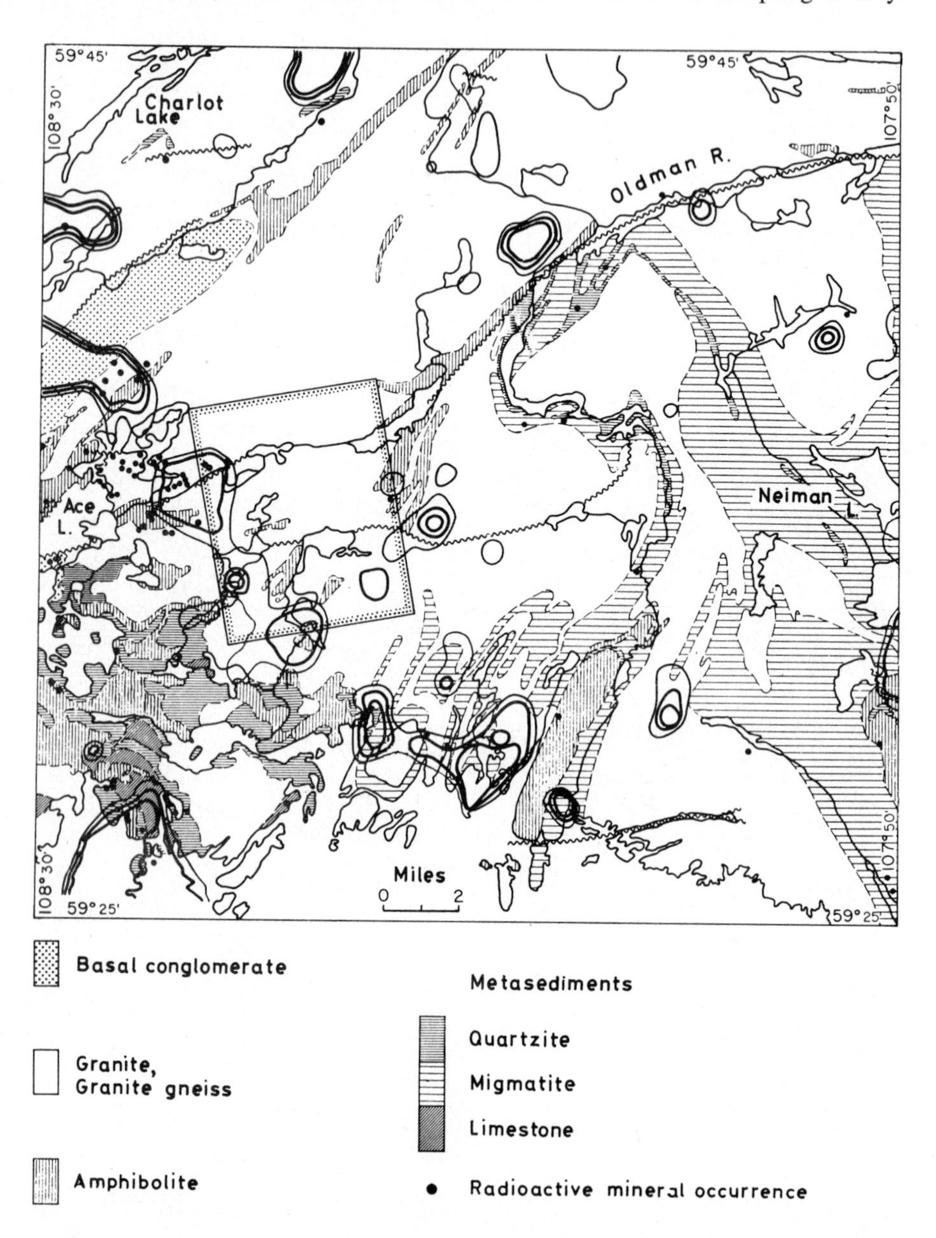

Fig. 16 Uranium concentration ratios of stream waters, Beaverlodge area, Saskatchewan. Sample density, 1 per square mile; contours: lowest, 2·1; intermediate, 3·1; highest, 4·5

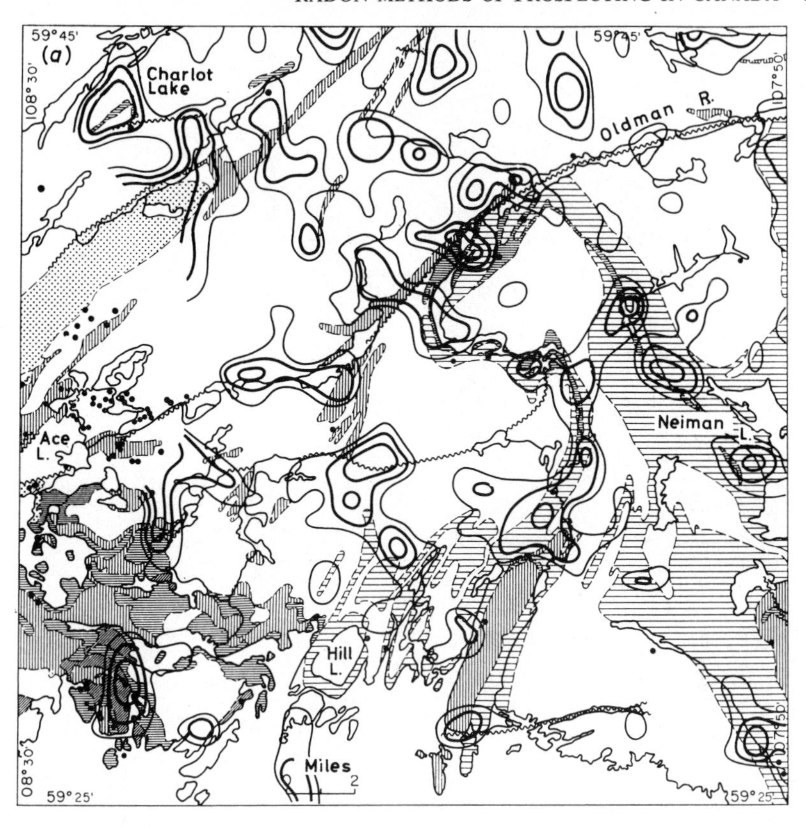
59°45'
(a)
108° 30'
Charlot
Lake
59°45'
107°50'
Oldman R.
Ace
L.
Neiman
L.
Hill
L.
Miles
2
108°30'
59°25'
107°50'
59°25'

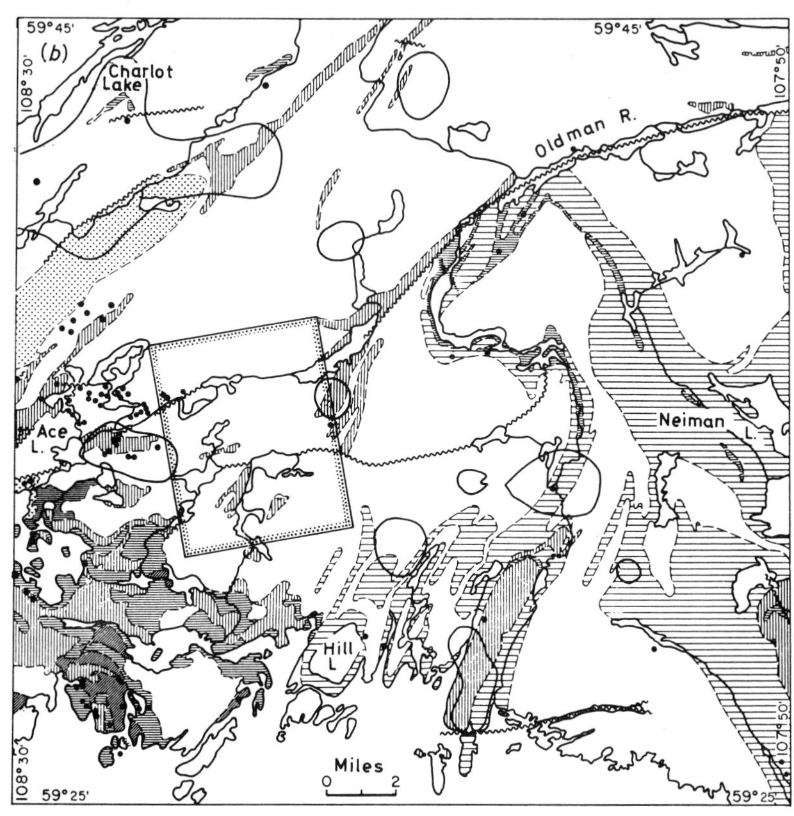
59°45'
(b)
108° 30'
Charlot
Lake
59°45'
107°50'
Oldman R.
Ace
L.
Neiman
L.
Hill
L
Miles
0
2
108°30'
59°25'
107°50'
59°25'

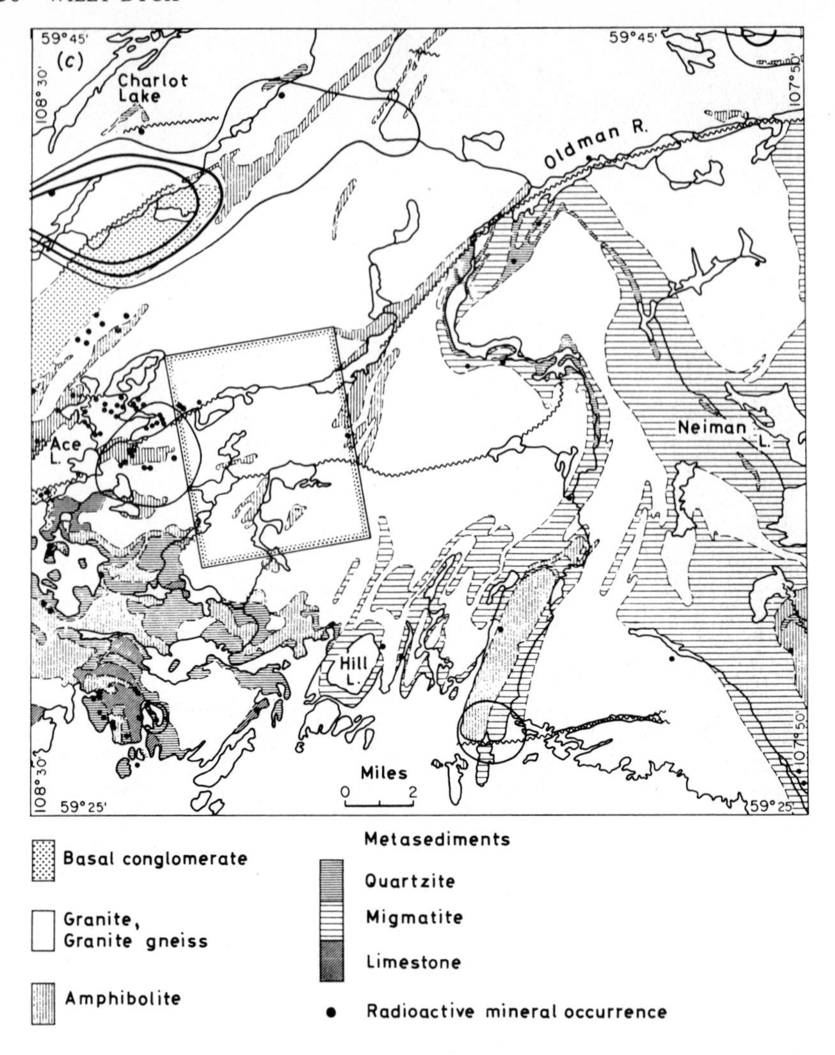

Fig. 17 Radon content of lake waters, Beaverlodge area, Saskatchewan. Sample density, 1·3 per square mile; contours: lowest, 2·5; intermediate, 5·2; highest, 10·6 pCi/l (*a*); sample density, 1 per 5 square miles; contours: lowest, 5·2; intermediate, 11; highest, 23 pCi/l (*b*); sample density, 1 per 12 square miles; contours: lowest, 5·2; intermediate, 11; highest, 23 pCi/l (*c*)

Because only one-quarter of the lake sites from the regional survey were measured for depth, which is the most important variable for radon in lakes, ratio plots are not more meaningful than raw data plots. Instead, the computer was asked to select a predetermined number of samples uniformly from the area and to plot the radon and uranium results to see what effect sample density had on the element distribution (Figs. 17 and 18). In Fig. 17 the radon distribution at 13 (the original density), 2 and 1 sample per 10 square miles are shown, and in Fig. 18 the uranium distributions at the corresponding sampling densities are given. In all cases the contour levels were based on the lognormal distribution of the most dense sampling. The actual values of the contours are given in the captions, the first or lowest contour being

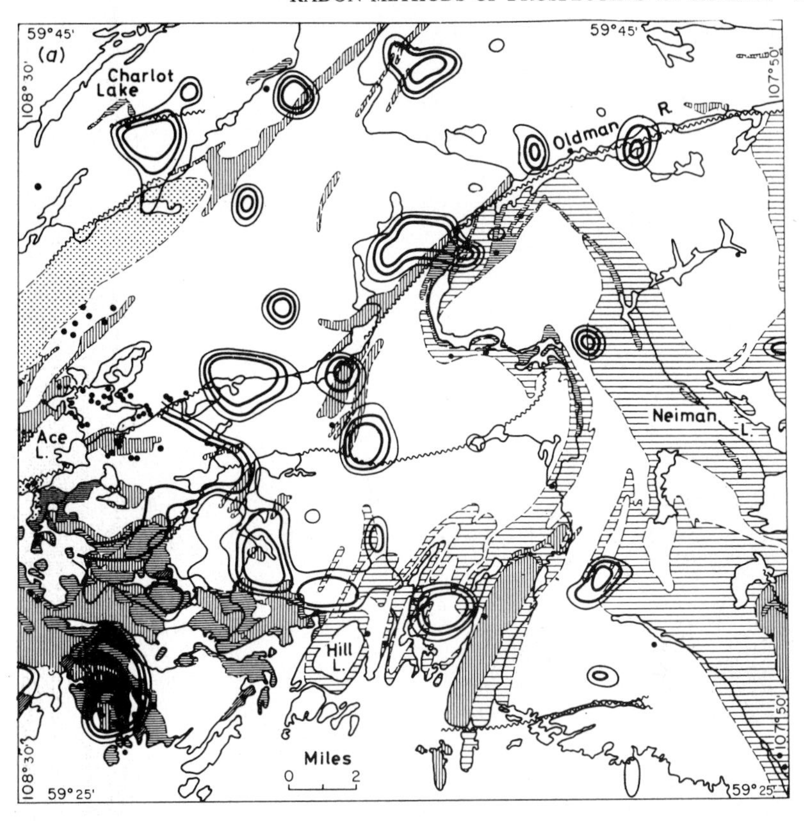
59°45'
(a)
108°30'
Charlot Lake
59°45'
107°50'
Oldman R.
Neiman L.
Ace L.
Hill L.
Miles
0 2
108°30'
59°25'
107°50'
59°25'

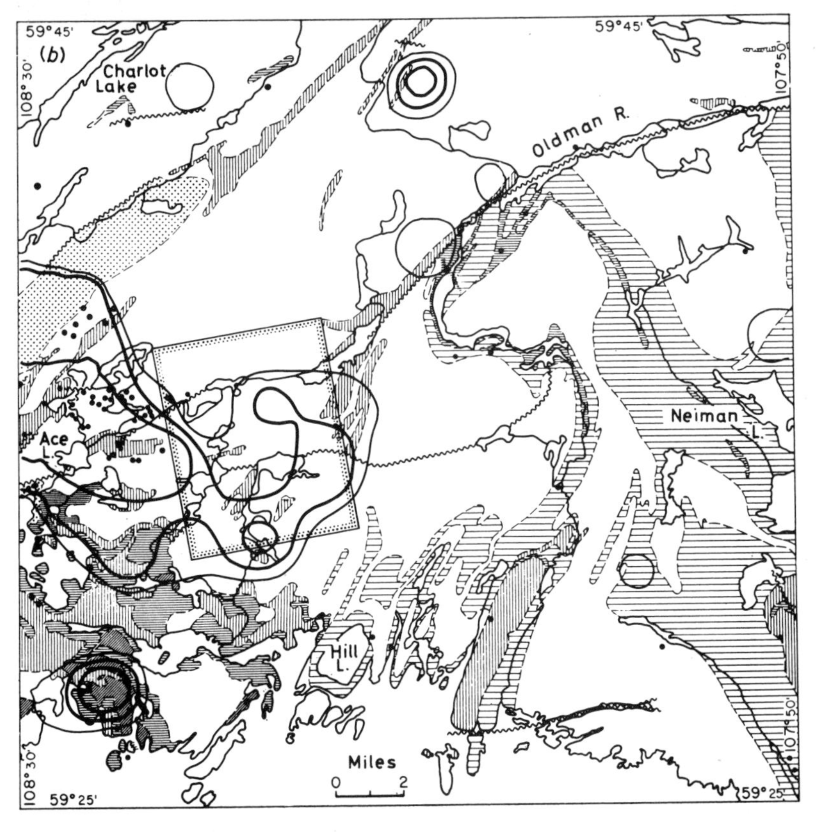
59°45'
(b)
108°30'
Charlot Lake
59°45'
107°50'
Oldman R.
Neiman L.
Ace L.
Hill L.
Miles
0 2
108°30'
59°25'
107°50'
59°25'

equal to the mean plus one-half sigma, the next mean plus one sigma, the third-mean plus one and one-half sigma, and the fourth mean plus two sigma. Had a serious prospector followed up on the radon or uranium highs, he would have discovered the Beaverlodge uranium deposit starting with as

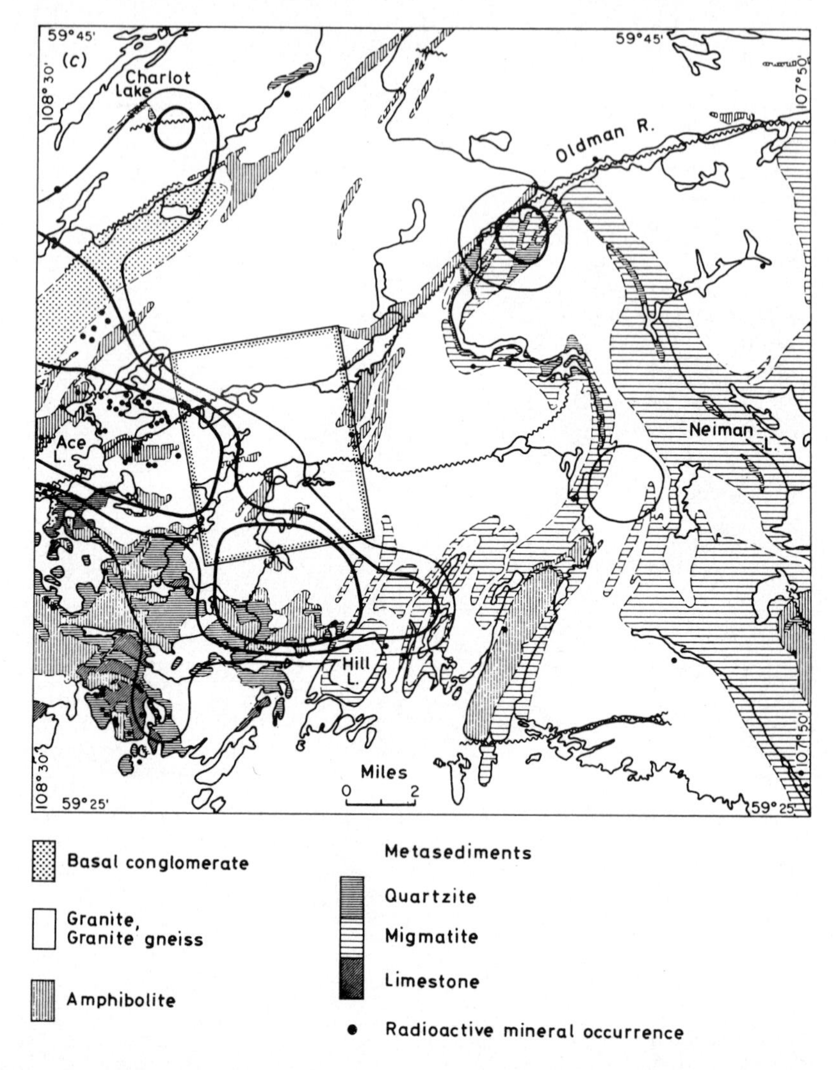

Fig. 18 Uranium content of lake waters, Beaverlodge area, Saskatchewan. Sample density, 1·3 per square mile; contours: lowest, 0·7; intermediate, 1·0; highest, 1·4 ppb (*a*); sample density, 1 per 5 square miles; contours: lowest, 0·7; intermediate, 1·0; highest, 1·4 ppb (*b*); sample density, 1 per 12 square miles; contours: lowest, 0·7; intermediate, 1·0; highest, 1·4 ppb (*c*)

few as 50 samples for the initial regional survey. The one sample per 10 square miles selection was repeated with another set of samples. The coincidence of highs is greater than chance alone would produce. The big problem, however—how to decide on what constitutes an anomaly—is very difficult and sometimes impossible. The only sure guide is to follow up on values which rise noticeably above the background in an area.

Conclusions

The development of portable radon counters has made possible the detection of radon in soils and water in the field. Although radon in soil gases has been shown to outline radioactive zones, the method has found limited application for uranium prospecting in the Canadian Shield because of poor soil conditions and/or abundant outcrop. Because of the great abundance of small lakes and streams in the Canadian Shield, radon in surface waters has been a useful tracer for uranium mineralization on a detailed as well as on a regional scale.

Computer contouring and multiple-regression analysis aid in locating and evaluating uranium showings.

Factors which most affect the radon content of water in the Beaverlodge area are distance between sample and sediment, temperature of the water and alkalinity; size of water body, elevation, flow, level and turbulence also affect the radon content noticeably. Except for alkalinity and elevation, these variables lower the radon content of water. These two variables enter indirectly via the lithology of the region; calcite and uranium are found together in the lower-lying southwest corner of the study area, reversing the positive elevation–radon correlation observed for the area as a whole.

Acknowledgment

The author would like to thank the following persons of the Geochemistry Section of the Geological Survey of Canada for assisting him in the various aspects of this work: J. J. Lynch and his staff for analysing the samples—in particular, J. C. Pelchat for the many radon analyses; J. D. Hobbs and Dr. R. G. Garrett for assistance in the use and modification of the computer programs; C. C. Durham and A. S. Dass for carrying out certain phases of the Beaverlodge field work. G. Meilleur and his staff of the Geological Survey of Canada Instrument Shop are due thanks for the fabrication of the many parts for the soil probes and radon detection chambers.

The author would also like to express his appreciation for the assistance and information received from Dr. E. E. N. Smith and other members of Nuclear Eldorado, Inc., concerning the Beaverlodge field work area and the setting up of the field laboratory and the accommodation of the field crews.

References

1. La radioactivité de l'atmosphère et du sol. *Le Radium*, **1,** no. 4 1904, 2–5.
2. Baranov V. I. *Radiometry* (Moskow: Akademy Nauk SSSR, 1956), 391 p. (Russian text)
3. Blake D. A. W. Oldman River map-area, Saskatchewan. *Mem. geol. Surv. Can.* 279, 1955, 52 p.
4. Bowie S. H. U., Ball T. K. and Ostle D. Geochemical methods in the detection of hidden uranium deposits. In *Geochemical exploration* (Montreal: CIM, 1971), 103–11. (*Proc. 3rd Int. geochem. Symp., Toronto 1970*) (*CIM spec. vol.* 11)
5. Christie A. M. Goldfields–Martin lake map-area, Saskatchewan. *Mem. geol. Surv. Can.* 269, 1953, 126 p.
6. Dyck W. *et al.* Comparison of regional geochemical uranium exploration methods in the Beaverlodge area, Saskatchewan. In *Geochemical exploration* (Montreal: CIM, 1971), 132–50. (*Proc. 3rd Int. geochem. Symp., Toronto 1970*) (*CIM spec. vol.* 11)
7. Dyck W. Field and laboratory methods used by the Geological Survey of Canada in geochemical surveys. No. 10. Radon determination apparatus for geochemical prospecting for uranium. *Pap. geol. Surv. Can.* 68–21, 1969, 30 p.

8. Dyck W. Development of uranium exploration methods using radon. *Pap. geol. Surv. Can.* 69–46, 1969, 26 p.
9. Dyck W. and Smith A. Y. Use of radon-222 in surface waters for uranium geochemical prospecting. *Can. Min. J.*, **89,** April 1968, 100–3.
10. Dyck W. Radon-222 emanations from a uranium deposit. *Econ. Geol.*, **63,** 1968, 288–9.
11. Garrett R. G. A program for the rapid screening of multivariate data from the earth sciences and remote sensing. *Northwest Univ. Rep.* no. 12, 1967, 98 p. (*NASA grant* NGR–14–007–027)
12. Grimbert A. Evolution et perspectives de la prospection géochimique des gîtes uranifères. In *Geochemical exploration* (Montreal: CIM, 1971), 21–3. (*Proc. 3rd Int. geochem. Symp., Toronto*) *1970* (*CIM spec. vol.* 11)
13. Grimbert A. and Loriod R. La prospection géochimique detaillée de l'uranium. *Pap. geol. Surv. Can.* 66–54, 1967, 167–71.
14. Higgins F. B. Jr. *et al.* Methods for determining radon-222 and radium-226. *J. Am. Water Works Ass.*, **53,** no. 1 1961, 63–74.
15. Israël H. and Israël G. W. A new method of continuous measurement of radon (Rn^{222}) and thoron (Rn^{220}) in the atmosphere. *Tellus*, **18**, 1966, 557–61.
16. Israël H. and Horbert M. Tracing atmospheric eddy mass transfer by means of natural radioactivity. *J. geophys. Res.*, **75**, 1970, 2291–7.
17. Krcmar B. Personal communication, 1970.
18. Krumbein W. C. *et al.* Whirlpool, a computer program for sorting out independent variables by sequential multiple linear regression. *Tech. Rep.* no. 14, ONR no. 389–135, 1964. (Northwestern University, Evanston, Illinois)
19. Macdonald J. A. An orientation study of the uranium distribution in lake waters, Beaverlodge district, Saskatchewan. *Colo. Sch. Mines Q.*, **64**, 1969, 357–76.
20. Morse R. H. Comparison of geochemical prospecting methods using radium with those using radon and uranium. In *Geochemical exploration* (Montreal: CIM, 1971), 215–30. (*Proc. 3rd Int. geochem. Symp., Toronto 1970*) (*CIM spec. vol.* 11)
21. Novikov G. F. and Kapkov Yu. H. *Radioactive methods of prospecting* (Leningrad: NEDRA, 1965), 760 p. (Russian text)
22. Partington J. R. Discovery of radon. *Nature, Lond.*, **179**, 1957, p. 912.
23. Peacock J. D. and Williamson R. Radon determination as a prospecting technique. *Trans. Instn Min. Metall.*, **71**, Nov. 1961, 75–85.
24. Robertson J. A. Geology of Township 149, Township 150. *Geol. Rep. Ont. Dep. Mines* 57, 1968, 162 p.
25. Rogers A. S. Physical behavior and geological control of radon in mountain streams. *Bull. U.S. geol. Surv.* 1052–E, 1958, 187–211.
26. Rogers A. S. and Tanner A. B. Physical behavior of radon. *Rep. U.S. geol. Surv.* TEI–620, 1956, 349–51.
27. Rose A. W. and Suhr N. H. Major element content as means of allowing for background variation in stream-sediment geochemical exploration. In *Geochemical exploration* (Montreal: CIM, 1971), 587–93. (*Proc. 3rd Int. geochem. Symp., Toronto 1970*) (*CIM spec. vol.* 11)
28. Rose A. W. Dahlberg E. C. and Keith M. L. A multiple regression technique for adjusting background values in stream sediment geochemistry. *Econ. Geol.*, **65**, 1970, 156–65.
29. Ruzicka V. Personal communication, Geological Survey of Canada, 1969.
30. Satterly J. and Elworthy R. T. Mineral springs of Canada; part 1, the radioactivity of some Canadian mineral springs. *Bull. Can. Mines Br.* 16, 1917, 55 p.
31. Smith A. Y. and Dyck W. The application of radon methods to geochemical exploration for uranium. Paper presented at Annual General Meeting of the Canadian Institute of Mining and Metallurgy, Montreal, 1969. Abstract in *CIM Bull.*, **62**, 1969, p. 215.
32. Stevens D. N., Rouse G. E. and De Voto R. H. Radon-222 in soil gas: three uranium exploration case histories in the Western United States. In *Geochemical exploration* (Montreal: CIM, 1971), 258–64. (*Proc. 3rd Int. geochem. Symp., Toronto 1970*) (*CIM spec. vol.* 11)
33. Tanner A. B. Radon migration in the ground—a review. In *The natural radiation environment* (Chicago, Ill.: University of Chicago Press, 1964), 161–90.

34. Tremblay L. P. Geology of the Beaverlodge mining area, Saskatchewan. *Mem. geol. Surv. Can.* 367, 1968, 468 p.
35. Wennervirta H. and Kauranen P. Radon measurement in uranium prospecting. *Bull. Commn géol. Finl.*, **30**, no. 188, 1960, 23–40.

DISCUSSION

Professor M. Dall'Aglio enquired as to the spread of the hydrogeochemical aureoles of radon in stream waters. His own experience had shown that in running waters radon aureoles were as little as a few hundreds of metres.

The author replied that he was unable to give the length of the radon dispersion in streams waters; but, in one case, samples were taken from an interconnected stream and lake a few hundred feet apart: the stream sample was very active, but the lake sample was not. He suspected that that stream sample came from a nearby spring and, thus, was not supported by radium in the stream sediment. When compared with uranium, radium moved very little in the surface environment. As evidence of that they picked 100 samples having the highest anomalous uranium values, determined their radium content and found their mean U:Ra ratio to be 8—indicating a great degree of disequilibrium. From that it was concluded that uranium was much more mobile.

A. Grimbert asked if any follow-up work had been done around the anomalies found. If the choice had to be made between analysing for radon or radium or uranium, which would be considered to be the more significant?

The author said that no attempt had been made to trace the sources of the radon at Beaverlodge, but, after a regional survey of the Gatineau Hills, carried out in 1968, they had in 1969 sampled some of the lakes in more detail. The source of the lake water radon was shown to be some pegmatite dykes on the uphill side of the general topographic slope. Similar downhill pegmatites did not give rise to anomalies. That example suggested that groundwater movement carried the radon from the dykes into the lake. With regard to the second question, he would analyse surface waters for uranium for reconnaissance work, and for radon for detailed work. At Beaverlodge radon pinpointed the mineralization more accurately than uranium in waters.

R. N. Aitken asked the author if he took measurements of wind speed, barometric pressure and water turbulence when measurements in water were being taken. What allowance was made numerically for those in the computer program?

The author said that at Beaverlodge the streams sampled were given relative 1, 2, 3 values for turbulence and flow; and the output from the computer showed that those were significant observations. Lakes, however, with zero flow and no turbulence could not be fitted into that scheme.

In reply to a request by G. R. Ryan for more details on the effects of speed, turbulence and other factors on the content of uranium in stream water, the author replied that both with radon and uranium, the larger the stream the lower was their content. That was to be expected, as large volumes of water tended to clean themselves as they moved along the bed of the river. The organic content of the stream bed had a very marked effect on the uranium content of the sediment; and that was the most important factor governing between 20 and 30% of the variability in uranium content in bottom sediments. Turbulence and flow rate gave an inverse correlation significant at the 95% level of confidence.

In reply to B. O. Kvendbo, the author said that the most favourable soils for radon measurement were those which were dry and porous. The fewer boulders they contained, the better, but those were generally present in Canada.

N. R. Newson said that his company had built and operated for two field seasons a radon gas emanometer apparatus for measuring radon dissolved in groundwaters. Holes were made with a 1- to 1½-in auger, and those filled up with water, which was then collected. A fixed volume of air was bubbled through each sample and the air was then treated in much the same way as the soil air technique. Daily production in difficult soil conditions on a 100 ft × 400 ft grid was about 40 samples.

Orientation surveys over relatively undisturbed known deposits had shown that the method had very good resolution; and reproducibility was about on a par with soil sampling. In fact, radon survey results virtually duplicated soil uranium sampling results in terms of outlining anomalous areas, and they thought that the two methods were largely interchangeable in their area. Generally, radon work was done in top priority areas, because the results were immediately available, and soil sampling in areas of lower priority. They had come to rely on the radon groundwater method as one of their main tools because it worked in their particular set of circumstances. Obviously, it had to be used intelligently—as did any other method.

Dr. H. D. Fuchs said that in Canada soil sampling and radon determination could only be performed by means of air transport. He wondered if it were justified under those conditions to undertake expensive sampling when a detailed airborne spectrometer survey might give even more information.

The author commented that soil sampling and radon determinations were not cheaper than an airborne spectrometer survey, but they came in at a more detailed stage of work when the presence of an orebody, possibly covered by overburden, had been inferred from other types of regional survey.

P. H. Dodd said that he would like to make a comment about the possible use of ^{222}Rn as an airborne prospecting technique. There were too many ambient parameters related to atmospheric radon concentrations to make its measurement a prospecting tool. Such factors included local and distant barometric history for several days; wind direction and velocity; moisture in soils; rainfall; electrical storms, etc. Traversing over 'barren' or water-saturated soil would yield low values. Inversion layers would yield a blanket concentration: the altitude of the blanket depended on temperatures and air density differences. Intersections of stream valleys might produce Rn clouds. Dust clouds with ^{214}Bi and ^{222}Rn had been measured at 1400 ft above North African desert areas. It would be some time before basic atmospheric research on ^{222}Rn and ^{220}Rn and their daughter elements could provide a basis for interpreting atmospheric measurements in terms of mineral deposits.

Dr. A. G. Darnley agreed with Mr. Dodd's comments on the limitations of atmospheric radon monitoring as an exploration method. If the other airborne radiometric methods did not exist, then it would merit consideration; but they did exist. Atmospheric radon was too ephemeral to be worth chasing; however, equipment for monitoring radon in the air was worth developing in order to provide corrections for gamma survey measurements.

It should be noted that the author's uranium-in-streams and uranium-in-lakes results agreed quite well with the airborne radiometric map of the same area. Details of the cost of collecting samples at Beaverlodge would be welcomed.

The author said that at Beaverlodge the cost of collecting a stream sample was $Can.50.00 and a lake sample cost $Can.30.00. In all some 500 stream and 700 lake sites were visited.

In reply to Dr. Darnley's comment that the same area was flown with a gamma spectrometer in about 7½ h, at a cost of $Can.15 000.00, which was quite close to the cost of the uranium–radon survey, the author said that the airborne gamma-spectrometry contour plot was quite similar to the uranium–radon pattern at much lower sampling densities. In fact, 50 lake waters (about one day of work at a total cost of about $Can. 1500) outlined the major uranium showing quite well.

Dr. C. T. Meyertons asked if there were any information to suggest that there might be variations in radon content of soils because of lithology. There had been indications that that might cause problems in the interpretation of anomalies as a result of the excessive release of radon in the more porous units and retention in the denser, less permeable units. J. M. Miller said that he and his colleagues had experience of variable porosity in Scotland: there, hills were frequently capped by impermeable layers of peat, which were absent on their slopes, where the overburden was sandy. When uranium mineralization in the bedrock passed from a sandy to a peat-covered area, its radon anomaly was cut off due to the decrease in porosity. The radon-in-soil values in those circumstances were, therefore, heavily influenced by the porosity of the overburden.

Dr. M. Davis said that radon might perhaps enjoy wider use as an aid to uranium prospecting if there were more understanding as to how it moved in ground, air and waters. That might make a valuable subject for university research. The author drew attention to recent work by Morse,* who had concluded that the radium in the stream sediments was the major contributor of radon to the stream water. That appeared to be generally true in the Beaverlodge area also, although in some cases radon was contributed by springs.

Mr. Dodd suggested that concentrations of ^{222}Rn in soil gas conformed to the formula

$$^{222}\text{Rn (pCi/l)} = \frac{^{226}\text{Ra g/g} \times E \times P}{\varphi} \times 10^3$$

where E was the emanation coefficient, P the density of soil, g/cm^3, and φ fractional porosity.

Variations in any factor might cause an apparent anomaly: each should, therefore, be evaluated when an anomaly was being assessed. In spite of those problems he favoured the technique, but it should be interpreted with a model of possible systems and interrelated factors in mind.

In reply to a question by H. Müller regarding the use of correction factors to eliminate changes due to temperature, air pressure, etc., the author said that in soil gas measurement he relied on the variation from one site to the next with a view to determining which sites were relatively high in comparison with their neighbours. The general level fluctuated considerably from day to day, but the pattern remained the same. If a radon survey continued over several days, it would be best to set up a test strip of, say, 6 holes and correct the day-to-day values by reference to the mean variation of the test strip.

*Morse R. H. The surficial geochemistry of radium, radon and uranium near Bancroft, Ontario, with applications to prospecting for uranium. Thesis, Queen's University, Kingston, Ontario, 1970.

Borehole logging techniques for uranium exploration and evaluation

P. H. Dodd B.S., M.S.

U.S. Atomic Energy Commission, Grand Junction, Colorado, U.S.A.

D. H. Eschliman B.S.

Lucius Pitkin, Inc., Western Uranium Project, USAEC, Grand Junction, Colorado, U.S.A.

550.822.3 : 550.37 : 553.495.2

Synopsis

Boreholes, totalling tens of millions of feet, are geophysically logged each year to support exploration for uranium. The basic programme consists of natural gross gamma-ray, resistivity and spontaneous-potential (SP) logs. The electric logs provide qualitative identification and correlation of lithology; the natural gamma-ray log may be calibrated in equivalent uranium, and manual or computer analysis provides a major source of grade and thickness data for ore-reserve estimation.

The principles of calibration and analysis of the gross gamma-ray log are reviewed, with emphasis on corrections for non-standard conditions, including formation fluid and increased equivalent atomic number (Z) effect of uranium. Equivalent uranium concentrations can be determined with precision, but disequilibrium remains a problem.

In-hole spectrometry is being tested, applications including identification of the anomalous isotope, resolution of low-intensity radium–uranium haloes, indication of geochemical environments and changes, evaluation of low-grade resources, evaluation of mixed U–Th ores, and investigation of disequilibrium. The Compton/photo log is based on photoelectric absorption and may be developed to measure uranium directly.

Supplementary logs are being adopted to provide additional information: they include drift surveying, temperature, sonic, gamma–gamma density, neutron–neutron and caliper. The last three provide quantitative correction parameters for basic logs and data for estimation of tonnage factors and porosity and evaluation of water and mining problems. The qualitative *n–n* log indicates lithology and may be substituted for the resistance log in air-filled or cased holes.

Promising experimental logs include pulsed neutron activation, for direct determination of uranium, and a sensitive magnetic susceptibility log, now being field-tested to indicate alteration and heavy mineral concentrations and, possibly, to distinguish crystalline rock units.

Industry recognizes the need for more and better low-cost subsurface information; this ensures a continuing effort to develop sophisticated logging methods to guide exploration and to evaluate uranium resources.

Borehole logging techniques have come to represent an important phase of most uranium exploration programmes, and many of the techniques are also applied to the evaluation of ore reserves. Logging techniques which were impossible 25 years ago, when natural gamma-ray logging for uranium was first tried in the United States, are now routinely employed by industry to obtain most of the basic lithologic and stratigraphic information used by the geologist to guide exploration and to provide the ore-reserves engineer with quantitative sample data. Improved equipment, together with imaginative development and an increasing awareness on the part of the mining industry of the value of and need for more and better information from the borehole, ensures continuing improvements and reliance on logging.

Some of the technical advantages and economic incentives for geophysical logging are noted below. (1) High-cost coring can be largely replaced with non-core drilling at one-fourth or less of the cost (this is especially true for exploration of sandstone-type deposits). (2) Logs provide information when core recovery is poor or impossible (this includes poorly consolidated sediments or badly fractured rocks). (3) Data can be obtained from holes drilled previously for other purposes (including cased holes). (4) Multi-purpose logging probes provide several types of information simultaneously and rapidly, and records are immediately available for qualitative evaluation. (5) Continuous analogue or frequent digital records permit re-sampling, if necessary, for statistical and economic evaluation. (6) Geophysical logs are consistent and unbiased by personal observation or experience. (7) Digital or digitized records can be analysed objectively and rapidly by computers. (8) The volume of 'sample' sensed by the probe is generally larger, and therefore statistically more reliable, than cores or cuttings. (9) Delays and costs of sampling and laboratory analysis are reduced. (10) The bulk physical properties of undisturbed rocks are often measured more readily and reliably by geophysical logs than by laboratory tests of small and possibly disturbed samples.

Logging programmes and applications

Logging programmes for uranium exploration and evaluation are becoming more varied and sophisticated.

BASIC LOGGING PROGRAMME FOR URANIUM

The standard gamma and electric log programme for uranium was adopted by industry in 1955 when the combined Gamma-R log was introduced.[1] Today, a basic programme generally consists of natural gross-count gamma ray, single-point resistance or 16-in resistivity (R), and spontaneous-potential (SP) measurements made with one 'trip' in the hole.

Resistance and resistivity log

Because of the simplicity of equipment, the single-point resistance log has become the standard electric log for use to define qualitatively lithology and stratigraphy and for detailed correlation of thin rock units and facies (Fig. 1). The single electrode is not satisfactory for determining the true resistivity of the 'beds' and has a small effective penetration beyond the wall of the hole. Nevertheless, the variations in resistance at the electrode are sufficient for the detection of small qualitative differences in the resistivity of thin lithologic units, and the single-point resistance log provides ready identification of sandstone and fine-grained clay–siltstone–shale because of the characteristic distribution and resistivity of interstitial water.

The short normal resistivity log, either 16-, 32- or 64-in, or some combination of these, is being field-tested by several companies, and it may eventually replace the single electrode. As yet, methods have not been developed by the uranium industry to exploit advantages of greater penetration and quantitative analysis; it remains essentially a qualitative lithologic and correlation log. The 32- and 64-in spacings do not resolve thin beds as well as single-point resistance or 16-in resistivity; beds less than 1 ft in thickness may be significant in uranium geology.

Spontaneous-potential (SP) log

The SP log has been gaining popularity with the industry, but there is little

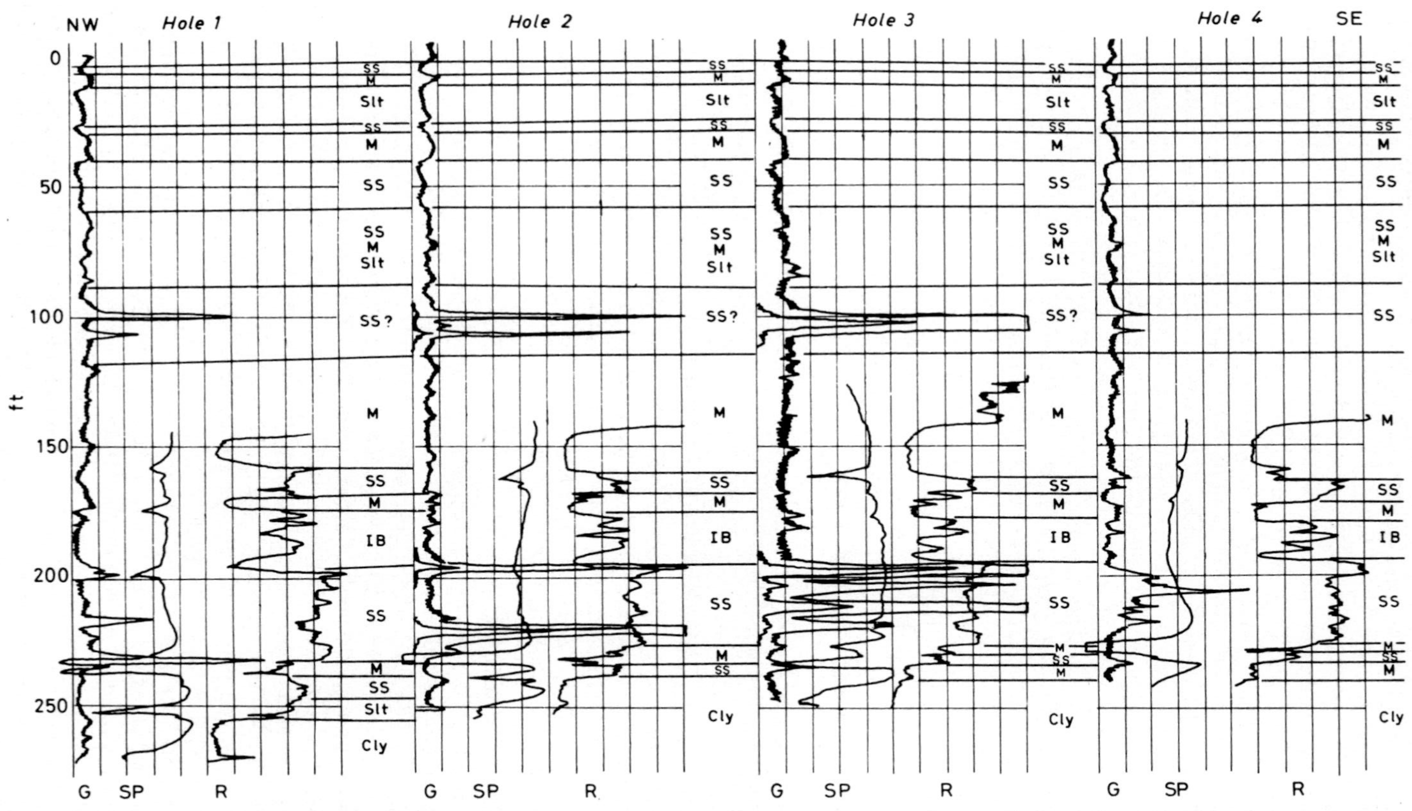

Fig. 1 Identification and correlation of lithologic units and mineralized zones across roll front by use of basic programme of gamma-ray, resistance and SP logs

understanding of the basic principles involved. Some of the fundamentals of the SP measurement and the variables influencing the SP curve have been reviewed recently by Hallenburg.[2] Mostly, the SP log is used by the uranium industry to confirm the lithology and correlations derived from the resistance log (Fig. 1). Recent investigations, however, to relate the SP effects, especially the redox potentials, to the environments of uranium roll fronts appear promising.[3] Better SP logs are being obtained in the deeper holes and they could be greatly improved by adequate conditioning of the holes prior to logging.

These electric logs are limited to uncased holes filled with water or water-based muds.

Gamma-ray log

The sensitive gross gamma log detects small variations in total gamma activity characteristic of lithologic composition and can be used for stratigraphic correlations and identification of lithologic units in the absence of anomalous radioactivity (Fig. 1). The gamma log can be used in air-filled, oil-based muds, foam or cased holes. Properly designed equipment having linear response to broad ranges of gamma intensity can be used to evaluate radioactive ores and can be calibrated in units of equivalent uranium, thorium or potassium.

Deep holes which intersect ore are routinely drift-surveyed with a directional log to compute the true depth and location of anomalies.

The basic logging programme can be used to identify roll fronts, as suggested by Adler and Sharp[4] (the cross-section in Fig. 1 illustrates such a case). The upper and lower limbs are indicated by small gamma anomalies near the upper and lower contacts of the host sandstone (hole 1). The limbs thicken and the intensity increases in hole 2; the 'ore roll' fills the central zone of the sandstone in hole 3; and the broad slightly anomalous zone in hole 4 indicates proximity to the front, probably on the 'barren exterior' side.

SUPPLEMENTARY LOGS FOR URANIUM EXPLORATION

Supplementary logs now include caliper, gamma–gamma density–porosity or combined density–caliper, and, occasionally, neutron–neutron, temperature or differential temperature, and sonic.

Caliper log

The caliper log is used to correct the response of other logs. It is too often omitted on the erroneous assumption that the diameter of the hole can be estimated with sufficient accuracy. Precise and accurate diameter measurements are absolute requirements for quantitative analysis of nuclear logs.

Gamma–gamma density log

The simple gamma–gamma density log developed by the petroleum industry[5] and applied to hydrology and coal cannot be used to measure the density or porosity of uranium ore host rocks because of high-intensity natural gamma-ray anomalies.[6] Dodd and Droullard[7] demonstrated the feasibility of the short–long spaced difference technique which overcomes the problem. Density information is used in several ways: (1) to obtain tonnage factors to compute reserves; (2) to indicate the competence of rocks which affects mining methods and costs; (3) to determine porosity and variations in cementation; and (4) to determine interstitial water of the formation, which modifies response of nuclear logs and indicates possible water problems in mining.

Neutron–neutron log

The short-spaced neutron–neutron log was introduced to uranium logging by

Dodd and Droullard[8] in 1964. Its acceptance by the U.S. uranium industry is now increasing. Recent developments and some applications are discussed here.

Temperature and differential temperature logs are offered to the uranium industry by several logging service companies in the U.S. The primary application is identification of aquifers and of interconnected aquifers.

EXPERIMENTAL LOGS

A number of logs for uranium applications are in various stages of development. These include the 'selective natural gamma' or Compton/photo ratio log, suggested by Czubek,[9] the natural gamma spectral or KUT log,[10] induced polarization (IP), Eh and pH[11] and magnetic susceptibility, which are now being field-tested but need further development to become routinely operational. Some preliminary results with a sensitive magnetic susceptibility probe are discussed here.

The neutron activation log for uranium is in the earliest stages of development. A neutron–fission neutron log for uranium was suggested by Dodd[12] in 1961, and the feasibility of the pulsed neutron version of a fission neutron log was experimentally demonstrated by Czubek in 1969.[13] To date, experiments with uranium X-ray fluorescence have been disappointing, but they deserve future investigation for special situations.

Interesting and potentially valuable as these experimental logs are, the fact remains that natural gross gamma logs constitute the major tool of the uranium industry today throughout the world. Tens of millions of feet of uranium exploration holes are logged by natural gamma each year—some 20 000 000–30 000 000 ft in the U.S. alone. Therefore, some of the principles of this log and its analysis are reviewed and the results of some recent investigations to improve this standard log for uranium are presented.

The quantitative gamma-ray log

REVIEW OF BASIC PHYSICAL PRINCIPLES

Origin and characteristics of gamma rays

Gamma rays of characteristic energies are emitted by the conversion ^{40}K to ^{40}A and at certain steps in the ^{232}Th, ^{235}U and ^{238}U decay series. Fig. 2, a generalized diagram of these naturally radioactive isotopes, indicates the gamma-ray emitters of most significance to uranium logging methods.

The primary energy of the gamma photon is a function of the energy imbalance created by the alpha- or beta-particle decay of each isotope. The primary energy of the gamma is, therefore, characteristic of a specific isotopic decay. The energy of this electromagnetic photon is commonly stated in units of 10^3–10^6 electron volts (keV or MeV) or in terms of its electromagnetic wavelength or frequency. The gamma photon is without mass or charge and, therefore, can penetrate significant thicknesses of earth materials; the length of penetration is a function of the energy of the photon, the bulk density of the medium, which is proportional to the electron density, and the chemical composition of the medium which can be described by the equivalent atomic number of the medium (Zeq).

Reactions of gamma rays

The natural gamma rays react with the geologic medium in two principal ways: (1) Compton scattering, in which a collision with an electron occurs and some energy is imparted to the electron and lost from the gamma photon according to the same laws as govern the collisions between billiard balls; and (2) photo-

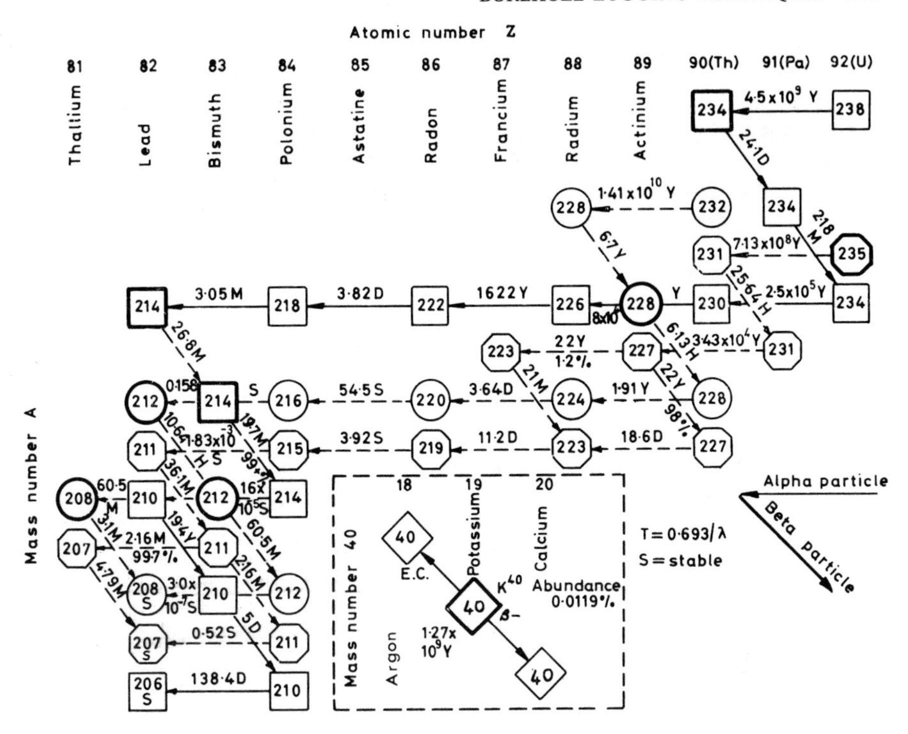

Fig. 2 Diagram of the natural decay series of uranium, thorium and potassium. Isotopes in heavier blocks emit gamma rays used for field and laboratory geophysical measurements

electric absorption, in which the total energy of the gamma photon is expended in a collision with an electron which overcomes the binding energy of the electron and imparts any remaining energy to it in the form of kinetic motion. A third reaction, pair production, is improbable because of the few natural high-energy gamma rays and the generally low atomic number of rock-forming elements (low Z).

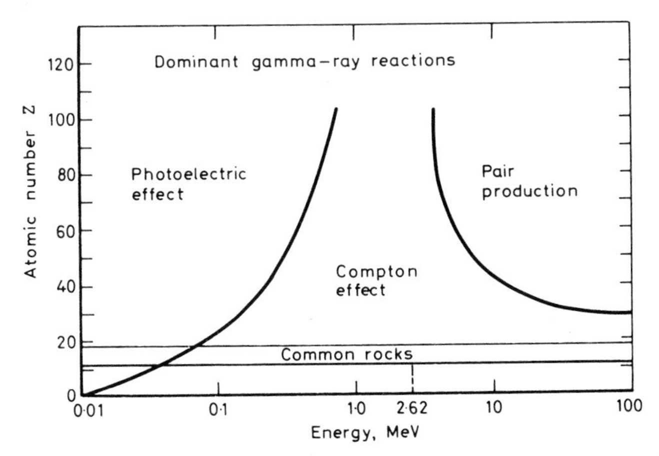

Fig. 3 Type of gamma-ray reaction depends on photon energy and atomic number. Equivalent atomic numbers of common rocks, maximum natural gamma energy and dominant reactions are indicated

The dominant gamma-ray reactions related to the energy of the gamma ray and atomic number of the medium are shown in Fig. 3. It should be emphasized that most gamma rays have undergone multiple scattering in the rock, with attendant losses in energy, before being intercepted by a detector in the logging probe; few unscattered gamma rays which retain diagnostic energies of emission remain to be measured in the borehole. Compton scatter and photoelectric absorption significantly influence both gross count and spectral logging methods.

PRINCIPLES OF CALIBRATION AND ANALYSIS

Sample volume concept

The concept of the effective sample volume and factors governing its size and shape are fundamental to any design of a gamma-ray logging system, the calibration methods and log analysis. For practical purposes the effective sample volume may be taken as a spheroid, which contains the material contributing 98 or 99% of the gamma rays intercepted by the detector.

Fig. 4(*a*) is a simplified diagram of an equatorial section of such a spheroid surrounding a detector. The length of the arrows represents the several primary energies of the gamma photons and the probable mean distance travelled before scattering reduces the energy below the minimum which will penetrate the

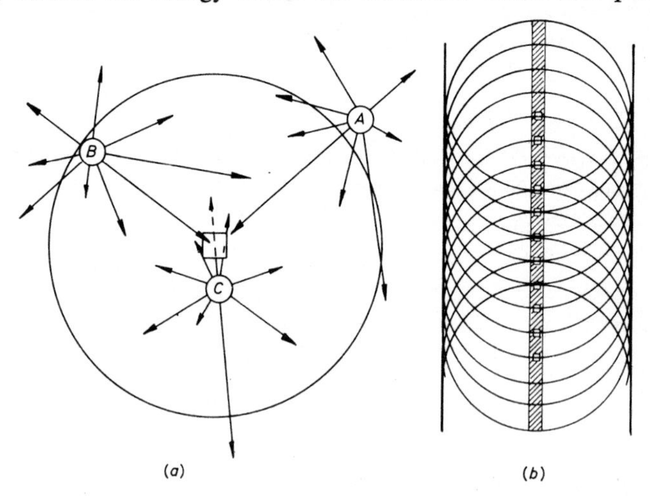

Fig. 4 Effective sample volume and cylinder of sampled volume

detector or the point where total absorption occurs. The probability of detecting a natural gamma ray originating at point *A* outside the sample volume is negligible, whereas a photon originating at *B*, just within the sample, has a small but significant probability of arriving at the detector with sufficient energy to penetrate the detector and to be recorded. There is a high probability that gamma rays originating at *C* may be detected because there is less chance of absorption by the medium and less chance of multiple scattering, which lowers the energy below a detection threshold value, and proximity to the detector increases the solid angle occupied by the detector, thereby increasing its chances of occupying the path of the photon. The degree of flattening or depression of the spheroid in its polar regions depends on the absorption by the probe along its axis.

Dynamic logging, or a series of closely spaced static measurements, moves the detector up the hole, as indicated in Fig. 4(*b*), and the spheroid of effective sample volume is translated into a cylinder having the same radius and a height measured along the axis of the hole.

The radius of the spheroid of effective sample volume depends on (1) the maximum primary energy of gamma rays emitted by the source; (2) the density of the medium (high density increases the probability of scattering per unit length); (3) the chemical composition, Zeq of the medium; and (4) the minimum energy accepted into the detector and counting circuitry. On the assumption of a nearly constant primary energy spectrum from uranium ore minerals, the radius is, therefore, controlled by the total absorption by mineral grains and interstitial fluid surrounding the hole; absorption by the borehole media, such as air, water, mud, casing or drill rods; absorption by the probe materials, such as probe shell, filter and detector housing; and in adequate equipment, a capability of always measuring the minimum energy of gamma rays penetrating the detector. Because of the strong influence of density, the radius of sample is measured in gramme-centimetres and, for a constant energy response by the probe and electronics, the length varies with density of the formation and borehole media. Most gross counting probes used for uranium logging have an effective sample radius of about 75–120 g-cm; methods or probes which accept only high-energy photons have a more restricted radius.

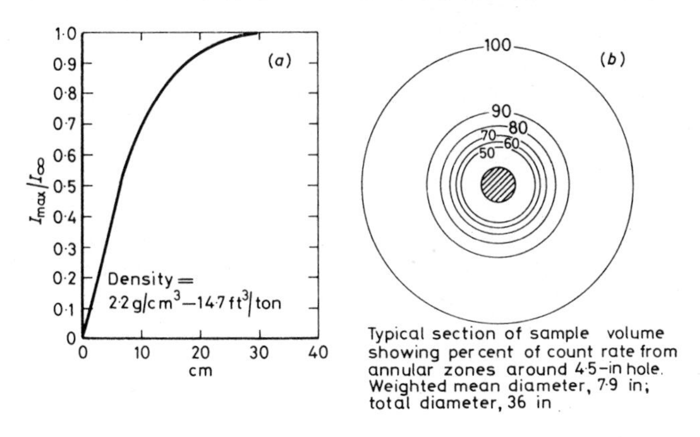

Fig. 5 Relative influence of thickness and annular zones within sample volume

In Fig. 5(*a*) the count rate (gamma intensity) is related to the thickness of cylindrical layers having an effective infinite radius around a typical probe. The gamma intensity increases with thickness until an effective infinite thickness is achieved at about 40 cm in material having a density of 2·2 g/cm³; the effective sample radius for this probe is, therefore, about 88 g-cm. This radius, typical of gross count systems used for uranium, is large in comparison to cores or cuttings from most exploration drill-holes.

Material close to the probe contributes a large part of the gamma flux measured by the detector, and all parts of the sample volume do not have equal influence (Fig. 5(*b*)). In an air-filled hole 11·4 cm (4·5 in) in diameter and a formation having a density of 14·7 ft³/ton (2·2 g/cm³), about 50% of the measured gamma intensity comes from material within 6·6 cm (2·6 in) of the hole and 90% is contributed by an annular zone within 18 cm (7·1 in) of the hole.

Calibration theory and methods

The effective sample volume concept is fundamental to the calibration theory for natural gamma-ray logs. *Only when the sample volume is completely filled with homogeneous material is the count rate proportional to equivalent grade.* This condition is rarely, if ever, met in exploration holes, and 'calibration' obtained by relating count rate to grade alone, especially grade of core samples, is not

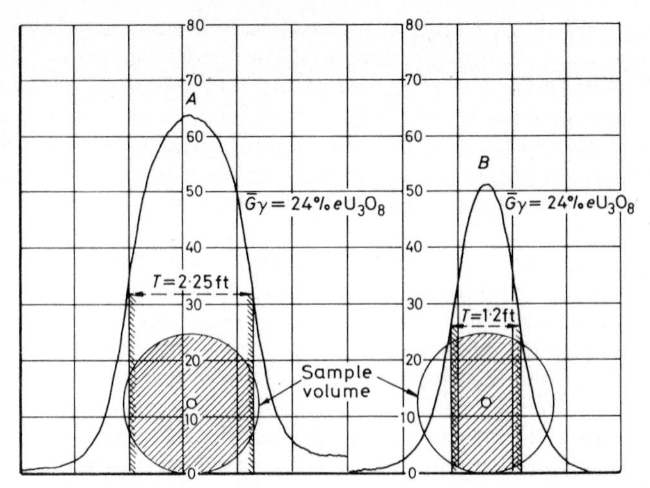

Fig. 6 Gamma intensity, in count rate, depends on thickness as well as grade

rigorously correct. Fig. 6 illustrates this effect. Analogue records from two model holes containing the same grade indicate different maximum count rates because the 'ore' zones are not the same thickness and do not fill the sample volume equally.

Empirical model studies by Droullard and Dodd[14] indicated that, if the detector is moved up the hole at a nearly constant rate, the area under the curve is proportional to the product of the average equivalent grade and the thickness. Scott *et al.*[15] mathematically demonstrated this principle for standard conditions of calibration. In practice, the equation has been expressed as

$$\bar{G}\gamma T = KA \tag{1}$$

where $\bar{G}\gamma$ is the average radiometric grade of uranium mineralization, expressed in per cent equivalent U_3O_8 by weight, T is the thickness of the mineralized zone, ft, K is a constant of proportionality determined by specific instrument calibration, and A is the corrected area under the gamma-ray log curve. The value of the calibration constant K is dependent on the value of $\bar{G}\gamma T$ assigned to the calibration model, the sensitivity of the detector, the units of thickness T, and the units of area A. Fig. 7 illustrates this linear relationship and is based on six full-scale model holes, each having a different grade and thickness, including a multi-grade and multi-layered model. The fit includes any errors of sampling, assaying of material in the model holes, non-uniformity in moisture and Zeq, errors of logging and of integrating the area under the curve; the maximum deviation of any point from a line drawn through the average and the origin is 3%.

Calibration of a logging system for uranium is theoretically possible by use of equation 1, and cored samples from exploration drilling, assuming that (1) the equivalent uranium grade ($\bar{G}\gamma$) is correct; (2) a sufficiently large number of anomalous intervals have been adequately sampled (good core recovery) to ensure that the grade of the cored samples is statistically representative of the material surrounding the holes; and (3) all intervals either have the same standard conditions of borehole diameter and fluid filling, the same Zeq and interstitial fluid, or adequate data are available to correct any variations to standard conditions. These prerequisites for calibration are seldom completely fulfilled by exploration holes, and, at best, require many ore intersections and additional measurements for correction factors to be obtained and for the establishment of the statistical reliability of sampling and analyses.

For these reasons it has been found expedient to substitute a few carefully constructed, well sampled and carefully analysed model holes for the many uncertainties of exploration drill-holes. Potential errors are minimized by crushing and blending materials and selecting standard conditions similar to typical boreholes. Standard conditions include hole diameter and fluid, and formation

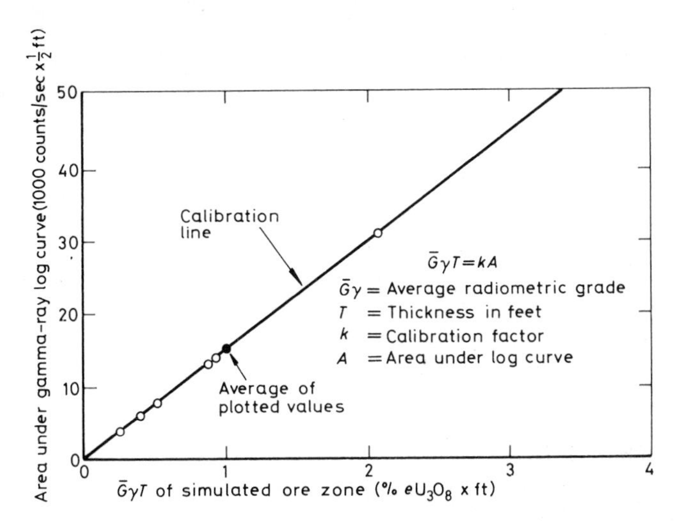

Fig. 7 Relationship of area under log trace, integrated count rate, to grade–thickness product: this proportionality is the basis for calibration

parameters of density, Zeq, and interstitial fluid (per cent moisture). The models should exceed effective infinite diameter and the thickness of zones used for calibration should, for several practical reasons, exceed the diameter of the effective sample volume of any detectors which may be used. Thick zones reduce the percentage error in thickness measurements, the influence of non-homogeneity and the chance for atypical material being concentrated near the edge of the hole, and experimental errors by providing more representative samples of the deflection or count rate. Also, if the effective sample volume is filled in each of several measurements, it is possible to make use of the unique experimental situation of an infinite thickness of homogeneous material in which response of the logging system is proportional to concentration.

Theoretically, it is possible to obtain a reliable calibration from a single model for which all the standard conditions are reliably and accurately known. It is recommended, however, that two or more models, or one model containing two calibration zones, well separated by low-activity (background) material, should be used. The gamma intensity of the calibration zones should differ by a factor of about ten and the intensities should be representative of those likely to be encountered in logging exploration holes (for example, about 0·2 and 2·0% equivalent uranium). Two or more zones provide a check on the accuracy of the models, permit checking response capabilities of a logging system in a typical range of intensities and provide data for determining system dead time[14] and for TWOPIT calibration techniques developed by Crew and Berkoff.[16]

Theory and practice of log analysis

Quantitative analysis of the gross gamma log is based on equation 1. Manual and mechanical integration methods were described by Droullard and Dodd[14] and by Scott *et al.*,[15] and methods for automated analyses by computers were described by Scott.[17] Recently, Coulomb *et al.*[18] presented similar sophisticated techniques.

When the manual method is being used the $\bar{G}\gamma T$ of the zone to be analysed is determined by numerical integration of the response at regular small intervals. The units to be integrated depend on the logging system. These may be the deflection, in inches, millimetres, per cent of full scale or count rate. In the case of a digital system the units may be the number of pulses accumulated in a regular period of time while logging through a small regular interval in the hole. These may be recorded on a printer, on perforated or magnetic tape or, after digital to analogue conversion, on a strip chart. The average equivalent grade, $\bar{G}\gamma$, is calculated by dividing the grade–thickness product, $\bar{G}\gamma T$, by T, the thickness of the zone.

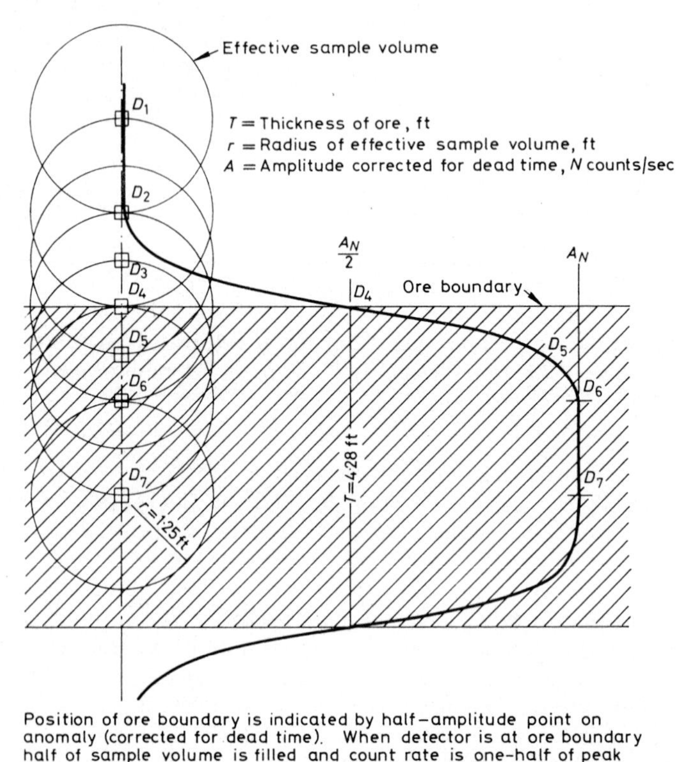

Fig. 8 'Ore' boundary is indicated by half-amplitude point. When the detector is at the boundary half of the sample volume is occupied by ore. Radiation detected beyond the ore develops the 'tail areas' of the anomaly

An empirical method, based on the concept of the effective sample volume, is used to determine the thickness, T, of a mineralized zone from the log (Fig. 8). The zone to be analysed will cause no response in the detector until the detector is one radius of the sample volume or less from the zone boundary. As the detector further approaches the zone, the sample volume is increasingly occupied by material from the zone. If the zone is at least one radius in thickness, when the detector is at the zone boundary half of the volume is filled. On the assumption of a random distribution of radioisotopes within the zone, and a thickness equal to or exceeding the diameter of the sample volume, the gamma intensity (count rate) is half of the intensity measured when the sample volume is entirely within the zone and filled with anomalous material. For this simple case the half amplitude (half of the maximum intensity or count rate) defines the boundary. If the thickness of the anomalous zone is less than the diameter of the sample volume, the maximum intensity will be somewhat less than twice the intensity at the boundary, and picking a boundary at half amplitude causes an over-estimation of thickness T. This error in T for thin zones does not affect the value of $\bar{G}\gamma T$, which represents the quantity of radioactive material; the resulting over-estimation of T causes a slight 'dilution' by adjacent material and under-estimation of the average grade. For practical mining and reserve calculation, however, this error in thickness can be tolerated or ignored because thicknesses of less than 15–18 in can seldom be extracted economically without similar dilution.

As was explained by Scott *et al.*,[15] the sum of the log deflections at the zone boundaries may be multiplied by a 'tail factor' for determination of the area of the anomaly extending beyond the boundary: this eliminates the need to integrate from background to background. The 'tail factor' technique overcomes the problem of distinguishing the point of departure from background on the log and also offers the opportunity to exclude, for economic evaluation, certain 'low-grade' zones adjacent to ore zones. The value of the tail factor is related to the radius of the sample volume and the value may vary within small limits as the density and atomic number of the rock and borehole change.

The analysis by computer is also predicated on equation 1 and the sample volume concept. In an infinitely thick, homogeneous zone the gamma intensity will be uniform and all log deflections, in units of true counts per second, N (counts/sec corrected for dead time or other non-linearity), will be equal, and any one can be represented by $\bar{N}$. If the intensity is recorded or read from the log at regular intervals representing regular intervals of T in the hole, such as $\frac{1}{2}$ ft ($T/2$), the number of values will be $2T$ and

$$A = 2T\bar{N} \tag{2}$$

From equations 1 and 2 it follows that

$$\bar{G}\gamma T = K2T\bar{N} \tag{3}$$

$$\bar{G}\gamma = 2K\bar{N} \tag{4}$$

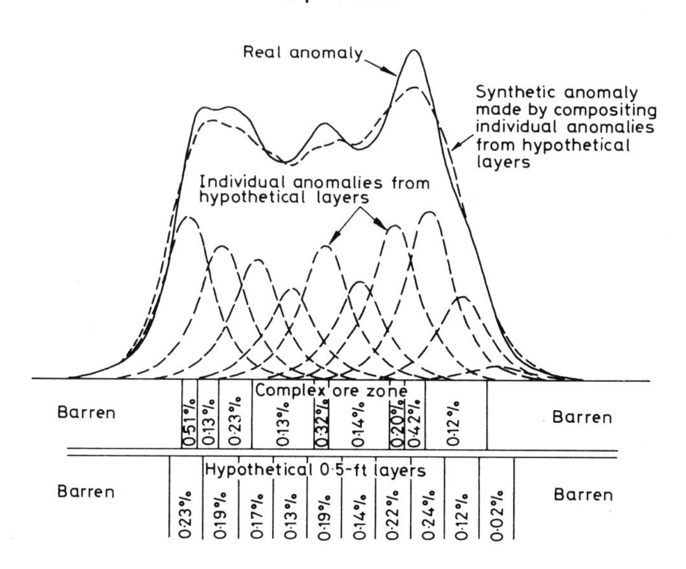

Fig. 9 GAMLOG generates synthetic anomalies for hypothetical layers and adjusts grade of these layers until composite anomaly matches log

Therefore, the grade of an anomaly representing a homogeneous, infinitely thick zone is represented by the product of the count rate multiplied by an appropriate calibration factor K, and the number of intervals per unit thickness T.

By use of equation 4 the GAMLOG program makes a first approximation of the grades of ½-ft layers of an anomaly. Weighting factors for adjacent intervals have been determined from model studies. These are used to develop a composited total anomaly, which is compared with the real log anomaly. The grades assigned to individual layers are adjusted by an iterative process until the composited anomaly closely matches the log anomaly, as is shown in Fig. 9. The output tabulates the depth or elevation and the 'best' grade assigned to each hypothetical layer. GAMLOG data are used in a number of other computer programs to calculate reserves, to evaluate mining plans and to estimate production at various market prices or production costs.[19,20]

With subsequent modifications and the addition of more subroutines the GAMLOG computer program developed by Scott[17] has been adopted by the U.S. Atomic Energy Commission, at least three service companies and a number of mining companies to analyse gamma logs for ore-reserve estimation.

COMMENTS ON LOGGING EQUIPMENT

Log analysis, based on equation 1, assumes short- and long-term stability of the system, sufficient sensitivity to provide reliable counting statistics, linear response of the logging system and, equally important, that the standard conditions in the calibration models are duplicated in the exploration drill-holes.

Table 1 Sensitivity and relative counting error of typical gamma-ray logging probes used for uranium exploration*

Detector	Probe shell Metal g/cm²	$\frac{\bar{N}}{100 \text{ ppm } eU_3O_8}$	$\sigma = \sqrt{\bar{N}}$	% Error† $\frac{\sigma}{N} \times 100$
3 G–M tube	Brass 1·4	19	4·26	22·9
7 G–M tube	Fe 2·5	48	6·93	14·4
½ in × ½ in NaI	Fe 2·5	56	7·48	13·4
½ in × ½ in NaI	Al 0·8	73	8·58	11·7
½ in × 1½ in NaI	Fe 2·5	143	11·96	8·4
½ in × 1½ in NaI	Al 0·8	184	13·56	7·4
¾ in × 1 in NaI	Fe 1·2	226	15·03	6·7
1¼ in × 1¾ in NaI	Fe 2·5	640	25·30	3·9
1¼ in × 1¾ in NaI	Al 0·8	803	28·34	3·5
1¼ in × 3 in NaI	Filter 8·2	266	16·31	6·1

*Response is for 1-sec counting period in an infinite thickness of 0·01% eU_3O_8.
†For single sample point.

Several basically different approaches to gamma logging systems are successfully used in the recent generation of equipment. Detailed description of the various sophisticated techniques is best left to the instrumentation engineer. Adequate gross-count probes, highly adaptable standard NIM pulse processing and recording modules or complete vehicle-mounted systems are available from a number of manufacturers. Except for short system dead time, modern logging systems are linear at all rates up to 10^5 counts/sec or more. NaI(Tl) scintillation detectors have been adopted almost to the exclusion of G–M detectors. When operated near the centre of the gain plateau,[14] with acceptable phototubes, high voltage supplies, amplifiers and lower discriminators, the plateau is sufficiently flat and long that a gain shift of four will result in a change in count rate of no more than 2 or 3%. Normal minor gain shifts, caused mostly by the temperature of the phototube, or small fluctuations in other components result in almost undetectable changes in count rate and the requirement for short- and long-term stability is met.

The system dead time of most recent equipment is less than 10 μsec. Techniques to provide short dead time, even through 5000 ft or more of logging cable, include (1) placing the discriminator and rate meter or scaler in the probe and transmitting only dc logic signals on the cable; (2) prescaling in the probe; (3) derandomizing with buffer storage in the probe; and (4) excellent matching to the cable by use of specially designed cable driving circuits (the latter also permits pulse-height spectral measurements by standard NIM modules at the surface). Short and nearly constant dead time, some systems having less than 1 or 2 μsec, and nanosecond pulse processing and counting electronics result in a capability for counting well in excess of 100000 random pulses per second. This has led to adoption of larger and more sensitive NaI(Tl) crystals and the requirement for reliable counting statistics is more than adequately met (see Table 1 for typical examples).

Multiple measurements within an anomaly further reduce the statistical errors of counting random events and adequate sensitivity permits the selection of short time constants or short gate times (1–2 sec) without significant error. Logging speeds of 5–10 ft/min in the anomaly will not override the short time constant and permit digital recording at intervals of 0·2 ft.

TECHNIQUES AND CORRECTION FACTORS FOR QUANTITATIVE LOGGING AND ANALYSIS

The calibration and log analysis theory assumes stability of the system, a linear response to a wide range of gamma intensities, adequate sensitivity for reliable measurements and duplication in the borehole of the standard physical conditions in the calibration model. Great emphasis has been placed on the need for stability, linearity and sensitivity, which has resulted in a modern generation of commercially available equipment that satisfies these requirements. Adequate equipment having been achieved, it is now possible to evaluate the possible errors and correction techniques to meet all of the above assumptions, including corrections for non-standard conditions.

Corrections for system dead time

Modern gamma-ray logging systems have short but measurable periods of dead time. This is the time, generally measured in microseconds, which the system requires to detect a photon and to process the resulting pulse. During this period the instrument will not record additional photons incident on the detector. The random nature of gamma emission ensures, even at low intensities, that, occasionally, two or more photons will arrive at the detector nearly simultaneously. Only the first arrival will be measured and additional photons incident during the dead time will be lost. As gamma intensity increases, the probability of loss due to dead time increases rapidly until the number of photons lost is greater than the number counted; when the system is totally and continuously occupied in processing pulses it is saturated and the system will not respond to an increased gamma intensity. Fig. 10 illustrates the non-linear response caused by the system dead time of typical gamma-ray logging units. Even a moderate dead time, e.g. 5 μsec, requires a 20% correction at an observed rate of 40000 counts/sec, and this increases to a 50% correction at an observed rate of 65000 counts/sec. Some recent digital instrumentation includes automatic dead time correction by use of some type of live timing in which the counting period is extended by the amount of time the system is dead. Without this type of com-

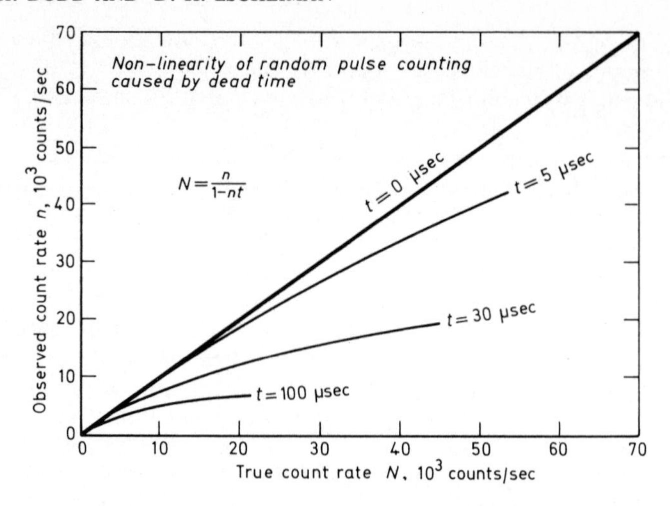

Fig. 10 System dead time results in significant non-linear response at high count rates

pensation each numerical value of observed count rate, n, should be statistically corrected by use of the standard equation

$$N = \frac{n}{1-nt} \tag{5}$$

where N is the true rate and t is dead time. Because this is at best a statistical correction, it becomes increasingly inaccurate for large values of n, and corrections in excess of 50% are undesirable. Therefore, because of uncertainties in determining the absolute value of t, variation of t with count rate found in some equipment and the statistical errors in large corrections, it is necessary to optimize the sensitivity of the detector and the system dead time. The errors caused by uncertainty in dead time correction at high rates may well exceed errors of counting statistics.

Corrections for non-standard borehole conditions

To minimize corrections the standard conditions of calibration in the model holes should approximate common field logging conditions. Any departure from standard conditions in the borehole environment requires a correction. Variations in diameter, hole fluid, drill rods or casing require significant corrections.

Effects of borehole diameter and fluid Corrections for changes in diameter of air-filled holes in the range 2–8 in are insignificant if the probe has been calibrated in a typical 4½-in air-filled model hole. When the hole is filled with water or natural mud (low Zeq), however, correction for non-standard diameter becomes more significant, as is shown by the correction curves for typical probes presented in Fig. 11. Factors which influence the hole-size correction include (1) the diameter of the probe, and, hence, the g/cm² of surrounding absorber; (2) the location of the probe in the hole, centralized or decentralized (the latter is common practice and generally requires smaller corrections); (3) the density and Zeq of the borehole fluid; (4) the sensitivity of the detector to various energies; and (5) the minimum threshold energy counted by the system. If the detector is operated on the gain plateau in the gross-count mode, and all detected gamma rays are counted, the energy threshold of the probe is determined by the material and thickness of the probe shell (Z and g/cm²) (this has been treated theoretically by Czubek[21]). The threshold values can be approximated by empirical tests by use of low mono-energetic sources; the threshold for typical probes is in the range 40–60 keV.

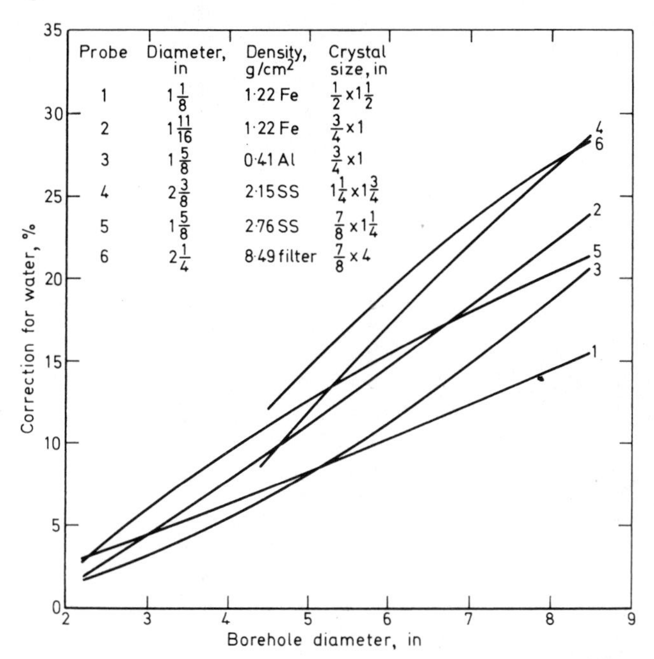

Fig. 11 Examples of correction for diameter of water-filled holes for typical probes

The low atomic number (Zeq) of water and natural muds lowers the mean energy of gamma rays by Compton scattering while absorbing only the lowest energies. The thick-walled and high Z probe shells absorb more of the low-energy photons than the thin-walled shells and, therefore, thick-walled probes are more influenced by the decrease in energy caused by the water and mud and require larger corrections.

The Zeq of the formation increases with the uranium content, which, in turn, increases the mean energy of the gamma-ray spectrum. This effect partially compensates for the decrease in energy caused by borehole water and mud. Data for recent empirical experiments in model holes are presented in Fig. 12; these indicate a decrease in the value of hole-size corrections with increasing grade. The magnitude of the effect is dependent on the energy sensitivity of the probe. Additional work should be done to evaluate the problem and devise operational correction techniques.

The curves relating correction to diameter are nearly linear for typical probes over the normal range of hole size. Correction factors, empirically derived from models, are fitted by simple power or low-order polynomial equations, and the coefficients can be entered into computer programs for log analysis. The corrections are significant and the diameter of critical portions of holes (anomalous zones) should be determined by caliper logging.

Effects of casing Similar curves should be developed for each type of probe to correct for absorption by casing or drill rods. These corrections may be large (Fig. 13) and drillers' records should be carefully searched to determine the correct thickness (check for more than one 'string' in the hole).

Non-standard conditions in the formation (*ore zone*)
Non-standard conditions encountered in the formation which modify the natural gamma photon intensity and energy are less easily determined than

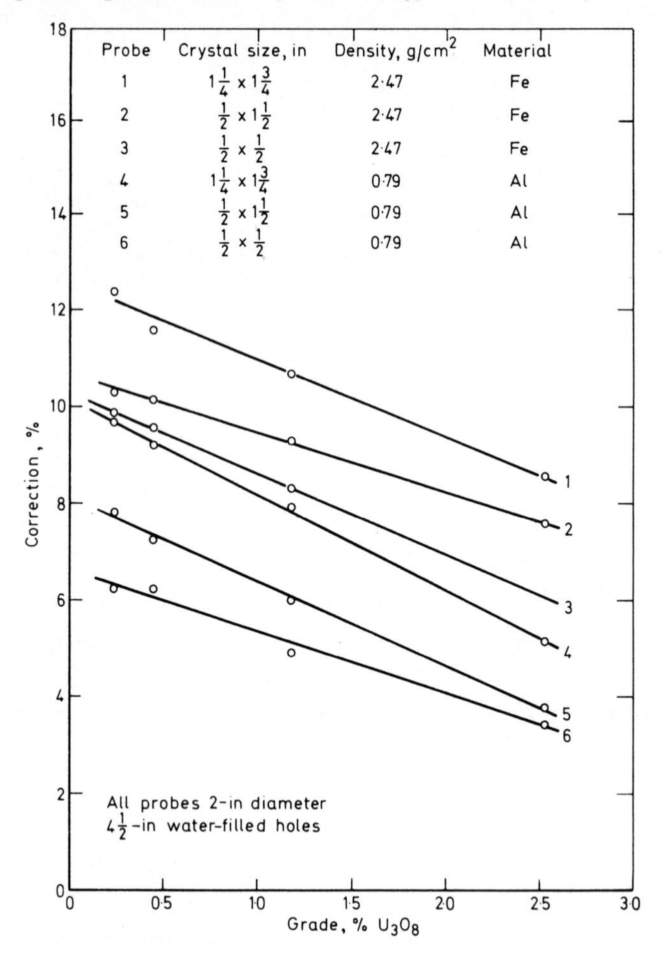

Fig. 12 Experimental data indicate influence of ore grade, Zeq, on correction for water in 4½-in diameter holes for typical probes. Hole-size correction decreases with grade

borehole conditions, and the cumulative effects are difficult to isolate and evaluate. These include interstitial fluids and variations in composition, particularly unusual abundances of high atomic number elements, which affect the Zeq of the formation.

Correction for interstitial water Free water in the formation increases absorption without contributing gamma photons. The interstitial fluid, in effect, dilutes the concentration of the radioactive minerals in terms of dry weight per cent (this effect is illustrated by the two logs in Fig. 14). Except for the 36% by weight of water in model *P–W*, models *P–D* and *P–W* contain the same materials, which assayed 0·50 and 0·49% eU_3O_8, respectively. The log of the water-saturated 'formation' in *P–W*, however, has a distinctly smaller area under the curve. An 'uncorrected' analysis of the water-saturated *P–W* zone indicated a grade of 0·36% U_3O_8, a discrepancy in grade of 0·13% eU_3O_8 or an error of 26·5%. By use of the relationship

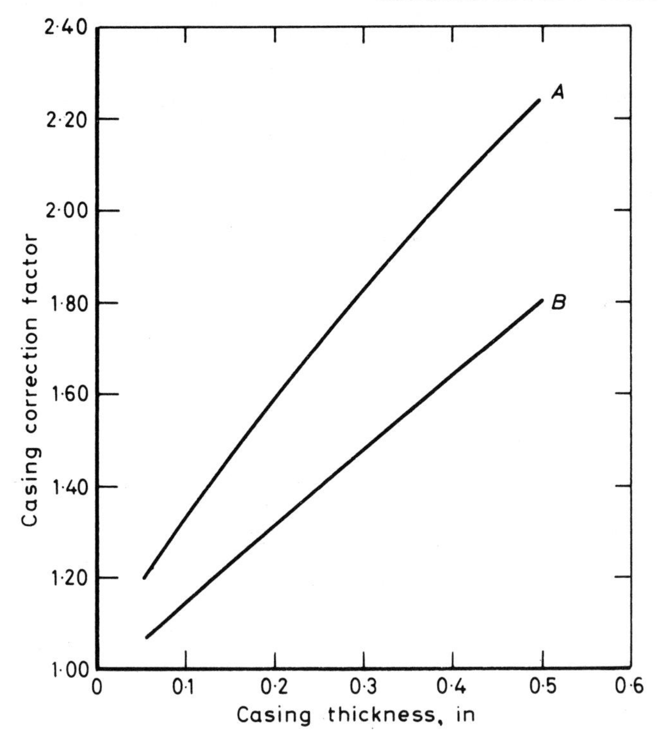

Fig. 13 Examples of correction factors for wall thickness of casing for typical uranium logging probes (probe *A*: steel shell, 2·0-in outer diameter; wall, 0·065 in thick; NaI detector, $\frac{1}{2}$ in × $\frac{1}{2}$ in; probe *B*: brass shell, 1·0-in outer diameter; wall, 0·065 in thick; G–M detector, Amprex 912 NB)

$$\text{Free water correction factor} = \frac{100 - \%H_2O \text{ in calibration model}}{100 - \%H_2O \text{ in formation}} \quad (5)$$

the 'corrected' grade of *P–W* becomes 0·495 % eU_3O_8, which agrees well with the 0·49 % grade of the model. This empirical correction appears adequate for normal variations of interstitial water from 8 to 22 %. The free water below the water-table can be closely approximated from the gamma-density log[7] or determined from the neutron–neutron log, independent of saturation.

Correction of atomic number, Zeq, effect Differences in composition, Zeq, of the calibration model and of the formation are more difficult to evaluate. Danes[22] described the need for a chemical correction factor in gamma–gamma density logging. The problem affects natural gross gamma logging as well. Fig. 3 shows the fields of probable gamma-ray reaction in terms of the photon energy and atomic number (Z) of the medium. The Zeq of common rocks ranges from about 12 to 16.[23]

Multiple Compton scattering in an infinite thickness of common rocks modifies the primary spectrum to one which is heavily weighted in the energy region below 200 keV. This has been treated theoretically by Czubek and Lenda.[24] Fig. 15 includes two spectra obtained by a typical logging probe—one of a ^{226}Ra needle source through air, the other of an infinite zone of uranium ore (^{226}Ra) in a borehole. Compton scatter in the rock medium degrades the photo peaks and enhances the Compton 'fill-in', particularly in the energy region below 0·6 MeV. This extreme scattering is one of the major limitations in downhole gamma spectrometry. Fig. 16 shows the energy distribution, in cumulative per cent, which was measured in-hole by a thin-walled aluminium probe (0·41 g/cm²) and a small (½ in × ½ in) low-energy sensitive NaI detector. More than half of the photons detected fall in the low-energy region in which changes in Zeq will affect the ratio of Compton scatter to photoelectric absorption.

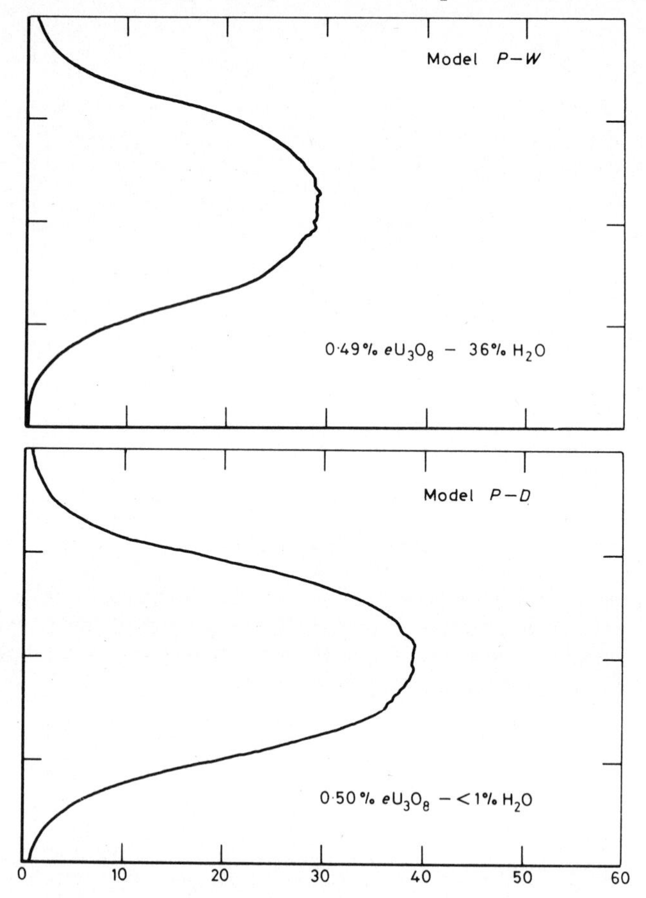

Fig. 14 Logs of model holes show attenuating effect of interstitial water in saturated and dry zones of nearly equal grade

Czubek and Guitton[6] pointed out that addition of 'ore grade' concentrations of uranium (Z = 92) will significantly increase the Zeq of host rocks. This modifies the spectrum of scattered gamma quanta by enhancing photoelectric absorption. The low-energy photons which predominate in common rocks are increasingly and selectively removed by photoelectric absorption with an increase in Zeq of the formation. Consequently, if the probe responds most efficiently to low-energy photons (< 400 keV), the measured intensity (counts/sec) is a function of the Zeq of the formation. Hence, the number of gamma photons are not directly proportional to the concentration of uranium in secular equilibrium.

The Z effect of uranium on a typical probe which is sensitive to energies of < 400 keV is demonstrated by measurements made in two 'ore zones', *A* and *B*, containing 0·24 and 2·54% eU_3O_8, respectively. Measurements shown in Table 2 were made in two parts of the spectrum, below and above 400 keV, by use of a ¾ in × ¾ in NaI(Tl) detector in a thin stainless steel probe shell (1·23 g/cm²).

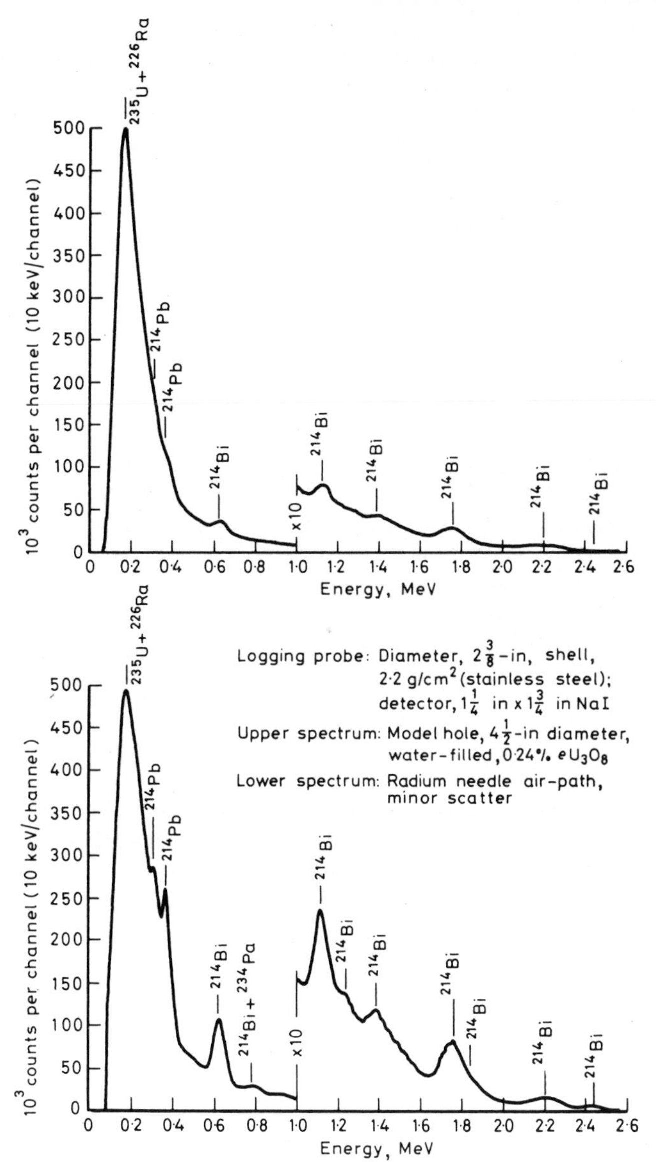

Fig. 15 Spectra measured by typical logging probe show fill-in and loss of primary photo peaks caused by Compton scattering in borehole environment

The Σ count measured in the energy region above 400 keV is directly proportional to the grade; the ratio of the counts nearly equals the ratio of the grades. Because of photoelectric absorption of photons in the energy region below 400 keV, the Σ count in 'high-grade' zone *B* is deficient and the Σ count ratio is much smaller than the grade ratio.

The magnitude of errors caused by the Z effect of uranium concentration for typical low-energy sensitive probes is shown in Fig. 17. The error is greatest in probes most sensitive to low-energy photons by reason of small crystals and thin low Z probe shells. The error increases with increasing grade of the 'ore'. Errors at 2·5% U_3O_8 range from about 18 to 24%, depending on the low-energy sensitivity of the probe.

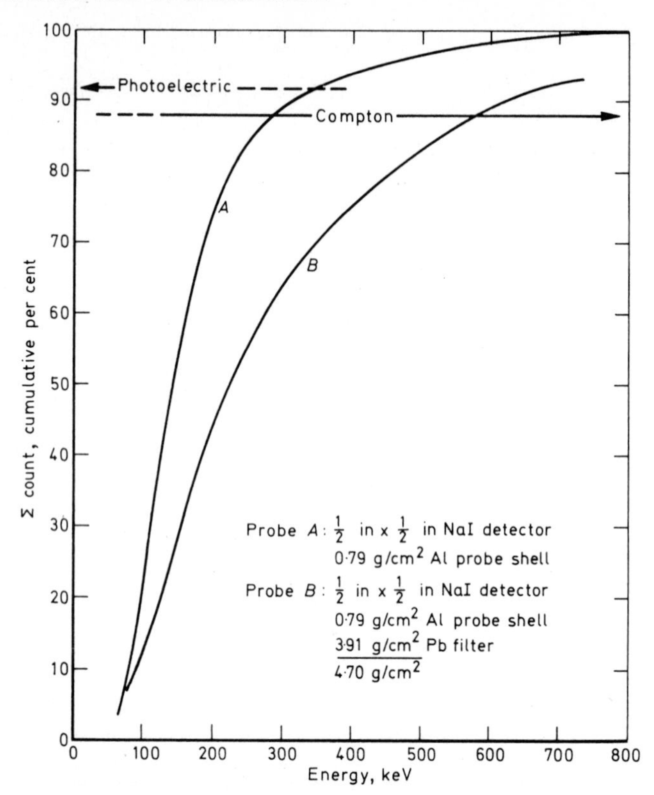

Fig. 16 Energy distribution measured in borehole by thin-walled probe and detectors sensitive to low energy is concentrated in the region below 200 keV

The experimental data for each probe can be fitted to a third-order polynomial expression, which might be taken as the 'correction curve' for the Z effect of uranium. Unfortunately, this relationship is true only if the ore is in equilibrium. It does not necessarily follow that a high measured intensity indicates high concentration of uranium and high Zeq.

Czubek and Lenda[24] measured the dependence of the total intensity detected from media of differing Zeq, for various levels of discrimination. To eliminate the problem of non-standard Zeq values in the formation it is necessary to exclude low-energy gamma rays. In the energy region above 400 keV the measured intensity is proportional to grade in equivalent uranium (see Table 2).

The ultra-stable electronics required to maintain a constant level of discrimination are frequently unreliable and costly. The addition of a physical filter around the detector will exclude most of the low-energy photons and will reliably provide the necessary discrimination. A graded filter of lead, cadmium and brass has been used successfully. The graded filter eliminates X-rays generated in the primary lead filter. The filter, because of the dominantly low-energy spectrum in the borehole, will reduce the count rate by 70–80%; however, the desired

Table 2 Response in uranium ore models measured above and below 400 keV discriminator

Zone	*A*	*B*	*A/B*
Grade, % eU_3O_8	0·24	2·54	0·954
Σ Count <400 keV	262 281	1 956 177	0·134
Σ Count >400 keV	25 553	271 946	0·940

sensitivity (counts/sec) can be regained by increasing the size of the NaI(Tl) crystal.

Curve *F* in Fig. 17 shows the linear response achieved by use of a $\frac{3}{4}$ in × 2 in crystal and a graded filter consisting of 0·140 in of lead, 0·062 in of cadmium and 0·035 in of brass, and a stainless steel probe shell with a resulting total density of about 7·25 g/cm². Similar filtered probes have recently been adopted by industry.

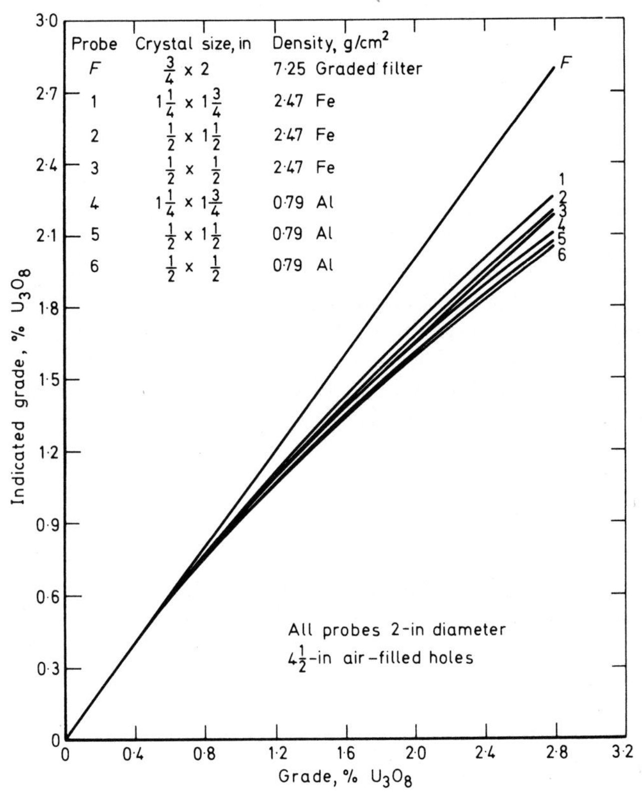

Fig. 17 Experimental measurements with typical uranium logging probes sensitive to low energy show non-linear response caused by increasing Zeq of high-grade ore. Z effect of uranium does not influence adequately filtered probe

Correction for disequilibrium The discussion to this point has dealt with calibration and analysis of the gross-count gamma-ray log in units of equivalent uranium. The average *equivalent* grade of a mineralized zone can generally be determined with an accuracy of $\pm 5\%$ at the 100 ppm level, or at lesser concentrations if counting periods are extended.

In spite of its generally successful application to uranium exploration, the gross gamma log has three serious limitations: (1) disequilibrium between uranium and its major gamma-ray emitting daughters is not uncommon—this can lead to serious errors in estimating the true concentration of uranium; (2) at concentrations of less than 100 ppm the contribution from normal rock levels of potassium and thorium becomes significant; and (3) analysis of combined uranium–thorium ores must be predicated on an estimated ratio.

Three types of disequilibrium in the uranium series may cause significant errors. Most serious is disequilibrium between ^{238}U and ^{226}Ra, which, in extreme cases, may approach total lack of either isotope and the time to reestablish equilibrium is controlled by ^{234}U ($T/2 = 2{\cdot}5 \times 10^5$ years) (for additional discussion see Rosholt[25]). Disequilibrium between ^{226}Ra and ^{222}Rn, and, consequently, the major gamma-ray emitters ^{214}Pb and ^{214}Bi, is generally more subtle. It is difficult to evaluate the degree of emanation, diffusion or transport of the Rn by atmospheric pumping in unsaturated formations and dry holes or transport by rapid flow of groundwater.[26] Disequilibrium between ^{238}U and ^{234}U and progeny is apparently caused by auto-oxidation and selective leaching of ^{234}U. This type of disequilibrium is relatively unimportant in ore evaluation, but it is of concern to genetic studies.

A knowledge of the type, distribution and degree of disequilibrium significantly aids economic evaluation and mine planning, which may be based, to a large extent, on gross gamma log data. It is also valuable to the exploration geologist. Disequilibrium information indicates geologically recent leaching or secondary enrichments; it may indicate the direction of movement of solution fronts common to many sandstone-type deposits; and disequilibrium is one of the criteria suggested by King and Austin[27] to indicate relative position on a roll deposit.

Chemical and sealed-can gamma-only assays described by Scott and Dodd[28] provide information on which to base corrections for U/Ra disequilibrium. The amount of in-growth after sealing is also a relative indicator of the emanation coefficient of the ore. Several logging techniques to detect or compensate for disequilibrium are being tried which involve various forms of energy discrimination; they must be classed as experimental and are not routinely used by industry.

Natural gamma-ray spectral logging

INTRODUCTION

The theoretical capability of NaI(Tl) scintillation detectors to obtain gamma energy spectra in uranium exploration boreholes was recognized in 1952.[29] Development of practical operational methods in the succeeding 15 years was inhibited by lack of adequate instrumentation. Some of the technical problems of borehole spectrometry were reviewed by Dodd and co-workers;[10] many of these problems have been solved by the recent generation of instrumentation.

APPLICATIONS TO EXPLORATION AND EVALUATION

Identification of gamma-ray haloes

In the search for uranium there is an obvious need to distinguish between gamma-ray anomalies caused by uranium, thorium or potassium; qualitative identification is the simplest application of in-hole spectrometry. Radioactive haloes are important guides in uranium exploration, and detection of low-level variations in uranium (radium) is important. Without spectral analysis it is often difficult to recognize minor, but significant, enrichment or depletion of perhaps 2–5 ppm eU_3O_8 if variations in gamma background caused by potassium and thorium are of similar magnitude.

Environmental changes

Geological–geochemical environments may be ‘fingerprinted’ by characteristic ratios or changing U:Th:K ratios. Changes in the relative ratios can be measured with spectral logging. Qualitative recognition of environmental change does not require determination of absolute concentration, but there is a need for high sensitivity and precision.

Low-grade resources

It is advisable to distinguish between sources of gamma rays in the evaluation of low-grade uranium resources containing 100 ppm or less. Normal intrinsic concentrations of potassium and thorium may contribute 20–30% or more to the gamma intensity.

Disequilibrium

Perhaps the most important application of energy discrimination methods are those which, under favourable conditions, can detect U–Ra–Bi disequilibrium. Lyubavin and Ovchinnikov[30] proposed a method based on the relative fraction of gamma radiation of uranium and its decay products prior to radium and the gamma radiation from the total uranium decay series. Photons emitted by ^{234}Th, ^{235}U and U X-rays, which are closely related to ^{238}U, contribute to the low-energy region of the spectrum (<200 keV), even though the photo peaks are obscured by the Compton continuum. Photons emitted by the decay of ^{214}Bi dominate the spectrum of uranium ore and account for nearly all of the spectrum above 200 keV. A shift in the ratio of hard to soft photons is, therefore, presumed to represent a proportional shift in the U:Ra ratio. Blake[31] reported encouraging results from logging ore zones with a filtered and unfiltered detector to measure a ratio of hard gamma rays to the total spectrum. Electronic discrimination, if adequately stabilized, will improve this differentiation between low- and high-energy bands of the natural in-hole spectrum. However, the Zeq effect at moderate grades will tend to increase the hard/soft ratio much as would a high-radium disequilibrium. Therefore, the H/S log is limited to zones of normal background and low-grade ore.

NATURAL SELECTIVE GAMMA–GAMMA (COMPTON/PHOTOELECTRIC) LOGGING

Theoretical basis

In 1968 Czubek[9] followed up his previous suggestion[23] with additional theoretical and experimental data to show that small quantities of uranium in common rocks will increase Zeq sufficiently to provide a possible method for measurement of the uranium concentration. He indicated, for example, that at 300 keV a change in uranium concentration from 0·025 to 2·5% changes the ratio of the photoelectric to Compton absorption coefficients, μ_{ph}/μ_c, from 0·0075 to 0·1.

Method of measurement

By measuring intensities I_1 and I_2 at two energy intervals E_2–E_1 and E_4–E_3, where the low-energy band, E_2–E_1, is strongly influenced by photoelectric absorption and the high-energy band, E_4–E_3, is not, it is possible to relate the ratio I_1/I_2, designated the P parameter, to the concentration of uranium, irrespective of equilibrium.

Experimental results

An example of the experimental results obtained by Czubek by use of the energy intervals 100–150 keV and 300–700 keV and a $\frac{1}{2}$ in × 1 in NaI(Tl) detector is presented in Table 3. Similar experimental results, but with slightly different

Table 3 Experimental values of *P*

	I_1, counts/min	I_2, counts/min		
%U	(100–150 keV)	(300–700 keV)	*P*	*P**
0·063	16 785·8	5 075·9	0·30239	1·0000
0·516	140 378·2	49 069·5	0·34955	1·1559
0·88	215 098·8	83 441·9	0·38792	1·2828

**P* normalized to 0·32039.

energy windows, have been obtained by Droullard[32] from the model holes at the Grand Junction Office of the U.S. Atomic Energy Commission. The instrument system employed automatic gain stabilization, single-channel analysers and three automatic digital scalers controlled from the same crystal clock time base. The third scaler automatically provided the ratio P, which was recorded on a strip chart simultaneously with I_2; also, the values of I_1, I_2 and P, measured in each 0·1 ft of hole length, were recorded by the digital printer.

Applications

It is nearly impossible to maintain or exactly to reset single-channel analysers to precisely the same window width and energy position each time a hole is logged. It is much easier to obtain relative values for each log where P is proportional to variations in uranium concentration but without a reproducible constant calibration value to relate P to absolute concentrations. Until the specialized instrument modules are developed, the uncalibrated Compton/photo log can be used in conjunction with a gross gamma log to indicate relative changes in grade and disequilibrium. When practical on-site calibration procedures can be developed, it will be possible to measure directly the absolute grade of uranium ore above some threshold value, possibly as low as 0·1 or 0·05%.

In spite of these limitations, Compton/photo (selective natural gamma) logs are being experimentally tested in the field, and improved instrumentation is being developed. In 1970 Czubek[33] also proposed a theoretical basis for interpretation of this log, and one of the companies which provide uranium logging services in the United States is developing a computer program to analyse Compton/photo test logs.

Neutron–neutron log

INTRODUCTION

The neutron–neutron log is being tested by the uranium industry as a supplemental log, but it is not yet routinely included in many programmes. The neutron log, which responds principally to the hydrogen in the environment, was developed by the petroleum industry to measure porosity in saturated rocks.[34,35,36] It can be used successfully in air-filled holes and holes drilled with foam or oil-based muds, as well as in water- or mud-filled holes of moderate diameter; in addition, a semi-quantitative response can be obtained through casing. The neutron–gamma log, which has been used successfully for petroleum and hydrology, is of doubtful value for uranium exploration because of anomalous natural gamma radioactivity.

The neutron–neutron log was introduced to the uranium industry in 1964.[8] Because of low sensitivity or sensitivity of available neutron detectors to gamma rays in simple systems, the method was not adopted. Development of the ^{3}H detector and well matched preamplifiers now makes the method practical.

THEORETICAL PRINCIPLES

A fast neutron emitted by one of the isotopic sources, such as PuBe or AmBe, will be slowed down by elastic collision until it is in thermal equilibrium with its surroundings or undergoes inelastic collision and capture by a nucleus of an atom. Capture of the thermal neutron is usually accompanied by emission of high-energy gamma rays, which is the principle of the neutron–gamma log. Hydrogen is very effective in moderating the neutron. (The number of collisions required to thermalize a neutron increases with the atomic number.) The probability of capture is inversely proportional to the energy of the neutron and increases with atomic number of the target atom.

Neutron–neutron logs are of two types and depend on detecting thermal and/or epithermal neutrons which have been moderated by hydrogen, most of which is contained in water. Measurements which include the thermal neutrons may be influenced by variable concentrations of elements having a high absorption cross-section to thermal neutrons. Detection of only the epithermal neutrons greatly minimizes spurious effects of thermal neutron absorbers;[37] thermal neutrons are excluded by a thin cadmium filter.

The neutron–neutron probe, in effect, responds to the amount of water per unit volume. The response can be calibrated in several ways. The fundamental calibration relates measured neutron intensity in counts per second to W, the grammes of equivalent water per cubic centimetre (g eH_2O/cm^3). More commonly, the response is related to per cent porosity for standard 'calibration' conditions, such as a limestone saturated with fresh water; corrections are required for other rocks or interstitial fluids.[37]

Most rock-forming minerals contain only a very small percentage of hydrogen by weight. It may be generally assumed that $eH_2O = H_2O$. The per cent moisture (H_2O) by weight is obtained by the relationship

$$\frac{W}{\rho_{bw}} \times 100 = \text{wt }\%\ H_2O \qquad (6)$$

where ρ_{bw} is the wet bulk density (ρ_{bw} may be determined from the gamma–gamma density log).

The response of the n-epithermal n probe, counts/sec/W, depends principally on the sensitivity of the detector, the neutron source strength and the spacing between source and detector. Corrections are required for non-standard hole size and for water, mud, mud cake and casing in the hole. The relative response of a decentralized short-spaced epithermal neutron probe and the effects of hole diameter, air- and water-filled, are illustrated in Fig. 18.

APPLICATION OF THE NEUTRON–NEUTRON LOG TO URANIUM EXPLORATION

The neutron–neutron (n–n) log supplements the standard gamma-ray, resistance and SP logging programme. It provides qualitative and quantitative information. Much quantitative information may be derived if the n–n log is used in conjunction with the gamma–gamma density and caliper log.

QUANTITATIVE ANALYSES

If the n–n probe is calibrated in counts/sec/W, as proposed, then weight per cent water is calculated from equation 6. The percentage of water by weight is required to correct the natural gross gamma-ray log to obtain eU_3O_8 per cent.

The parameter W can be used to determine the dry bulk density, ρ_d, from the in-place bulk density, ρ_{bw} from the relationship

$$\rho_{bw} - W = \rho_d \qquad (7)$$

The value of ρ_d may be used in several calculations. The tonnage factor, ft³/ton, used in calculating tons of ore from drilling, and, hence, the pounds of uranium, may be determined from equation 7 and conversion of units.

$$\frac{32{\cdot}03}{\rho_d} = \text{ft}^3/\text{ton} \qquad (8)$$

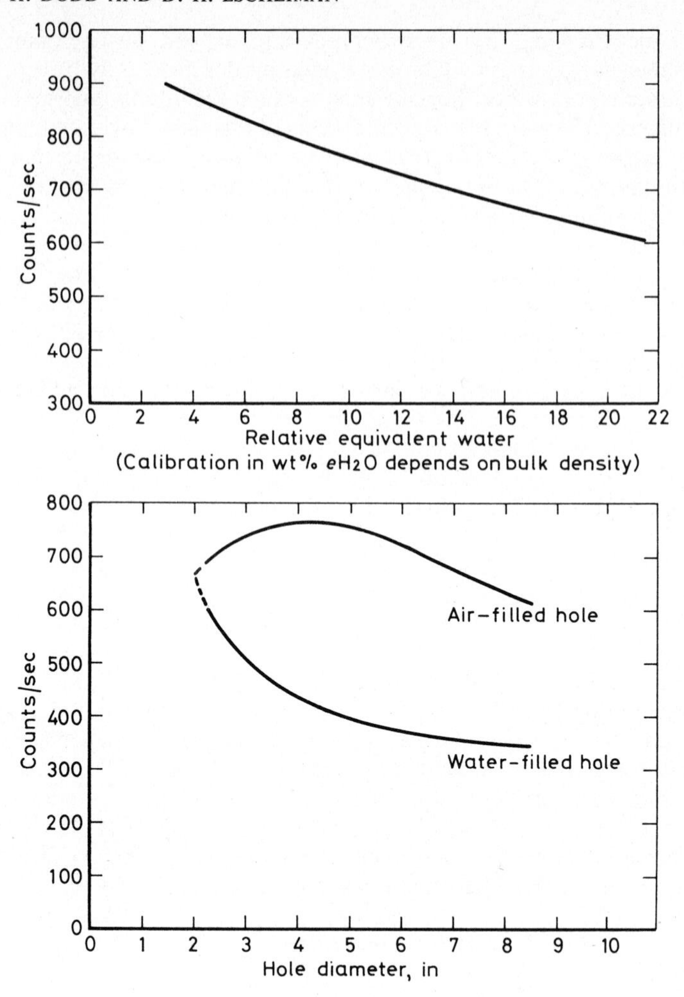

Fig. 18 Relative response of a short-spaced neutron-epithermal neutron probe

The relative or per cent porosity, φ, of unsaturated rock can be calculated by combining the *n–n* and gamma–gamma log. By estimating the grain density, ρ_g, from cuttings, φ is obtained from

$$\varphi = \frac{\rho_g - \rho_d}{\rho_g} \tag{9}$$

Porosity is a major parameter of rocks and can indicate the degree of sorting and relative permeability to the geologist.

The relative or percentage saturation may also be computed from φ, ρ_d and W. If

$$\text{Maximum } H_2O \text{ wt } \% = \frac{\varphi}{\rho_d + \frac{\varphi}{100}} \tag{10}$$

then

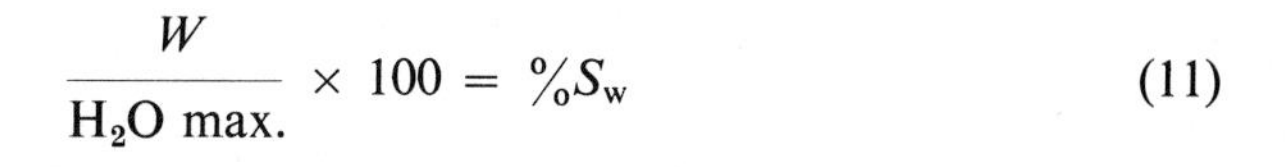

$$\frac{W}{H_2O\ \text{max.}} \times 100 = \%S_w \tag{11}$$

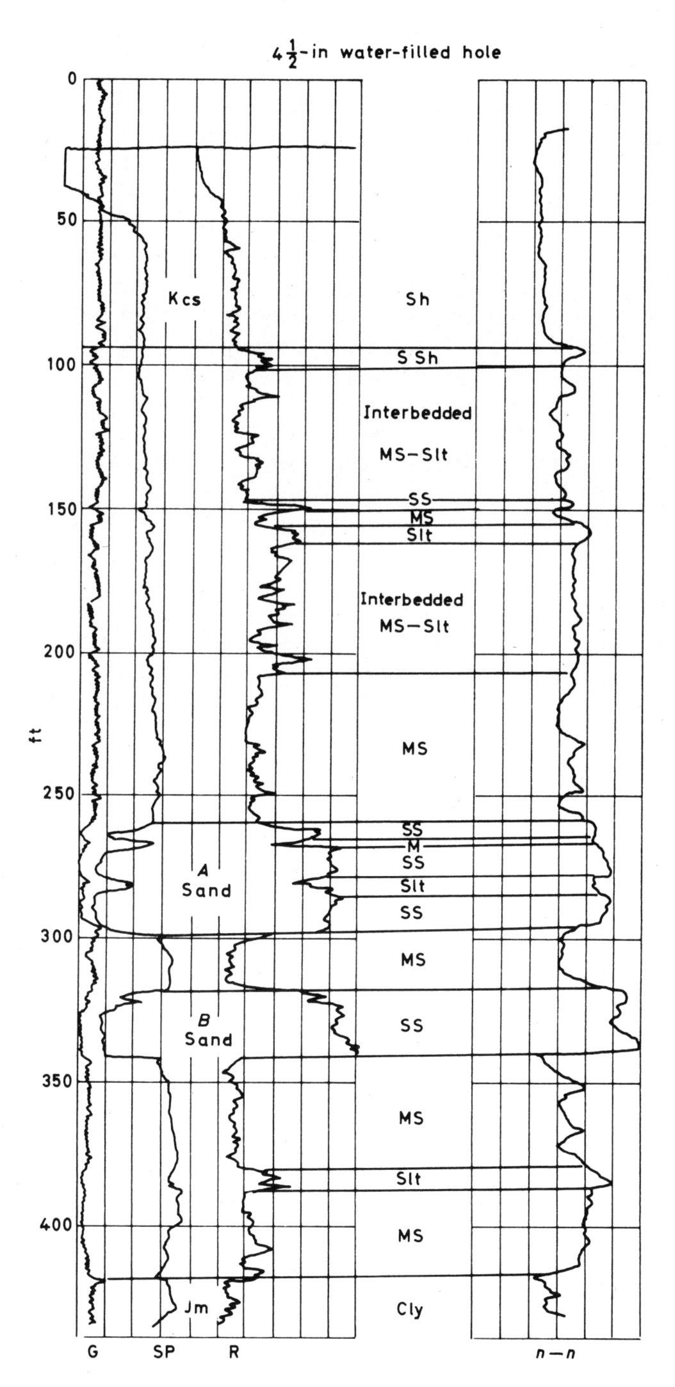

Fig. 19 The *n–n* log correlates well with single-point resistance log of typical continental sedimentary rocks

The value of S_w may be used for several purposes, including determination of φ from the gamma–gamma log for unsaturated conditions by substituting S_w for ρ_f, the fluid density, in the standard equation.

$$\varphi = \frac{\rho_g - \rho_b}{\rho_g - (S_w \times \text{Corr. } A/Z)} \tag{12}$$

The value of S_w is corrected for A/Z of water to obtain more accurate porosity measurements from the gamma–gamma density log.[7]

Qualitative lithologic information

The *n–n* log provides qualitative lithologic information similar to that obtained from the resistance or short normal resistivity log. Shale, mudstone and clay commonly have a greater porosity and contain more water than sandstone. The *n–n* log, therefore, indicates a sandstone by an increased neutron count rate, much as the resistance log indicates a sandstone by an increased resistance. The correlation between these logs is illustrated in Fig. 19.

The relationship between moisture content and porosity is commonly maintained in unsaturated rocks. Therefore, the *n–n* log can provide lithologic information above the water-table; the contrast between sandstone and mudstone is enhanced by air-filled holes. The *n–n* log becomes especially valuable for exploring arid regions where holes are drilled with air and resistance or SP logs are not available.

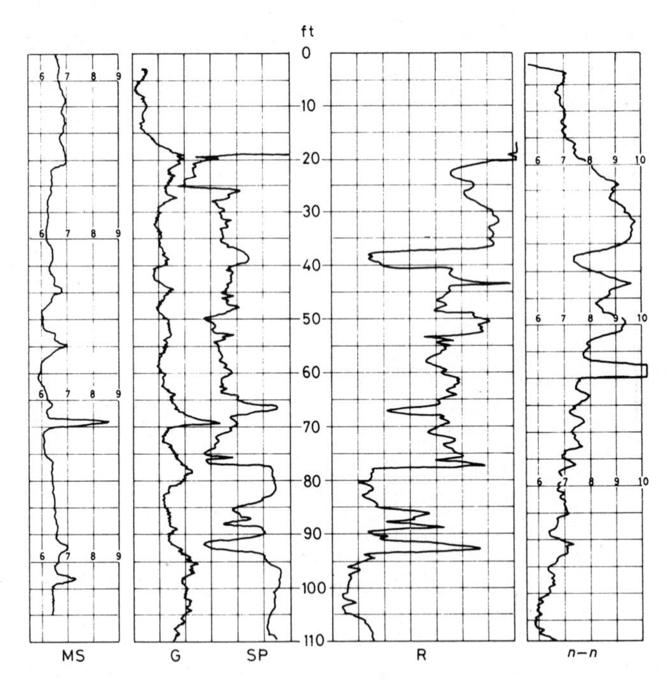

Fig. 20 Experimental high-sensitivity magnetic susceptibility log (*left*) indicates general low level with small local variations, which correlate with lithologic changes

The *n–n* and gamma logs can also be used to investigate cased holes which may have been drilled for other purposes.

Magnetic susceptibility log

INTRODUCTION

The inverse relationship between magnetic susceptibility and 'favourable' ground for uranium exploration has been described by Ellis and co-workers.[38] It is postulated that fluids related to uranium movement alter, to some extent, the intrinsic magnetic minerals in the host rock, thereby causing local negative anomalies. The values of magnetic susceptibility of normal sedimentary rocks are very low, in the range 5–300 $\times$ 10^{-6} cgs units. To measure small changes of perhaps 20–30% at such levels requires a very sensitive probe.

EXPERIMENTAL PROBE

An experimental sensitive magnetic susceptibility probe has been developed at the Grand Junction Office of the U.S. Atomic Energy Commission: this prototype will be used for preliminary field tests. The probe is 2¼ in in diameter and 47 in long. The probe shell is made from Delrin to minimize the magnetic influence from probe materials. The detector consists of an inductive coil in one branch of a balanced bridge, which is excited by a 1 kHz sine wave signal. The coil consists of two layers, a bifilar-wound layer over a single layer wound on a linear magnetic Ferrite core of high permeability. The probe is used with a standard three-conductor armoured logging cable and 30 V, dc, is supplied from the surface panel. The signal obtained varies from 0 to 10 V, dc.

EXPERIMENTAL RESULTS

The probe has recently been tested in a few holes in the field. The results are far from conclusive, but they are encouraging. Fig. 20 shows a magnetic susceptibility log obtained in a 4½-in water-filled hole in Tertiary continental clastic sediments. The average magnetic susceptibility for this section has not yet been determined, but is estimated to be about 60 $\times$ 10^{-6} cgs units or less. Small local anomalies occur on the log, generally at points of lithologic change, as indicated on the other logs. This log is from a 'red' and presumably 'oxidized' zone. A response some ten times greater was obtained in Tertiary rocks of another formation.

References

1. Broding R. A. and Rummerfield B. F. Simultaneous gamma ray and resistance logging as applied to uranium exploration. *Geophysics*, **20**, 1955, 841–59.
2. Hallenburg J. K. A résumé of spontaneous potential measurements. In *12th Annual logging symposium, Dallas 1971* (Houston, Texas: Society of Professional Well Log Analysts, 1971), I1–I17.
3. Hallenburg J. K. Personal communication.
4. Adler H. H. and Sharp B. J. Uranium ore rolls—occurrence, genesis and physical and chemical characteristics. In *Uranium districts of southeastern Utah: Guidebook to geology of Utah no. 21* (Salt Lake City: Utah Geological Society, 1967), 53–77.
5. Tittman J. and Wahl J. S. The physical foundations of formation density logging (gamma–gamma). *Geophysics*, **30**, 1965, 284–94.
6. Czubek J. A. and Guitton J. Les possibilités d'application de la méthode gamma–gamma à la détermination en place de la densité des minerais d'uranium. *France, CEA*-R2720, 1965, 32 p.
7. Dodd P. H. and Droullard R. F. A logging system and computer programme to determine rock density of uranium deposits. In *Radioisotope instruments in industry and geophysics* (Vienna: IAEA, 1966), vol. 2, 205–25.
8. Dodd P. H. and Droullard R. F. Some current concepts of nuclear borehole logging for uranium exploration and evaluation. Paper presented at 9th annual minerals symposium, AIME, Moab, Utah, 1964, 29 p.

9. Czubek J. A. Natural selective gamma logging. A new log of direct uranium determination. *Nukleonika*, **13**, 1968, 89–104.
10. Dodd P. H., Droullard R. F. and Lathan C. P. Borehole logging methods for exploration and evaluation of uranium deposits. In *Mining and groundwater geophysics/1967*, 401–15. (*Econ. Geol., Rep. geol. Surv. Can.* 26, 1969)
11. Pirson S. J. Environmental logging and mapping in the search for minerals. In *10th Annual logging symposium, Houston 1969* (Houston, Texas: Society of Professional Well Log Analysts, 1969), Il–147.
12. Dodd P. H. Neutron-neutron logging to assay U-235 in place. *U.S. A.E.C.*, 1961, unpublished.
13. Czubek J. A. Pulsed neutron method for uranium well logging. *Rep. Inst. nucl. Phys., Cracow* INP-732, 1971, 37 p.
14. Droullard R. F. and Dodd P. H. Gamma-ray logging techniques in uranium evaluation. In *Proc. 2nd U.N. Int. Conf. peaceful Uses atom. Energy* (Geneva: U.N., 1958), vol. 3, 46–53.
15. Scott J. H. *et al.* Quantitative interpretation of gamma-ray log. *Geophysics*, **26**, 1961, 182–91.
16. Crew M. E. and Berkoff E. W. TWOPIT, a different approach to calibration of gamma-ray logging equipment. *Log Analyst*, **11**, no. 6 1970, 26–32.
17. Scott J. H. Computer analysis of gamma-ray logs. *Geophysics*, **28**, 1963, 457–65.
18. Coulomb R. *et al.* Traitement par ordinateur des radiocarottages gamma dans les gisements uranifères. *France, CEA* N-1279, 1970, 29 p.
19. Grundy W. D. and Meehan R. J. Estimation of uranium ore reserves by statistical methods and a digital computer. *Mem. New Mex. St. Bur. Mines* 15, 1963, 234–43.
20. Patterson J. A., deVergie P. C. and Meehan R. J. Application of automatic data processing techniques to uranium ore reserve estimation and analysis. *Colo. Sch. Mines Q.*, **59**, 1964, 859–86.
21. Czubek J. A. Influence of borehole construction on the results of spectral gamma-logging. In *Nuclear techniques and mineral resources* (Vienna: IAEA, 1969), 37–53.
22. Danes Z. F. A chemical correction factor in gamma–gamma density logging. *J. geophys. Res.*, **65**, 1960, 2149–53.
23. Czubek J. A. Physical possibilities of gamma–gamma logging. In *Radioisotope instruments in industry and geophysics* (Vienna: IAEA, 1966), vol. 2, 249–75.
24. Czubek J. A. and Lenda A. Energy distribution of scattered gamma-rays in natural gamma-logging. In *Nuclear techniques and mineral resources* (Vienna: IAEA, 1969, 105–16.
25. Rosholt J. N. Jr. Natural radioactive disequilibrium of the uranium series. *Bull. U.S. geol. Surv.* 1084A, 1959, 30 p.
26. Tanner A. B. Physical behavior of radon. *Rep. U.S. geol. Surv.* TEI-700, 1957, 243–46.
27. King J. W. and Austin S. R. Some characteristics of roll-type uranium deposits at Gas Hills, Wyoming. *Min. Engng, N.Y.*, **18**, May 1966, 73–80.
28. Scott J. H. and Dodd P. H. Gamma-only assaying for disequilibrium corrections. *U.S. A.E.C.* RME-135, 1960, 20 p.
29. DiGiovani H. J., Graveson R. T. and Yoli A. H. A drill hole scintillation logging unit, type TU-5 A. *U.S. A.E.C.* NYO-4503, 1952, 24 p.
30. Lyubavin Yu. P. and Ovchinnikov A. K. Gamma radiation of uranium and its daughter products in radioactive ore bodies. *Vopor. rudn. geofiz.* no. 3, 1961, 87–94; *U.S. A.E.C.* trans. 6830.
31. Blake R. W. Personal communication, 1965.
32. Droullard R. F. Unpublished report, 1968.
33. Czubek J. A. Possibilités d'interprétation des diagraphies de radiocarottage gamma naturel sélectif. *France CEA* R-4061, 1970, 27 p.
34. Tittle C. W. Theory of neutron logging 1. *Geophysics*, **26**, 1961, 27–39.
35. Lebreton F. *et al.* Formation evaluation with nuclear and acoustic logs. In *4th Annual logging symposium, Oklahoma 1963* (Tulsa, Okla.: Society of Professional Well Log Analysts, 1963), IX-1–IX-19.
36. Czubek J. A. Neutron methods in geophysics. In *Nuclear techniques and mineral resources* (Vienna: IAEA, 1969), 3–20.

37. Tittman J. *et al.* The sidewall epithermal neutron porosity log. *J. Soc. Petrol. Engrs*, 1965, 1–20.
38. Ellis J. R., Austin S. R. and Droullard R. F. Magnetic susceptibility and geochemical relationship as uranium prospecting guides. *U.S. AEC*-RID-4, 1968, 1–21.

DISCUSSION

J. M. Miller commented that one answer to the problem of disequilibrium in borehole logging lay in the application of down-the-hole neutron activation. Information on progress in that field in the U.S.A. would be welcomed; he asked if it were thought that variations in sediment moisture content would present a serious problem.

Mr. Dodd said that progress in the U.S.A. in neutron activation for uranium was very small; two companies had investigative programmes. Variations in rock moisture would not present a serious problem. Czubek* proposed measuring the prompt epicadmium fission neutrons from ^{235}U or thermal neutrons from delayed fission of the products of ^{235}U fission. It appeared that pulses of 10^9 n/sec would be adequate to measure ordinary ore grade. The same pulsed neutron source used to induce fission of ^{235}U could be used to evaluate the moisture in the formation. If an isotopic neutron source were used, measurement of the thermal or epithermal neutrons prior to irradiation (by logging in the reverse direction) would evaluate the moisture.

L. Løvborg noted that it had been proposed that borehole logging be done by neutron activation, followed by counting the delayed neutrons resulting from fission ^{235}U. Would that technique be technically possible now that powerful neutron sources in the form of several milligrammes of ^{252}Cf could be purchased from the USAEC?

Mr. Dodd said that, until recently, that approach had not been seriously investigated because it required a very active source, probably 10^9–10^{10} n/sec, or greater, to achieve acceptable precision of analysis at ore grade, and more active sources would be needed to measure ppm levels in rocks. It was only recently that ^{3}He detectors, which had high sensitivity to neutrons and almost none to γ, had been developed; those would improve the possibilities for developing the method. He was concerned, however, at trusting the average logging operator with such a powerful source. Available ^{252}Cf sources could provide the necessary neutrons. Investigations being made by Kerr-McGee would be reported soon.†

R. J. Chico asked how many cored holes should be drilled to verify an orebody determined by means of gamma-log data, assuming a 'roll-front' type of orebody with two superimposed 'roll fronts', one in equilibrium and another out of equilibrium.

Mr. Dodd said that he would emphasize the importance of coring very early in the programme in order to correct, as far as possible, the effects of disequilibrium. Normally, not more than 10%, and usually 5%, of development drill-holes were cored in the western U.S.A., but that proportion depended largely on the complexity of the geology.

L. W. Koch enquired the moisture content of the ore material surrounding the AEC calibration holes. Were they sufficiently dry for field logging units to give conservative results under natural, water-saturated conditions?

Mr. Dodd said that they kept the calibration holes filled with water in order to maintain radon equilibrium, but standard conditions involved an air-filled hole. The calibration hole contained 12% moisture, which was fortunately about average for sandstones. A correction factor might be applied thus:

$$\frac{100 - \%\text{H}_2\text{O in environment}}{100 - \%\text{H}_2\text{O in model}}$$

That explained why they liked to do n–n logging in unusual environments.

Dr. U. Klinge asked the authors for their experiences of borehole logging in very cold areas, i.e. in or near the permafrost layer. Mr. Dodd replied that logging boreholes in arctic areas with permafrost conditions should be performed shortly after drilling. It was possible, however, to prevent the holes from freezing by filling them with a liquid having a low freezing point, e.g. kerosene.

*See reference 13 on page 274.
†*Californium-252 Progr.*, 1971, no. 7, 24–5; no. 8, 24.

B. L. Nielsen, with special reference to gamma spectrometry, which might be done by airborne or surface measurements, asked about possible disequilibrium in the decay series of uranium and thorium. What were the possible mechanisms of disequilibrium? For instance, was it due to leaching of the uranium? Was it due to the escape of radon or was it because of the age of mineralization, which had not been emplaced long enough to establish secular equilibrium? In other words, did one measure sometimes more or sometimes less uranium in relation to what was actually present in the rock?

Mr. Dodd replied that disequilibrium in the thorium decay series seldom presented a problem because of the relative chemical stability of thorium and thorium minerals, and the short half-lives of the decay products normally resulted in very minor disequilibrium, if any. All types of disequilibrium in the uranium series which Mr. Nielsen mentioned did occur, but the most common, particularly in soils and outcropping rocks, was loss of radon and differential leaching and removal of uranium; the first caused underestimation and the second overestimation of uranium concentration. The problem of underestimation caused by recently deposited 'new' uranium was less common, but it occurred occasionally.

P. L. Kitto asked if, during the data processing, there was continual monitoring of the calibration parameters or if they were considered essentially constant after the initial calibration.

Mr. Dodd replied that they liked to have the original calibration done at Grand Junction. A client would expect a logging company to check their equipment periodically (about once a week) while they were working for him by logging standard model holes located at Casper, Wyoming, Grants, New Mexico and George West, Texas. Those periodic checks provided updated values of factors for equipment.

Application of computers to the assessment of uranium deposits

R. Coulomb

Commissariat à l'Energie Atomique, Fontenay-aux-Roses, France

681.3:622.013.36:553.495.2

Synopsis

Computers are now in general use at the Commissariat à l'Energie Atomique (CEA) for all stages of uranium prospecting as well as for evaluating and mining uranium deposits. Mathematical models and algorithms have been developed for calculations which would be impossible by the use of human resources alone.

By means of sophisticated routines, data processing is under the control of mining engineers or geologists: data input, calculations and output are displayed either on digital plotters or on display tubes. For output, the process can be performed in a man–machine interactive mode, and it is therefore possible to achieve an optimal solution.

Hardware and software

The CEA has a powerful network of interconnected computers operated by the Département d'Informatique. The main computer, an IBM 360–91 (4000 K), on a multi-programming operating system (MVT), is situated at Saclay. An IBM 360–75 (800 K) controls the tele-processing network (SASP system).

Fig. 1 IBM 2780 terminal

Major terminals are located at the various centres: Fontenay-aux-Roses, Grenoble and Cadarache are equipped with 360–50 or 360–20, linked to Saclay by microwave circuits or multi-channel cables. These centres are generally equipped with graph plotters, display tubes, card readers, perforators and high-speed printers. Main users have medium terminals linked to Saclay by special lines (transmission speed, 4800 BIT/sec); thus, the Service de Minéralogie has an IBM 2780 terminal (Fig. 1) at Fontenay-aux-Roses. Hence, a job launched by Fontenay, included in the batch process at Saclay, can readily be sent and printed at Cadarache (800 km away), where GAM has an office responsible for overseas prospecting.

The Département d'Informatique makes all its facilities freely available. An important library of standard sub-routines is accessible to all users. A number of languages are available (FORTRAN H and G, COBOL, ALGOL, PL1), and for mining projects a whole range of special software has been devised by a group of specialists from the Direction des Productions.

Data storage and retrieval

GENERAL PRINCIPLES

Data storage is an essential but tedious task, particularly when, as at the CEA, it involves collecting information acquired over a period of years and, at the same time, coping with the daily arrival of a mass of new results.

The CEA has not attempted to design an ideal and unique data bank capable of meeting all problems, past, present and future. The data libraries are an evolving concept and are continually being adapted to cope with the expansion of work and to take account of technical progress. Specialized data files are being progressively integrated into the broader range. This is done periodically by suitable programs and creates no problems for a team familiar with programming techniques with a high-speed computer, with a large memory store, at its disposal.

Information is never punched twice manually as it is the function of the machine to carry out the task of re-punching. The program library can be used with any type of data files.

TYPE OF DATA

Data can be qualitative or quantitative. A geologist or petrologist provides highly valuable *qualitative* data. It is perfectly possible to store and use this information in a computer without coding it or changing its meaning. Geological language is not always precise, but it can be simplified and summarized by use of key-words and standardization of spelling and names (geographical locations, petrographic descriptions and nature of samples). The CEA have been recording this type of information in plain language over a long period.

Quantitative information is accepted in any form. Although it has been CEA policy to standardize data presentation and input format, new methods are not imposed on users: the changeover must be made by persuasion. All documents with suitable figures are punched on tapes by operators in free format in alpha-numeric sequences.

Several types of punching machines have been used (often secondhand) and any punch codes (IBM, Bull, ICT, Ferranti, etc.) are accepted. Punch equipment is progressively being replaced by a single type.

Non-continuous radiometric logging of drill-holes and face-scanning techniques are still commonly employed, but most CEA data comprise continuous recording radiation logs of drill-holes from developed deposits. The log traces must be digitized. Two systems are available, being used according to the complexity and length of the trace. (1) Point by point digitization on a plotter: an operator traces the curve with a pointer and the *xy* coordinates are automatically punched and stored on magnetic tape. At Saclay a Benson plotter linked to a PDP computer is used, but the bulk of series work is done by the Compagnie des Services Numériques. (2) Automatic scanner digitization: a

Fig. 2 Automatic scanner digitization

Belinogramme type of system is used for designs traced in Indian ink on translucent paper.

Several thousand gamma-ray logs have been digitized by these processes (Fig. 2).

Digital field equipment which provides a perforated tape is in course of development. This includes a mobile vehicle-borne operational unit, designed by the Electronics Section, which has now been working for some months.

CHECKING, CORRECTION AND STORAGE

Because the collection of data at CEA is highly decentralized and apparently

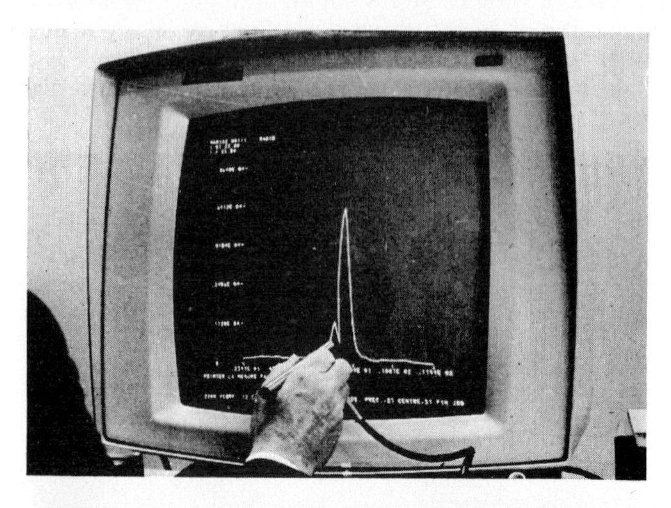

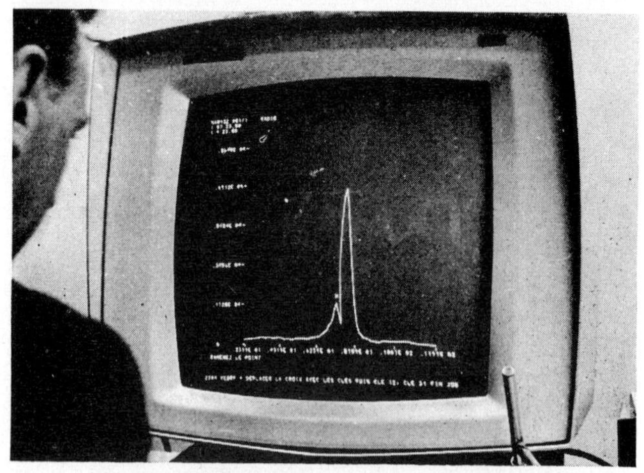

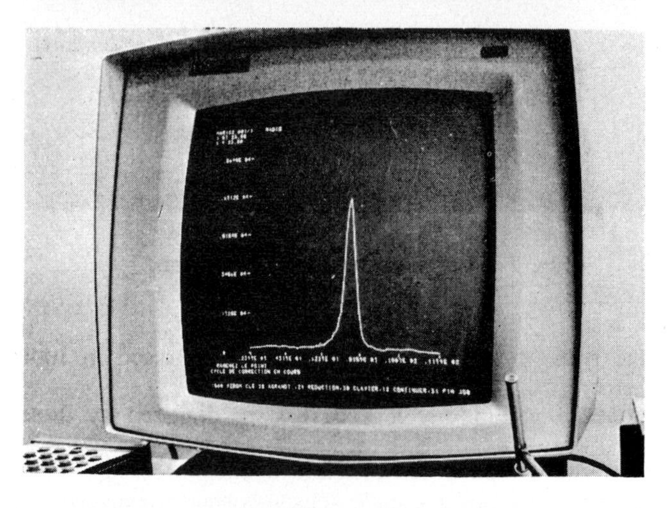

Fig. 3 IBM 2250 graphic display unit

uncoordinated, it is necessary to standardize and check data prior to computing. Checks are carried out by a small number of staff with personal experience in the field of uranium prospecting. A preliminary check is carried out by computer routines. Codes and sequences are standardized and errors are identified by statistical or logical tests.

The synthesized data are then displayed on an IBM 2250 console. This unit includes a cathode-ray tube on which display can be generated by the computer: (*a*) the engineer interacts with a series of graphic display panels by use of a light-pen (by touching the face of the screen); (*b*) a typewriter keyboard and series of function buttons provide direct access to the computer for the introduction of missing data and the execution of specific programs.

All radiometric logs are checked in this way and corrected (Fig. 3). Surfaces, elevations and cross-sections are also displayed and examined by an engineer, assisted by a geologist. A microfilm is taken of each step on an IBM 2280. Within a few hours (representing only a few seconds of computer time) hundreds of well logs, representing thousands of elementary items of data, can be checked. In general, all such operations are carried out disc to disc, tape to disc or disc to tape. Long-term storage is usually on magnetic tape.

RETRIEVAL

Programs are used for access to all or part of the CEA data files. Retrieval can be effected by reference to a number or even by syntactical analysis of records containing qualitative information. This is particularly useful for the retrieval of rock samples in geochemical studies.

For ore-reserve estimation, data files are arranged by reference to orebodies. Direct selection from a list of names appearing on a 2250 display panel is possible with the use of a light-pen (Fig. 4). Similarly, it is possible to choose or eliminate specific data.

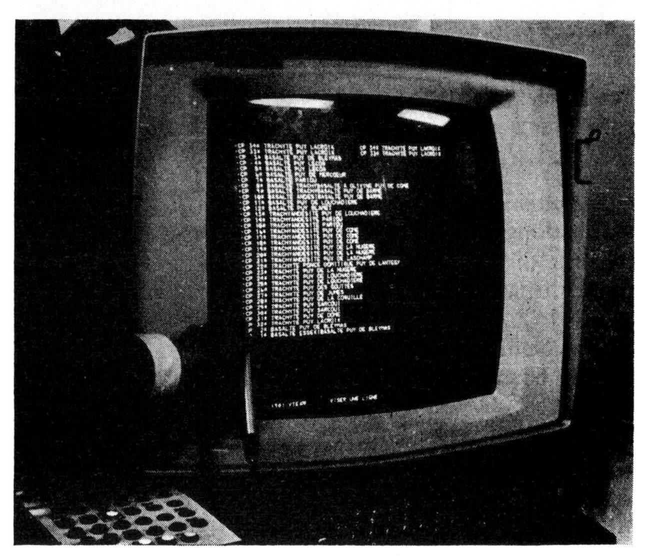

Fig. 4 Sample selection by light-pen

Uses

EXPLORATION

Computers and graph plotters are extensively used for automatic contouring. The theoretical and mathematical bases of these programs are trend surface

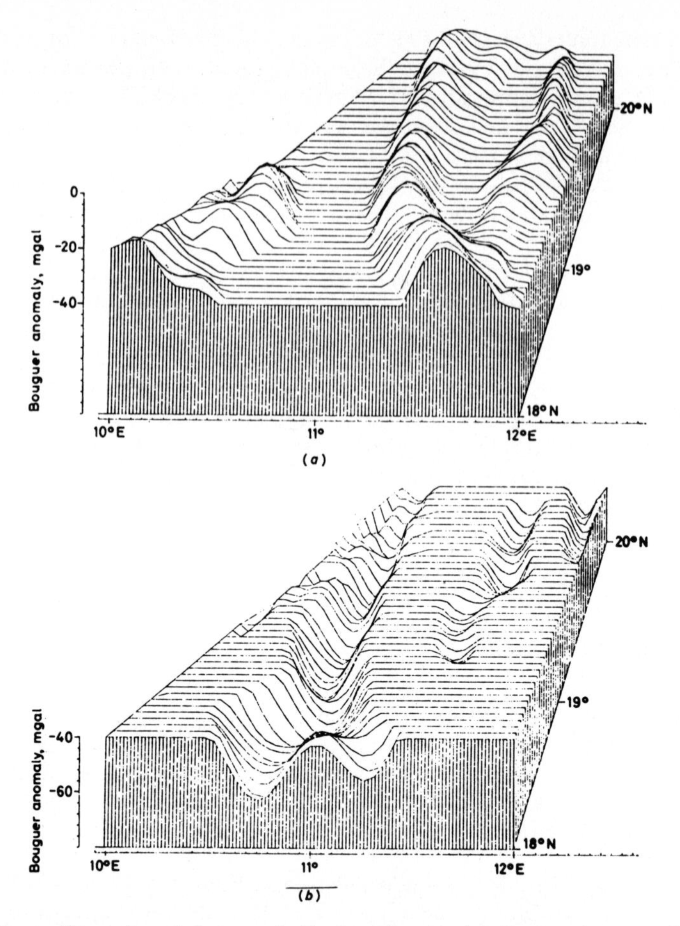

Fig. 5 Three-dimensional data and display forms: (*a*) Bouguer anomaly above −40 mgal; (*b*) Bouguer anomaly below −40 mgal. From Louis*

analysis or geostatistical methods (kriging, etc.). The results of many geophysical investigations, including gravity, magnetometric and aerial radiometric surveys, have been digitized and processed. Three-dimensional and perspective diagrams can be displayed, and parameters can be changed to provide different views (Figs. 5–10). The programs can also be used for the display and interpretation of geochemical prospecting data. Stratigraphic contour maps and cross-sections can be easily plotted by use of the results of resistivity logs.

ORE-RESERVE ESTIMATION

The estimation of ore reserves is usually carried out by employing gamma logs of surface drill-holes. A minimum number of core samples or cuttings is necessary for statistically significant results to be obtained.

Uranium is not always in equilibrium with radium, and gamma radiation detected by Geiger–Müller counter is mainly due to radium and its daughter products. Thus, the problem of estimating uranium reserves involves deduction of the distribution of radium content down the drill-hole from the total gamma log; determination of the distribution of uranium from knowledge of the

*Louis P. Contribution géophysique à la connaissance géologique du basin du Lac Tchad. Thesis, ORSTOM, Paris, 1970.

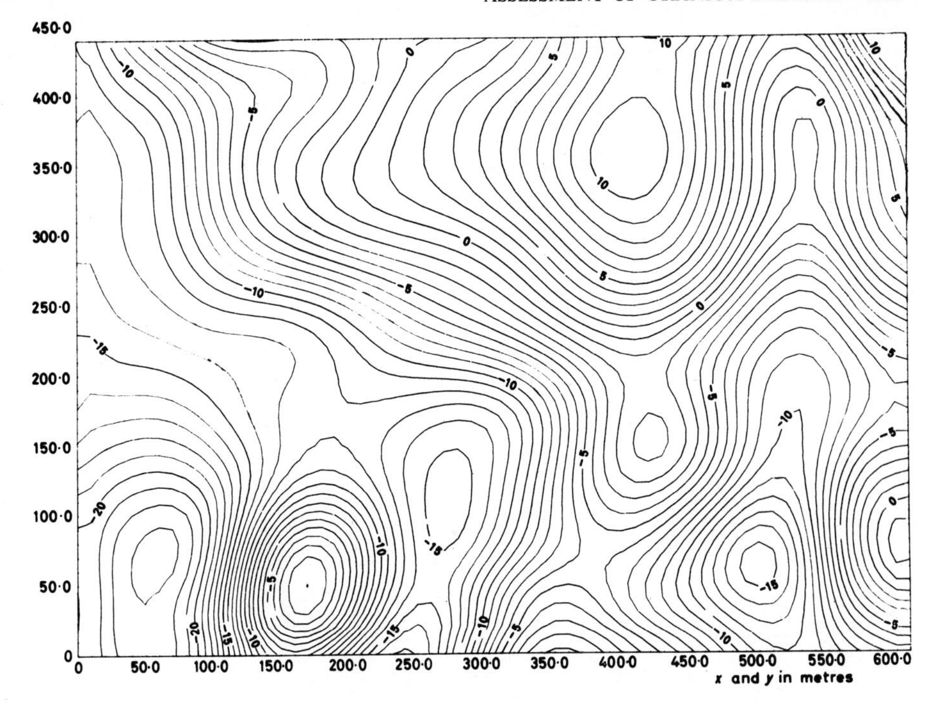

Fig. 6 Magnetometry: regional anomaly map

distribution of radium; extension of the information by further gamma-ray logging over the entire orebody in order to arrive at an overall estimate of grade and of the spatial distribution of the mineralization; and, finally, setting up a mining project based on exploitation criteria and economic considerations.

All the calculations involved in these operations are now undertaken by computers.

The 'deconvolution' principle

The radiometric log comprises total gamma activity down the drill-hole. The origin of the radioactivity is a combination of the influences of the radium content on the whole area, called a 'convolution'. 'Deconvolution' is the search for the radium content which caused the measured value of radioactivity. Normally, the problem is difficult to solve and it is assumed that the radium content is distributed in homogeneous thin layers. Finding the best fit for the radium concentration of these layers leads to the resolution of very large linear systems. The equations can, however, be solved by computer. When these calculations are performed, such factors as the characteristics of the bore tubing, the probes used and rock density are taken into account.

Comparison of the values for the radioactivity 'reconvoluted' (calculated from the start of the radium content arrangement found) and of the value measured allows a continuous check on the input data, assumptions and external conditions (characteristics of layers). Corrections can be carried out at any time (Fig. 11).

Regression analysis

The physico-chemical equilibrium of uranium/radium may be taken as a characteristic of a deposit (or a part of this deposit). Thus, there should be a good degree of correlation between the logarithms of the uranium and radium contents. It is therefore wasteful to analyse the uranium content of all the drill core from a deposit. The existence of a certain number of drill-holes for which there are both core and gamma logs enables the regression parameters to be calculated. From this it is possible to determine the distribution of uranium in boreholes for which only gamma logs are available. The calculations can be solved entirely automatically, but the results must undergo careful

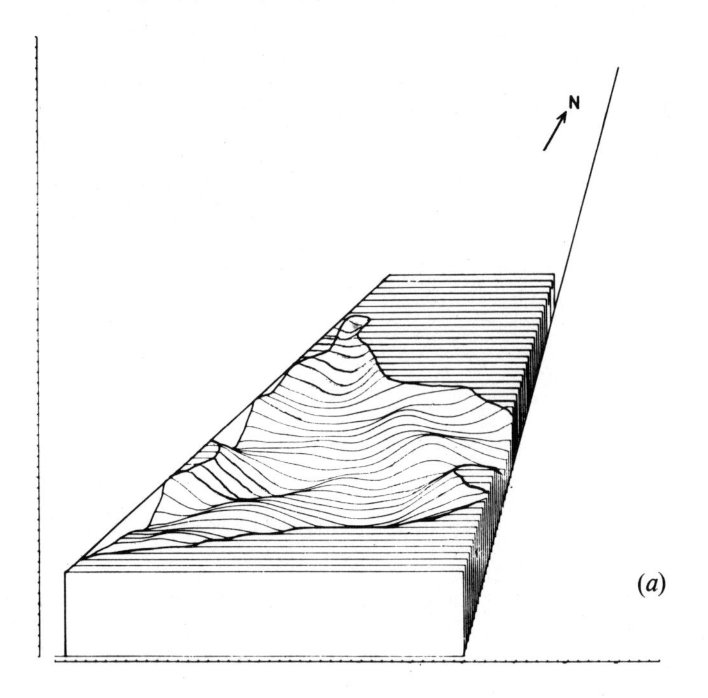

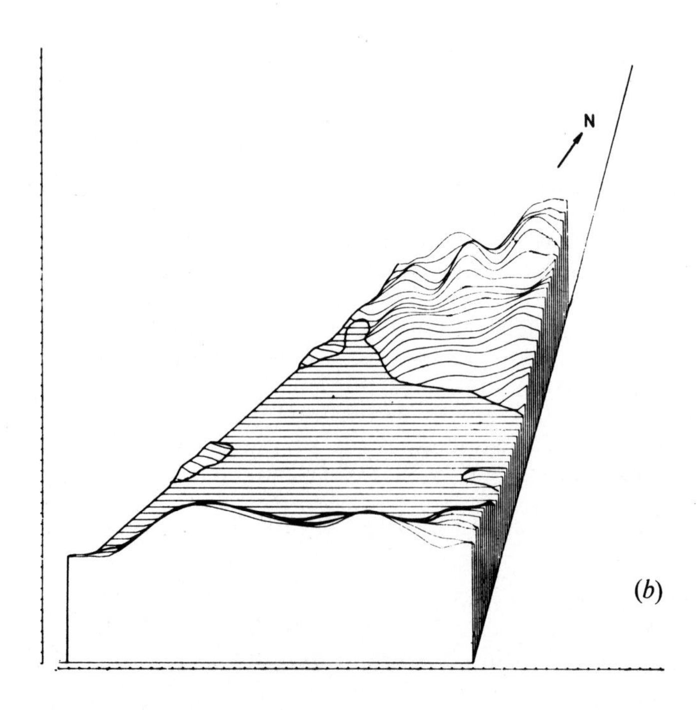

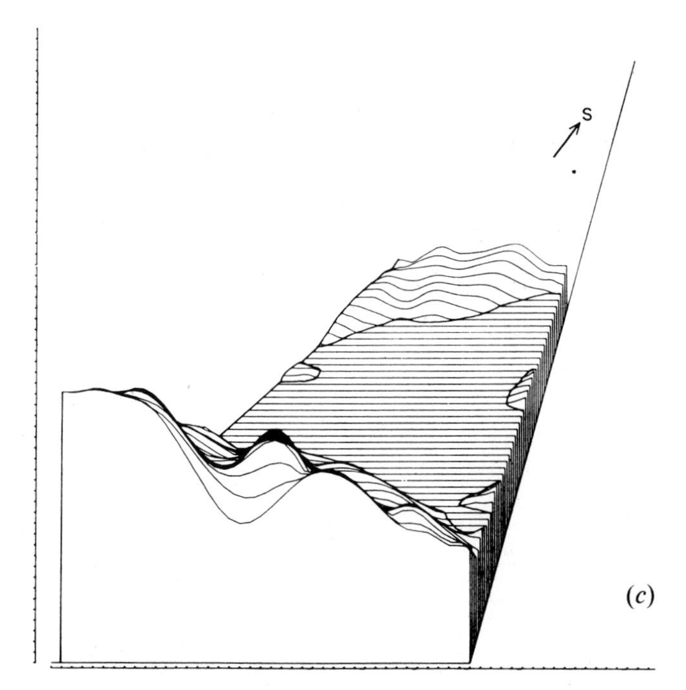

Fig. 7 Resistivity: three-dimensional data

scrutiny by a geologist to eliminate erroneous chemical contents and to check the assumption of a unique population. It should be noted that the proportion of borings sampled can be optimized.

Reduction of the number of core samples necessary to evaluate a deposit results in considerable economic gain, and the procedure results in a better knowledge of the actual values of the content of the orebody. The procedure described is used in the study of all the African deposits investigated by the CEA and several deposits in Metropolitan France. The advantages include speed, flexibility and accuracy.

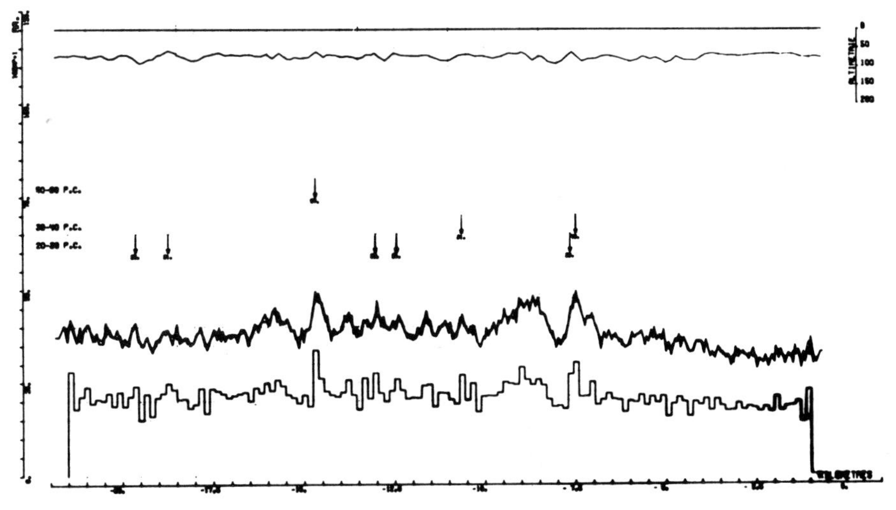

Fig. 8 Profile for each line flown to show count rate against distance

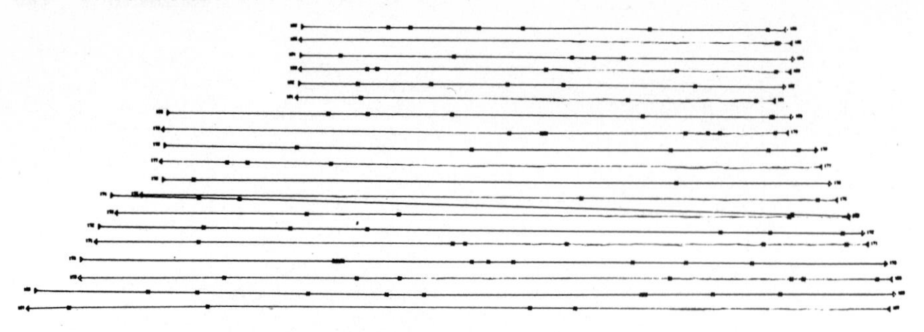

Fig. 9 Plots of flight lines superimposed on base map

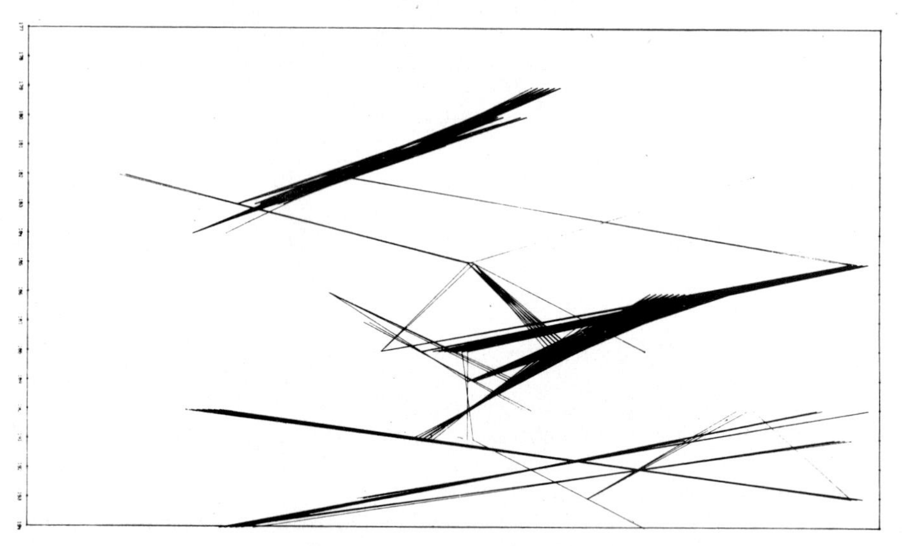

Fig. 10 Analysis of faults

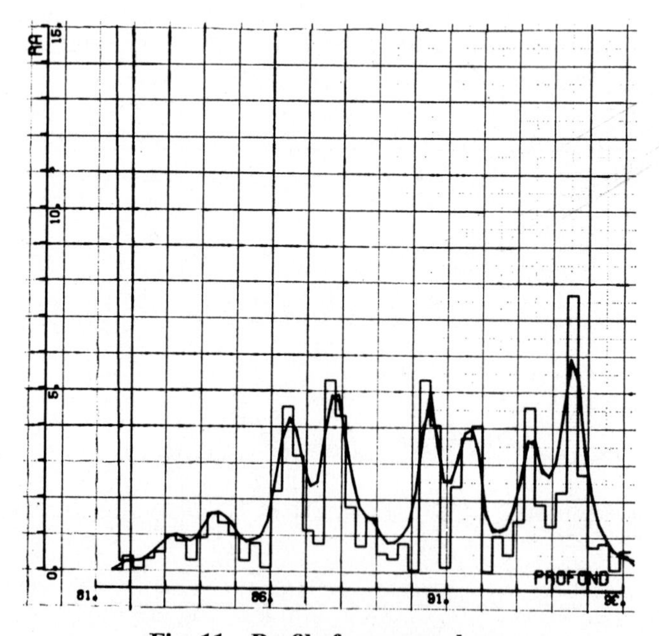

Fig. 11 Profile for gamma logs

Estimation procedure

Estimation involves the calculations of a weighted sum. The terms of the sum are the exploitable accumulation surveyed down the length of the boreholes: weighting is based on zones of influence of the boreholes. The program takes into account the conditions imposed by the mining engineers and applied to each drilling, and provides the following as output: (1) the tonnage of exploitable ore; (2) the average content of valuable elements in such ores; and (3) the average content of elements of significance to processing.

Sub-totals can also be given from which certain proportions can be calculated (e.g. the percentage of workable ore located above a certain level or the effect of the inclusion of a certain amount of undesirable elements).

Computer processing offers great facilities for calculating the variogram of a formation and for finding the geostatistical reference scheme. This may involve a lengthy statistical analysis—for example, analysis of variance, adjustment for statistical laws, correlation between metallic elements, etc.

Mining projects

The use of powerful information resources allows the investigator to go further than mere estimation. It makes possible the determination of the most probable site of the ores involved, with all their characteristics. Manual tests have clarified the system's philosophy, but the complete calculation of such large numerical models could only be processed by computers. Presently, the programs devised by the mining engineers allow a steady advance to optimum exploitation: for example, various models of open-pit mines extend to the complete automatic design, eventually in perspective.

Conclusions

The methods described have been able to achieve success because of complete collaboration between geologists, mining engineers and data-processing specialists. Visual presentation has greatly helped mutual understanding. Geologists have fully realized that computers offer an important aid, even if they do not replace the art of interpretation.

DISCUSSION

Dr. J. W. von Backström asked if the CEA programs were available on an exchange basis or by request or purchase. The author replied that they were not commercially available, but arrangements might be made to process data in Paris.

In reply to questions by B. L. Nielsen and Professor M. Dall'Aglio, the author stated that the curve follower could only deal with smooth curves, and that it was necessary to transfer to a transparent medium. The total cost of the computing equipment was about F10 000. The quantity of radium actually present in a narrow layer was not an indication of the overall grade. It was easier to compare that to the chemical results, since it was expressed as equivalent uranium, assuming a state of equilibrium.

Dr. F. Q. Barnes asked if, in plotting from the analogue curves derived from drill-holes, the spacing of the plotting points was varied or if they were at present spacings along the length coordinate. The author said that the number of plotting points varied along the distance coordinates.

In reply to a further question by Dr. Barnes, the author said that a standard width of 2 mm was taken on the plotter, but that that interval was adjustable within certain limits.

With regard to correction of the data by modification of the curve on the display screen, queried by D. L. Dowie, the author said that the geologist was the only person who could decide which cross-section he must investigate. It was very easy to correct or smooth data visually, although it might not be apparent from a numerical display. Data editing was very important.

Dr. A. G. Darnley agreed that editing digitally recorded data prior to computation was time-consuming, but it was also very necessary. The description of the use of a light-pen, with a graphical display, for the purpose of editing data sounded very attractive.

L. W. Koch said that he would like to utter a caution against the use of a cathode-tube display for correcting geological data. It might introduce unnecessary human bias and could be done better through built-in safeguards in the computer programs themselves. On the other hand, the cathode tube was a good place for introducing human judgment; but such manipulation should be recognized in the end-results. The author agreed that it was important to check for initial errors.

Uranium search in Australia, 1968: a case history

J. L. Montgomery

Saskatoon, Saskatchewan, Canada

550.8:553.495.2(94)

At the start of the Western Nuclear Mines search programme all known commercial deposits in Australia (at Rum Jungle, Alligator River and Mary Kathleen) were held by other uranium mining groups. There was a commitment with regard to uranium deposits in Queensland and no known Australian uranium deposits were available for purchase. Favourable ground adjacent to economic deposits was held by other uranium mining groups, but many areas with favourable geology were still open, and for much of the country there was no record of any uranium search having been undertaken. Nevertheless, there were geological maps outlining favourable lithologies.

A programme of uranium search was initiated with the use of the airborne scintillometer in areas of suitable topography in Queensland, Northern Territory and Western Australia, followed by ground examination, where possible, to evaluate air anomalies.

Design of search method

The object of the search was the location of an economic uranium deposit, but certain environmental factors which affected the design of the search method had to be considered. (1) Most of Australia has been exposed to surface weathering over a long period of time: hence, surface leaching of uranium is an ever-present possibility. Ground follow-up work, therefore, usually involves acquiring subsurface information at an early stage. (2) Many large economic sources of uranium have been located in coarse clastic formations—for example, the Witwatersrand, Blind River and Colorado Plateau deposits. (3) Known commercial deposits in Australia occur within relatively large areas of anomalous radioactivity. A difficulty with airborne search in the Mt. Isa area, for instance, is that anomalies may be 50 square miles in area. (4) Known Australian uranium deposits are concentrated in Precambrian rocks, particularly those of Archaean and Lower Proterozoic age. (5) The economics of follow-up work require an appreciable concentration of prospects, or one large prospect, to warrant the presence of crews and equipment in one area for some time. (6) Many companies stated their intention to commence uranium exploration and there was a need to get an effective programme under way.

The search method, therefore, was designed as an Australia-wide sampling procedure, and included an attempt to select samples over coarse clastic sediments of Precambrian age.

Scope of search method

A Piper Cub aircraft was chartered to fly the airborne scintillometer and a senior geologist over selected areas in the Northern Territory, Western Australia and, later, Queensland. 300 h of traverse was flown at an average hourly cost of $15.

As appropriate, a helicopter was chartered to locate anomalies on the ground and to land a senior geologist to carry out an evaluation. The anomalies examined were relocated from maps and photographs showing anomalies which had been located originally by the fixed-wing aircraft flown by the Company, by the Bureau of Mineral Resources or by other mining groups. 120 helicopter hours were flown.

Leases were pegged and other tenements were acquired over areas considered likely to have economic possibilities.

Procedure

The airborne instrument was used for traversing in the same way that a ground detector is used, namely to detect increases in radioactivity and then immediately to define the radioactivity as to area, intensity and spectral characteristics by flying at lower ground clearance and speed over the anomaly. The location of the anomaly was then recorded. The procedure for the helicopter work involved use of the airborne instrument to relocate the anomaly and to define its area, intensity and spectral characteristics. Landings were then made in chosen areas and an attempt made to evaluate the deposit with regard to the following criteria. *Geophysics*—(1) correlation with airborne anomaly; (2) intensity of radiation; (3) areal disposition of radiation; and (4) spectral characteristics of radiation. *Geology*—(1) stratigraphy, particularly permeability and environment of deposition; (2) petrology, particularly of igneous rocks; (3) mineralogy, particularly uranium minerals; (4) tectonics and structure, particularly relative ages; (5) geomorphology, particularly for age of exposure and superficial cover; and (6) economic geology, particularly (*a*) evidence for leaching or enrichment and transportation of materials and (*b*) correlation of radioactivity with a host rock or structure. *Economics*—(1) possible minimum tonnage of anomalous host rock; (2) possible continuity of anomalous host rock for mining; (3) type of follow-up work needed; (4) access and suitability for testing; and (5) possibility for concentration of uranium at economic levels for the method envisaged.

ADVANTAGES OF THE PROCEDURE

Work was started immediately with a method that was within the technical and administrative capability of the company. Small initial responses could be immediately investigated and developed into interesting anomalies by flying at lower ground clearance and speed.

Costs were low for anomalies sampled because the search was directed away from unfavourable geology, such as alluvial plains, salt pans, sand dunes and Mesozoic areas. Ground likely to be productive was acquired immediately and security of information was not a problem. Personnel could be trained in uranium search and follow-up work as necessary.

Mobility was excellent, and, although the sampling programme was Australia-wide, it was flexible enough to incorporate sampling at small isolated areas of favourable geology.

DISADVANTAGES OF THE PROCEDURE

There is no permanent tape record of the airborne survey results, and there is poor positional information for re-sampling or follow-up work because of the small map scales employed (1:250 000 or sometimes 1:500 000).

Psychological and physiological limitations arose from operator fatigue and difficult living conditions. Relative costs for transport of men and materials were high because of distance and mobility.

Results

An example of the type of information recorded in preliminary examinations of 500 radioactive occurrences is given in Table 1.

Table 1 Specimen record of preliminary examinations of airborne radiometric anomalies

Region	Counts times background Air	Ground	Sample	Remarks
Ashburton 11	3		30A	Hardey Sandstone, near Basement granite where Basal group crops out in a dome: fairly sharp anomalies, which, although not showing high counts, do reflect radioactive concentrations in clastic rocks. Ground prospecting recommended
Ashburton 12	4		30B	As above
Katherine 1			37	Sandstones with pebble beds; uniform count where examined. No anomaly located. Ground prospecting recommended
Katherine 2	6		45	Uranium minerals in fractures. Tuffs and volcanics of Upper Proterozoic *A.B.C.* prospect. Ground prospecting recommended
Katherine 3	3	3	56	No apparent cause. High counts over dolomitic argillite, light coloured. Uniform over large area. Nothing more recommended
Katherine 4	4	4	76	Edith River volcanics: landed and prospected thoroughly several areas of these volcanics, but found uniform counts over large areas. Nothing more recommended
Fletcher's Gully 1	3		87	Cenozoic cover: pilot could not land. Nothing more recommended
Fletcher's Gully 2	4		99	Depot Creek sandstone: pilot could not land. Flat plateau country with many trees. Ground follow-up when ground free. Recommended

Discussion of results

GEOLOGY

Stratigraphy

'Archaean' and 'Lower Proterozoic' are the names given, in most of the northern Australian area sampled, to strongly folded and faulted geosynclinal systems with volcanics and igneous intrusions. Age determinations are few. 'Upper Proterozoic' refers in most of the sampled area to relatively undisturbed sediments and volcanics with some dolerite intrusives but little evidence of granite intrusion. These formations are usually shallow-dipping sequences unconformably overlying the disturbed basement rocks.

Mineralization

Many anomalies were located over the Upper Proterozoic rocks, which are considered to be intracratonic, basin and shelf, mostly shallow-water sediments with much clastic material. None of the anomalies has yet developed into an economic deposit. Many were associated with sedimentary or volcanic horizons over very large areas, but ground counts and assays indicate concentrations of less than 100 ppm uranium in most instances. These relatively thin and relatively undisturbed, but much fractured, formations have presented easily leached exposures for much of post-Precambrian time.

Deposits in the South Alligator River and Rum Jungle areas are concentrated in Lower Proterozoic formations near overlying exposures of Upper Proterozoics, so they may have been exposed to weathering for relatively short periods of time. Elsewhere too, there are concentrations of anomalies where recently exposed basement rocks abut against mesas or tablelands of shallow-dipping cover. Examples are the new Nabarlek and Jim Jim finds.

In the Mt. Isa area uranium concentrations of ore grade are common on the surface. These deposits are usually associated with high concentrations of iron oxides, carbonates and or phosphorus, which should, theoretically, result in the uranium being less mobile, and it certainly presents treatment problems. It seems likely that uranium has been very mobile in most near-surface rocks and that subsurface information is needed for adequate assessment.

LOCATION OF TENEMENTS

One result of the sampling approach described has been the selection of areas for follow-up work. Mt. Isa contains numerous interesting prospects and a supply of experienced mining men is available there. *Authorities to prospect* and *leases* have been acquired there on the basis of the survey. The Bedford Downs area of Western Australia presents possibilities for developing large-tonnage operations and a *Temporary Reserve* has been acquired.

The Marble Bar and Ashburton regions have good potential for the occurrence of uranium deposits, but the sampling procedure did not reveal a concentration of prospects sufficient to justify the deployment of a follow-up crew. Consideration should certainly be given to uranium prospecting if other metalliferous investigations are undertaken in the area.

INSTRUMENTATION

The airborne scintillometer performed well, only three days being lost due to unserviceability, and only about 4 hours of committed flying time was wasted due to instrument failure.

TRAINING

Fifty hours of flying time (30 h fixed-wing and 20 h helicopter) was occupied in training people for uranium search. In the time available, however, some were unable to acquire either the navigational skills or the geophysical and geological understanding necessary for the effective application of the method described.

ANOMALIES

Anomalies could be classified as indicated below.

Granite Granite exhibited a generally high background, but several areas of granite or pegmatite also contained high activity above that background (2500–5000 counts/min).

Rhyolite Several areas of rhyolite (Edith River volcanics in the Katherine area, Bamboo Creek porphyry in the Pilbara region, Whitewater volcanics in the Kimberleys) displayed high levels of radioactivity (300–3000 counts/min).

Laterite Several areas of laterite produced count rates of 200–1300 counts/min. They comprised pisolitic, high-iron laterites of uniform radioactivity.

Argillite Black, purple-buff argillite, phyllite and slate produced counts in the range 200–1000, particularly in the Ashburton region and within the Speehwah Group in the Kimberleys. Similar shales gave rise to anomalies in the Arnhemland–Tungi and Roper River areas.

Sandstone and conglomerate Several areas of high radioactivity detected over coarse clastic rocks were carefully investigated. Despite reports of radioactivity associated with the Beatons Creek Conglomerate (Pilbara) and the Gardiner beds (Killi Killi Hills), count rates did not exceed 500 counts/min over these rocks. The Cheerybooka Conglomerate, however, was shown to contain anomalies in an anticlinal structure. An anomaly in the Bedford Downs Station area, near Halls Creek, was due to a pebbly bed which assayed 0·16% U_3O_8 from surface samples. Anomalously high radioactivity was also recorded in the Bangemall Conglomerate near Red Hill Station. Sandstone of the Beatons Creek Conglomerate Formation, the Speewah Group and the basal Hardey Sandstone, as well as sandstones of the Katherine River Group, were all demonstrated to be anomalously radioactive at some locations.

Fracture systems Samples taken from a fracture system at 16°25′S, 128°20′E contained 0·025% uranium, and anomalous radioactivity was shown to be associated with traces of manganese in breccia zones at 20°30′S, 120°15′E. Another radioactive high was detected in a fracture system containing quartz and copper secondary minerals. In the Mt. Isa–Cloncurry area anomalous count rates were almost invariably recorded over fracture systems exposed by mine workings.

Conclusions

The system employed for a rapid reconnaissance of potential uraniferous areas in Australia successfully discovered surface uranium enrichments in the Mt. Isa district which merited further investigation. A conglomerate horizon exhibiting anomalous surface radioactivity also justified the acquisition of ground for follow-up work.

The results of the survey reflect the high mobility of uranium in the weathering cycle under Australian conditions.

Bibliography

1. Brooks J. H. The uranium deposits of North-Western Queensland. *Publ. geol. Surv. Qd* no. 297, 1960, 50 p.
2. Brooks J. H. The occurrence of uranium in Queensland. In *Australian atomic energy symposium* (Melbourne: Melbourne University Press, 1958), 15–26.
3. Halligan R. and Daniels J. L. Precambrian geology of the Ashburton Valley region. *Rep. Dep. Mines W. Aust. 1963*, 1964, 88–96.
4. Heinrich E. W. *Mineralogy and geology of radioactive raw materials* (New York: McGraw-Hill, 1958) 614 p.
5. Hunty de la L. E. Report on Kimberley radioactive deposits. *Bull. geol. Surv. W. Aust.* no. 112, 1955, 41–5.

6. Low G. H. Copper deposits of Western Australia. *Miner. Resour. Bull. geol. Surv. W. Aust.* no. 8, 1963, 203 p.
7. MacLeod W. N. The geology and iron deposits of the Hamersley Range area, Western Australia. *Bull. geol. Surv. W. Aust.* no. 117, 1966, 170 p.
8. McKelvey V. E., Everhart D. L. and Garrels R. M. Origin of uranium deposits. *Econ. Geol. Fiftieth Anniversary Volume*, 1955, 464–533.
9. Morgan B. D. Uranium deposit of Pandanus Creek. In *Geology of Australian ore deposits 2nd edn* McAndrew J. and Madigan R. T. eds (Melbourne: Congress and Australasian IMM, 1965), 210–1. (*Publications 8th Commonw. Min. Metall. Congr., Australia N.Z., 1965*, vol. 1)
10. Noldart A. J. Radioactive anomalies, Pilbara goldfield. Unpublished records of ground investigations of airborne anomalies.
Noldart A. J. Summary progress report on a reconnaissance survey of portion of the Pilbara G. F., W. A. *Bull. geol. Surv. W. Aust.* no. 114, 1960, 134–49.
11. Noldart A. J. and Wyatt J. D. The geology of portion of the Pilbara goldfield. *Bull. geol. Surv. W. Aust.* no. 115, 1961, 199 p.
12. Prichard C. E. Uranium deposits of the South Alligator River. In *Geology of Australian ore deposits, 2nd edn* McAndrew J. and Madigan R. T. eds. (Melbourne: Congress and Australasian IMM, 1965), 207–9. (*Publications 8th Commonw. Min. Metall. Congr., Australia N.Z., 1965*, vol. 1)
13. Spratt R. N. Uranium deposits of Rum Jungle. In *Geology of Australian ore deposits, 2nd edn* McAndrew J. and Madigan R. T. eds (Melbourne: Congress and Australasian IMM, 1965), 201–6. (*Publications 8th Commonw. Min. Metall. Congr., Australia N.Z., 1965*, vol. 1)
14. Stewart J. R. An assessment of the search for uranium in Australia. In *Exploration and mining geology* Lawrence L. J. ed. (Melbourne: Congress and Australasian IMM, 1965), 343–51 (*Publications 8th Commonw. Min. Metall. Congr., Australia N.Z., 1965*, vol. 2)
15. Sullivan C. J. and Iten K. W. B. The geology and mineral resources of the Brock's Creek district, Northern Territory. *Miner. Resour. Bull. geol. Surv. W. Aust.* no. 12, 1952, 52 p.
16. Walpole B. P. The regional distribution of uranium occurrences Northern Territory. In *Australian atomic energy symposium* (Melbourne: Melbourne University Press, 1958), 6–14.
17. Walpole B. P. Westmoreland area uranium occurrences. *Rec. Aust. Bur. Miner. Resour. Geol. Geophysics.* 1957/40, unpublished.
18. *Western Australia Geological Survey Annual Reports, Bulletins and Miscellaneous Reports* relating to North Western Australia and Northern Territory.
19. Carter E. K., Brooks J. H. and Walker K. R. The Precambrian mineral belt of North-Western Queensland. *Miner. Resour. Bull. geol. Surv. W. Aust.* no. 51, 1961, 2 vols, 344 p.

Appendix

APPROXIMATE COSTS OF SURVEY—FULL YEAR

Piper Cub: 350 h at average cost of $15/h		$ 5 250
Helicopter: 120 h at average cost of $100/h		$12 000
Fuel for aircraft, positioned		$ 4 000
Personnel		$15 000
Maintenance and transport of personnel		$ 5 000
Head office, supervisory, communications, reports		$ 5000
	Total	$46 250
	(say,	$50 000)

DISCUSSION

Dr. F. Q. Barnes asked the author to indicate the results of drill sampling of the conglomerates in northern Australia which Western Nuclear Corporation discovered and which some Australian companies pegged. According to the Australian representatives who came to Canada, they considered that it was mineralization of Blind River type and would return good uranium values with deeper sampling. The author replied that the anomalies found were distributed along 150 miles of the conglomerate escarpment, but gamma-spectrometric determinations showed that they were nearly all due to thorium.

There was no improvement in uranium values with depth, but the area had been subjected to weathering since Precambrian times and the depth of weathering was very great.

Dr. R. G. Dobson said that the Rum Jungle shales contained pitchblende near the surface, but it had been suggested that, within the conglomerates, weathering had reached to depths of 700 ft. Recent drilling, however, had penetrated fresh conglomerate at no more than 170 ft. In any case, uranium did occur within weathered horizons. All recent work on Rum Jungle Crater Beds suggested that they contained no uranium, but new discoveries in the Kimberley Range suggested that that area had considerable potential. Generally speaking, the Central Australian sandstones were not very promising as they overlay high-grade metamorphic rocks. Rather than search there, it would be better to examine the possibilities of the Great Artesian Basin, where the Mesozoic carbonaceous and phosphatic sediments showed locally some highly anomalous radioactivity.

Ranger 1: a case history

G. R. Ryan B.A., F.I.M.M.
Geopeko Limited, Darwin, Northern Territory, Australia

550.8:553.495.2 (948.1)

Ranger 1 lies about 140 miles almost due east of Darwin, in the Northern Territory of Australia. Access is by partly formed earth road, which is only suitable for traffic between May and December in most years because of the monsoon. It can be reached by air to the Jabiru airstrip, which was constructed during 1970. Jabiru is the camp set up by Geopeko Limited in 1970 after the discovery, and lies about $1\frac{1}{2}$ miles from the prospect.

Geopeko Limited is the exploration subsidiary of Peko-Wallsend Limited, and acts as the operating company for a Joint Venture between Peko-Wallsend and the Electrolytic Zinc Co. of Australasia.

Discovery

In 1968 the Bureau of Mineral Resources published an updated version of their 1:500000 map of the Katherine–Darwin Province. On the older version the Nanambu Granite, which lies west of Ranger 1 and crops out at the surface in rare, very small exposures, was shown as Proterozoic in age, and it was assumed that the granite intruded the Lower Proterozoic sediments of the area. On the newer version the granite was shown as Archaean (and, hence, older than the sediments) as a result of age-dating.

Although outcrop is rare, photointerpretation by the Bureau indicated that the sediments were probably disposed concentrically about the granite, which now appeared to be a basement dome. This situation was directly analogous to Rum Jungle, and, accordingly, in 1968 Geopeko applied for an *Authority to Prospect* over part of the area.

From a reconnaissance undertaken in late 1968, before the monsoon arrived, it became apparent that surface exploration would be of little use without some overall guidelines since there was very little outcrop. Accordingly, the decision was taken to fly the area.

Flying was scheduled for late 1969, with spectrometer and magnetometer, and specifying Doppler control, as the flat, featureless terrain might have inhibited visual navigation. Unfortunately, commissioning of the Doppler equipment took time and, with the arrival of the monsoon, flying had to be postponed until mid-1970.

In the meantime Noranda (Australia) had located an area of high radiation on Geopeko ground while flying their adjacent area, and passed the information to the Joint Venture. Late in 1969 a Geopeko ground party located the anomaly and pegged two leases. A vehicle and materials for a small camp were left on site and secured against the worst of the monsoon, and trees were cleared for a helipad. Soil samples taken gave very high uranium values.

Evaluation

In March, 1970, four men were ferried in by light plane to Mudginberry Home-

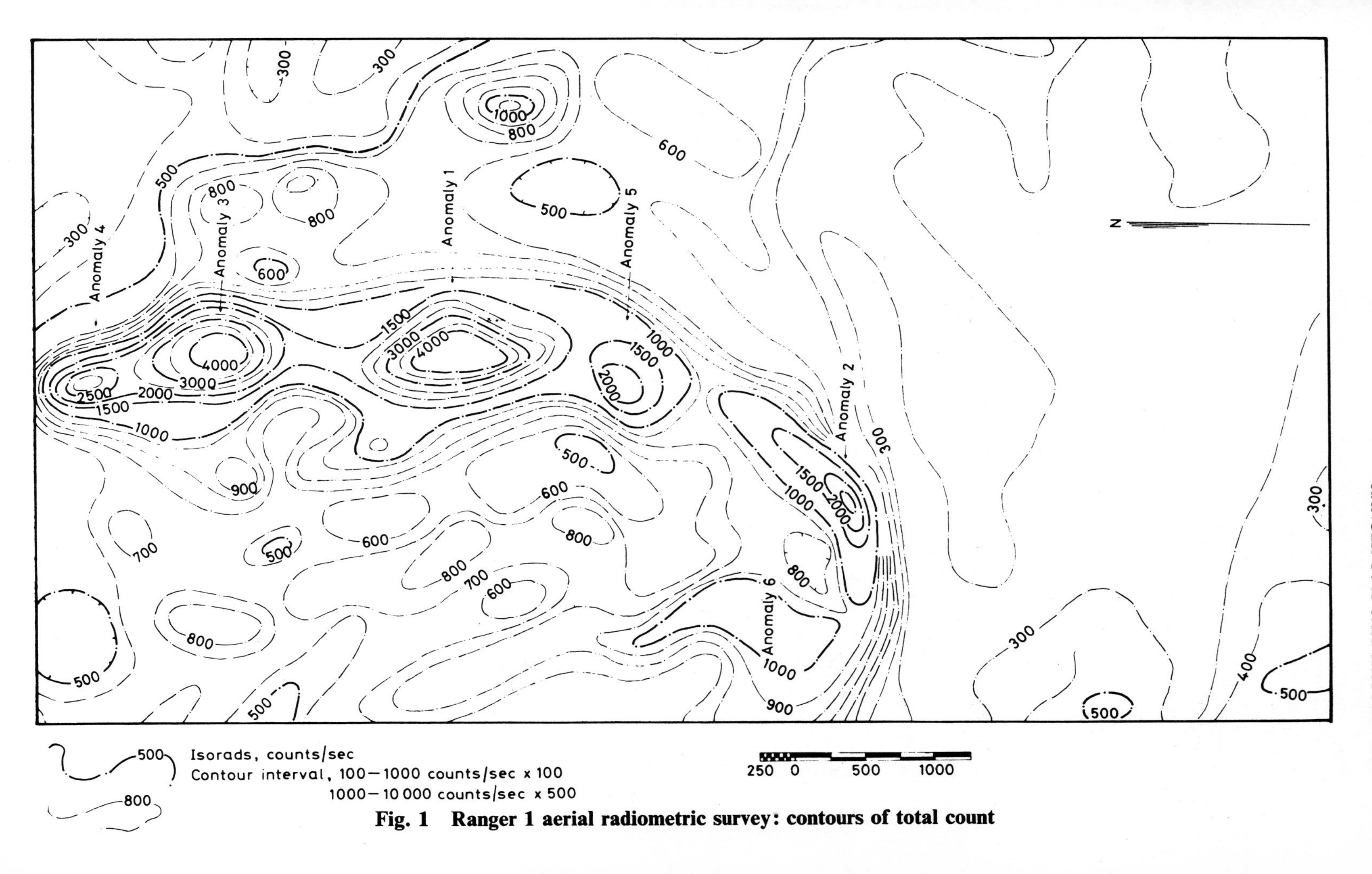

Fig. 1 Ranger 1 aerial radiometric survey: contours of total count

stead, eight miles to the north, and thence by helicopter. This party began surveying and clearing the initial grid over the anomaly, which was designated No. 1. A 2400-m north–south line was established, with 1200-m long cross lines at 100-m spacing.

The grid lines were read with a hand-held scintillometer, and revealed the presence of an anomaly over an area of about 160000 m^2 which attained values of up to 30 times background.

Ground reconnaissance on foot revealed other anomalous areas, and the grid was extended to 2000 m north and 2000 m south, and, ultimately, to 2700 m north and 3600 m south, airborne work revealing the presence of five anomalies in line over a distance of 6 km.

Road access became possible in late April, though with some interruptions, and extra personnel were brought in. By this stage it had become apparent that a mineral deposit of some significance was emerging. A semi-permanent camp was built with mess, workshop and married quarters. Field strength consisted of two geologists assisted by two technicians, with periodic assistance from a surveyor and a third geologist. A construction crew was responsible for the establishment of the camp.

The little outcrop available was mapped by use of plane table and alidade. Only one rock unit crops out, and this offered little help as it was itself not mineralized. It lies west of the anomalous areas, and dips east. A few occurrences of saleeite were, however, found in rocks of this type. Several geophysical methods were tried in case a useful tool should emerge, including S.P., magnetics, V.L.F. electromagnetic, and resistivity, but without success. A number of shallow auger holes revealed increasing radiation strength with depth, but did not reach fresh rock.

At this point a problem arose in that all of the anomalous area lay within the boundaries of a proposed National Park. It was essential that the size and potential of the area be demonstrated as rapidly as possible. Accordingly an Ingersoll-Rand Drillmaster was contracted, and rotary-percussion drilling was begun in late July, 1970. The first line of holes was put down at 50-m centres on section across the No. 1 anomaly to depths of around 100 m, revealing uranium mineralization over some 250 m in width and to the bottom of some holes. Holes were then stepped out north and south, and the No. 1 body was rapidly delineated. A similar programme was used to delineate No. 3 body, which is of about the same size in plan, with hole spacings at 100-m centres.

No. 2 body proved to be less predictable, and after three holes failed to encounter mineralization where expected, even though put down close to trenches revealing high-grade ore, the hammer drill was moved back to Nos. 1 and 3. Rain set in at this point, however, and the programme was concluded.

Two deposits of substantial size, and a third, probably smaller, but possibly richer, had now been delineated. The Joint Venture companies announced that a resource of the order of 70000 tons of contained U_3O_8 was suggested by drilling at that stage, and plans were laid for the development and ore-blocking stage.

Development

If work was to continue through the wet season, guaranteed access and comfortable accommodation were necessary. The prospect lies between two creeks, both of which flow to 10-ft depth and more during the rains. Furthermore, the water supply at the original camp site was not adequate for a larger establishment. A site for a landing ground was found, and, fortuitously, good-quality water in abundance was available nearby. The Jabiru camp, consisting of an air-conditioned office, bunkhouse, kitchen and mess, and married quarters, was brought on site in prefabricated units. A coreshed, powerhouse, workshop and store were built. The airstrip and roads, with necessary causeways and culverts, were put in. All this was accomplished in three months, despite the fact that most of the material had to be transported 2000 miles by road from southern centres. The complex was occupied just before the roads were permanently cut, though increasing rain had already caused a number of problems.

All supplies for a six-month period, other than perishable goods, were also on site.

On the technical side, diamond drilling replaced percussion drilling. Up to this point the geology of the deposit was poorly understood, except that some idea of the lithology was available from percussion cuttings. Accordingly, a series of nine inclined holes, three each on three sections 100 m apart, was planned to test what was assumed to be an easterly dipping body. It was found that the dip flattened to zero beneath the ore zone, and, subsequently, vertical holes were adopted. In the early stages a roller bit was used to drill the worst of the oxidized ground. In firmer ground a triple-tube HQ barrel was used, and, finally, NQ and BQ. Sludge samples were taken until full core was being obtained.

Subsequently, a technique of dry drilling, with an NX casing coring bit, and literally driving the casing into the soft muds in the top 10 m, gave much better recovery.

To eliminate freight and assay costs, core splitting and sample preparation is done on site. After pulverization, samples are radiometrically assayed by use of a Spectrometer Scaler, and ore-grade samples are forwarded for confirmatory assay at a Sydney laboratory.

All holes are probed by use of gamma logging, single-point resistance and self-potential, where possible, but it is commonly necessary to log inside the drill string, so that only the gamma log can be obtained. The logger was not designed for this purpose but was used principally as a lithological tool in the area. A low-sensitivity probe was purchased to cope with the ore, and a semi-quantitative log was obtained from the chart. Downhole assaying techniques are currently under investigation.

The ore is in equilibrium, and is low in thorium, so radiometric assaying is satisfactory, at least for primary ore. Equilibrium in secondary ore is now being evaluated.

For ore blocking the use of the CON-COR technique is being contemplated. Conventional rotary and percussion drilling is not satisfactory for three reasons. First, the holes make water, in some instances at rates up to 10000 gal/h: this inhibits the downhole hammer in particular, and makes sample catching uncertain. Secondly, the totally oxidized ground in the top 30 m is unstable and liable to collapse, particularly with the massive water flows encountered, so contaminating the samples from lower in the hole. Thirdly, it has been established that the walls of the hole may be contaminated locally by sample from below, giving a false downhole gamma profile.

CON-COR should obviate the latter two problems by sealing off the walls from circulating fluid. Non-core sampling also has its problems in that there is a considerable loss of values in the fines. With the high water returns the slime fraction tends to be carried away with overflow. Tests have shown that this fraction contains a significant amount of uranium. Sampling equipment is being designed to split the sample into two one-eighth fractions and collect the entire delivery, water and all, in bags large enough to prevent overflow. These will then be left to filter, thus retaining all the solids in the sample.

Until these techniques are perfected, ore blocking is being done by core-drilling.

In the initial core-drilling programme two vertical holes drilled alongside percussion holes showed that whereas the early percussion drilling gave results substantially below that obtained from core, the discrepancy was much less in later holes as sample collecting techniques improved. Nevertheless, the principal cross-section was drilled out with core-drilling on centres between the percussion holes so that a complete statistical comparison could be obtained between the two types of drilling. The results of this work are currently being evaluated.

Systematic ore blocking is now under way by use of vertical diamond drill holes on 50-m centres.

DISCUSSION

R. J. Chico asked (1) if the gamma logging was in equilibrium with results obtained via chemical assays; (2) what other metals were found in the deposit, and if there was evidence to support mineral zoning within the orebody; (3) for further information on the diamond drilling costs and the area of influence assigned to each hole for ore-reserve calculations; and (4) which methods were being considered for mining the orebody.

The author replied that good correlation between the chemical analyses of the diamond drill cores and the radiometric logs showed the deposit to be in equilibrium. In addition to uranium, they had done extensive chemical analyses of the cores for other elements, particularly rare earths and copper, lead and gold. Drilling costs were from $Aus.22 to $Aus.30 per metre, depending on size and depth of hole. Geostatistical methods indicated that a hole spacing of about 70 m would be adequate, but 50-m spacing had been adopted. The area of influence assigned to each hole was, therefore, 50 m × 50 m. He was unable to comment on the mining methods proposed.

The Nabarlek area, Arnhemland, Australia: a case history*

D. B. Tipper B.Sc.

G. Lawrence B.Sc., Ph.D.

Both of Hunting Geology and Geophysics, Ltd., Boreham Wood, Hertfordshire, England

550.8:553.495.2(948.1)

A summary is given of an airborne magnetic and gamma-ray spectrometric survey in Arnhemland, Australia, which directly resulted in the discovery of the rich Nabarlek uranium orebody, 170 miles east of Darwin. The survey was flown in early 1970 for Queensland Mines Pty., Ltd., by Hunting Geology and Geophysics, Ltd., and Allied Geophysics Pty., Ltd., their Australian Associates.

An airborne geophysical survey was selected as the most appropriate exploration method in terms of effectiveness, speed and cost for such a large area (a total of 20000 km of surveying was successfully flown). A gamma-ray spectrometer was used to detect anomalous radioactivity and to discriminate between potassium, uranium and thorium sources. Magnetic data were recorded simultaneously to assist in the understanding of geological structure.

Operational details

The geophysical and ancillary instrumentation was installed in a DC-3 aircraft, especially modified for low-level geophysical surveying. Cruising at 120 knots for up to 10 h, this aircraft remains the ideal flying platform for multi-system surveying of large areas.

The four-channel differential spectrometer was manufactured by Nuclear Enterprises, Ltd., Edinburgh, to Hunting's specification. Two thallium-activated sodium iodide detector crystals, each 6 in in diameter and 4 in thick, were used, although much greater crystal volumes are available as required.

Four energy channels were selected:

Channel	Channel limits	Characteristics
1	0·20–2·90 MeV	'Total' count rate
2	1·35–1·65 MeV	^{40}K peak at 1·46 MeV
3	1·65–2·30 MeV	^{214}Bi (uranium series) peak at 1·76 MeV
4	2·30–2·90 MeV	^{208}Tl (thorium series) peak at 2·62 MeV

The channel outputs were recorded as continuous analogue profiles, on ultraviolet sensitive paper, with no time-lag between the channels. The recording sensitivities were (counts/sec per in): channel 1, 750; channel 2, 50; channel 3, 15; and channel 4, 15.

The total-count data were recorded also on a twin-channel Moseley pen recorder. The magnetometer was a modified Gulf Mark III fluxgate, producing a continuous profile of total intensity at a scale of 60 gammas/in. A ground fluxgate magnetometer, at the survey base, monitored diurnal variations.

*In the original presentation this case history was illustrated by ten lantern slides. It is not possible to include these here, but prints can be obtained, on request, from the authors.

The magnetic and spectrometric data were related to ground features by a fiducial marking system, automatically synchronized with the shutter operation of a vertically mounted tracking camera.

The flight pattern comprised north–south flight lines spaced ¼ mile (400 m) apart, controlled by east–west tie-lines. The flying height, nominally 400 ft (120 m) above ground level, was measured continuously by radioaltimeter, and recorded on the ultraviolet and Moseley recorders.

Navigation of the aircraft was visual by use of 1:25000 and 1:50000 scale photomosaics. The actual flight paths were plotted in the field, on to mosaics, so that anomalies of interest could be followed up immediately on the ground by Queensland Mines staff. The remarkable Nabarlek anomaly was very quickly located in this way.

Geophysical results

The spectrometer responses to any radioelement will depend on the channel widths, and positions. With the settings used, a potassium source will affect channels 1 and 2 only, a uranium source will be recorded in channels 1, 2 and 3, and a thorium source will affect all channels. The counts in channels 2, 3 and 4, respectively, will approximate the following ratios for the various sources: potassium source, 100:0:0; uranium source, 48:48:4; and thorium source, 20:40:40.

The Nabarlek radiometric anomaly was particularly pronounced on two flight lines, the total-count anomaly being very intense, narrow and well defined. The spectrometer profiles provide much more information than the total-count rate profile. Fig. 1 shows the Nabarlek anomaly recorded with flying heights of 400 and 500 ft above ground level. At both heights the channel 3 (uranium) profile is off-scale, and channel ratios cannot be determined. The count rates at 500 ft, after subtracting the non-geological background, are (counts/sec) 1960 (channel 1), 58 (channel 2), at least 48 (channel 3) and 16 (channel 4). These values indicate a significant uranium content.

Also shown in Fig. 1 are the spectrometer profiles at heights of 800 and 1000 ft above ground. The anomaly remained conspicuous, and amplitude ratios (between channels) confirm a uraniferous source. The more important diagnostic ratio, that of channel 3 to channel 4, decreased with height from 5·6 at 600 ft to 4·3 at 800 ft and 3·8 at 1000 ft: this partly reflects the larger attenuation factor for the lower-energy radiation of channel 3.

The distribution of anomalous radioactivity is more clearly seen on an anomaly map, which shows the peak location and approximate width of each anomaly, together with amplitude values as recorded in each channel. This is not an interpretation map, of course, but a presentation of basic data.

The Nabarlek anomaly is conspicuous within a relatively featureless area. A second radiometric zone to the east, although of high total-count amplitude, has very different spectrometric characteristics that signify a thorium–potassium province, and the zone represents a large granite body.

The magnetic profile over the orebody shows it to occur within an area of relatively disturbed magnetic pattern comprising short-wavelength anomalies superimposed on one of much longer wavelength. The contours of total magnetic intensity have a basic interval of 10 gammas. The Nabarlek orebody is situated near the convergence of dominant positive and negative anomalies that are attributed to a dolerite sill. If the magnetization were wholly induced, then the pattern is reasonably consistent with a gentle anticline plunging easterly. There is, however, strong evidence in adjoining areas that the doleritic intrusives are

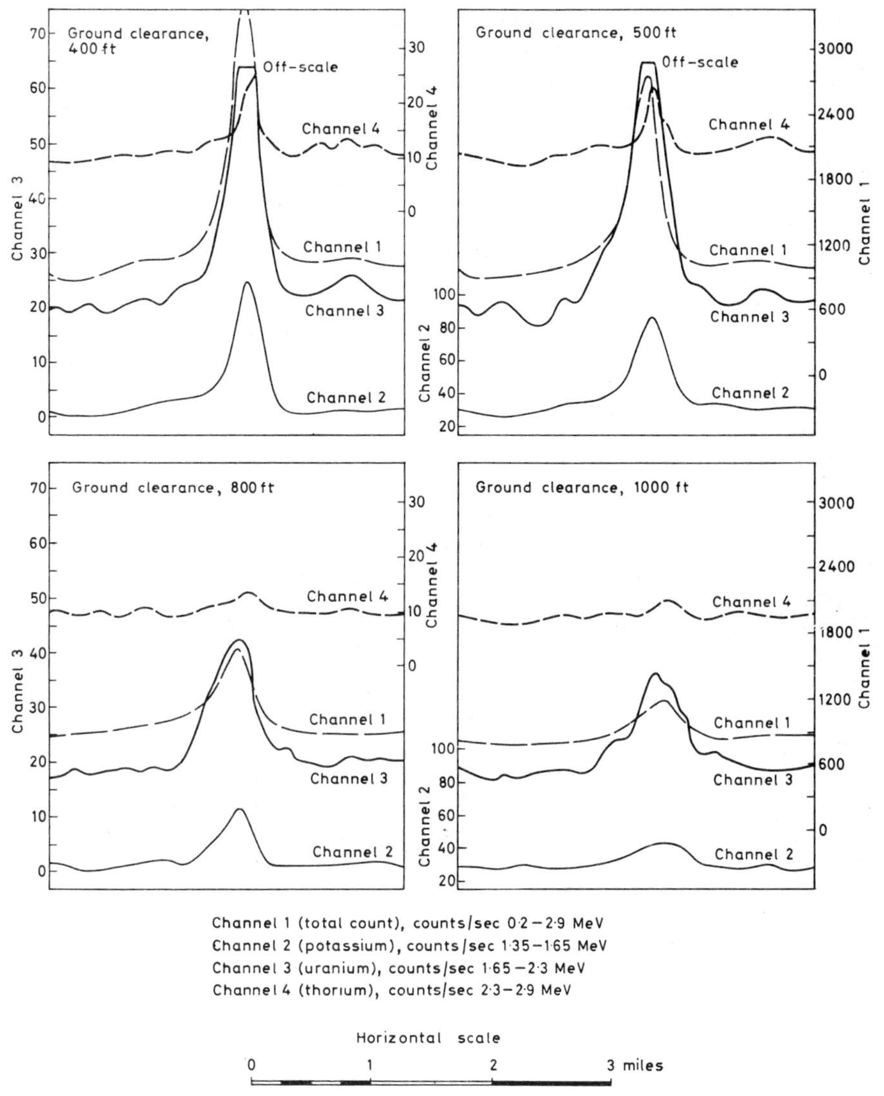

Fig. 1 Gamma-ray spectrometric profiles over Nabarlek orebody (different heights)

reversely magnetized. In this case, the fold would be synclinal, and this interpretation is consistent with the rather limited geological evidence.

Major faults with NNW and NNE directions are interpreted from the magnetic data to cut the syncline in several places. It seems probable that the Nabarlek deposit (striking 30° W of N) is controlled by a fracture.

The southern limb of the interpreted syncline may, from overall magnetic evidence, represent the line of a major ENE-trending fault, along which the dolerite may have been intruded. This is only tentatively proposed at this time; but, if correct, Nabarlek occurs at the intersection of two major fault systems.

Geological setting

The basement of the region is formed by low-grade metasediments of the Myra

Falls Metamorphics. These were originally believed to be Archaean in age because their metamorphic and tectonic style differ from an adjacent series of largely unaltered arenaceous sediments belonging to the Lower Proterozoic period. The flat-lying Upper Proterozoic Kombolgie Formation unconformably overlies the older rocks. The metamorphics and Lower Proterozoic rocks are intruded by granites and basic rocks—mainly dolerite. The Myra Falls Metamorphics are disposed in large east–west folds, whereas north–south fold axes characterize the Lower Proterozoic rocks to the west.

The major mapped faults are northwest–southeast and east–west, and affect rocks of Upper Proterozoic age and older.

The Nabarlek deposit lies in schists of the Myra Falls Metamorphics, close to their unconformity with the Upper Proterozoic Kombolgie Sandstone. The mineralized zone strikes 30° W of N and dips 45° NE. The main ore is a 12 ft wide band of massive pitchblende in the form of individual lenses totalling 830 ft in strike length. Patchily developed pitchblende extends up to 28 ft on either side. The overburden (Cenozoic laterite, sand, Kombolgie Sandstone rubble) is from 4 to 16 ft thick over the deposit, and it also shows a patchy development of pitchblende and secondary uranium minerals. The deposit may be limited in depth by the gently dipping, thick dolerite dyke shown on the magnetic pattern. The geological–geophysical relationships are interesting; and, in view of other nearby deposits, there is scope for further study of the geophysical data. The spatial association of uranium deposits with dolerite sills in the Myra Falls Metamorphics and Lower Proterozoic rocks is confirmed by new discoveries at Beatrice and Caramal (both of which are close to the unconformity with Kombolgie Sandstone). Similar occurrences have been reported at Jim Jim by Noranda and at Ranger 1 by Peko-Wallsend. The latter deposit is in sheared metasediments invaded by basic and pegmatitic dykes. The South Alligator River pitchblende deposits occur in shear zones associated with long northwest–southeast fractures, especially near to cross faults. Most are close to the Kombolgie unconformity.

The origin of many deposits remains unknown, although, overall, there is a suggestion of structural control, particularly along deep-seated faults. The spatial association with dolerite may be due to faults having acted as feeder channels for both igneous intrusion and mineralizing fluids.

The value of airborne spectrometric and magnetic surveying appears to have been well demonstrated. The spectrometric data have recorded anomalous radioactivity and distinguished between uraniferous sources and potassium–thorium granite outcrops. The extensive covering of superficial deposits and the spatial association of orebodies with doleritic sills and with faults indicate the particular value of simultaneous magnetic recording in this region to obtain basic geological information.

Acknowledgment

The cooperation of Queensland Mines in releasing the geophysical data presented is gratefully acknowledged.

DISCUSSION

Dr. G. Lawrence said that in northern Australia it seemed that the situation existed in which granitic rocks were eroded and the uranium was leached and deposited in environments favourable to fixation. Those favourable environments would seem to comprise a set of variable physico-chemical conditions that were likely to be more widespread than the number of known deposits suggested. The structural control of those rich deposits was more important than had been stated. Primary structural control was indicated by the fact that the original granitic masses (probably Archaean) were emplaced in structural 'highs', whereas the uranium deposits fringed those masses. The faults at South Alligator and Nabarlek were thought to be particularly important. The dolerite immediately to the south of Nabarlek might occupy a major east–west arcuate fault parallel to the other long arcuate faults shown on the East Alligator geological map. It was postulated that those faults (Caramal and possibly Beatrice lay along such faults) might be deep-seated, and they could have allowed migration of uranium from much deeper sources into the area.

Thus, it was envisaged that the groundwater of the province was rich in uranium from the erosion of granitic bodies. High-grade ore deposits had resulted not only from precipitation of the uranium but from additional concentration of deep-source primary material. The uranium would be precipitated where it met groundwater with an already high chemical potential of uranium. In that way deposits were localized at the coincidence of favourable environmental and structural sites, such as the intersection of the east–west and more north–south faults, as might have been the case at Nabarlek.

D. Ostle commented on the strong analogy between the geological environment at Nabarlek and that of the South Alligator River area. During investigation of the latter in the 1950s some workers, including the speaker, regarded the main mineralization control to have been doming, which exposed Archaean rocks and initiated deep-seated faulting. A sequence of igneous activity, represented mainly by granite and volcanic rocks, culminated in the introduction of uranium into faults associated with the major dislocations.

Doming of the basement could presumably give rise to the magnetometric anomalies described by Tipper and Lawrence. He asked if any control flights had been possible over the South Alligator area, as (assuming the above geological criteria to be valid) magnetometry could well be the most effective means of detecting favourable zones, even through a capping of Kombolgie Formation sediments.

Mr. Tipper said that other areas had been flown in northern Australia, but he could not reveal where they had flown. Although they did not fly the Alligator River area, they knew that it had been flown by magnetics and scintillometer, but not by spectrometer. There was little doubt that magnetics represented a useful tool in that part of the world.

Dr. W. Domzalski said that the very interesting example presented demonstrated two vital issues: first, that structural control constituted an all-important factor in the localization of uranium mineralization and, secondly, that magnetic data were of great value in the interpretation of structural conditions.

He would like to support that conclusion with another history, the whereabouts of which he was not in a position to reveal. In that instance the indication of 'structure' by magnetics was less dramatic than in the case under discussion, but no less important. Those familiar with magnetic surveys would appreciate the fact that even moderately magnetic rocks gave rise to mild magnetic field deformations revealed as 'patterns' over an area underlain by a given rock type. That was frequently referred to as the 'magnetic signature' of a rock. The two patterns had permitted the delineation of a contact between two different rock types, and that contact was a locus of uranium mineralization.

He was convinced that uranium exploration surveys, unless there were valid and strong reasons for a decision to the contrary, should include a magnetometer facility.

Mr. Tipper drew attention to a similar case in Central Africa, where uranium was associated with a conglomerate, which was magnetic and gave anomalies of 2–3 gammas. It was easier to trace that thin conglomerate band under overburden with magnetometry than with spectrometry.

Experimental survey with a portable gamma-ray spectrometer, Blind River area, Ontario: a case history

P. G. Killeen B.Sc., M.Sc., Ph.D.

C. M. Carmichael B.Sc., M.Sc., Ph.D.

Both of the University of Western Ontario, London, Ontario, Canada

550.8.002.5–182.3: 543.422.8: 553.495.2(713.7)

The present trend in uranium exploration is towards the increasing application of both ground and airborne gamma-ray spectrometers for the purpose of distinguishing between radiometric anomalies produced by thorium, uranium and potassium. The Blind River, Ontario, uranium mining area was chosen as a location for investigating the calibration and use of a portable three-channel gamma-ray spectrometer, and for developing a field survey technique suitable for uranium exploration.

The aim of the investigation was to determine (*a*) the radioelement content at selected stations across the exposed Huronian stratigraphic section of the Quirke Lake syncline and (*b*) the radioelement content over a relatively small area (e.g. a large outcrop) for detailed studies across and along strike. These geophysical observations were then compared with the geology. The instrument used was the Geological Survey of Canada's portable three-channel gamma-ray spectrometer.[1] The three energy channels were set to count gamma rays of 2·62 MeV (^{208}Tl) for thorium, 1·76 MeV (^{214}Bi) for uranium, and 1·46 MeV (^{40}K) for potassium. The energy width of each 'window' or channel of the gamma-ray spectrometer was set at 200 keV. A detailed discussion of the calibration of a gamma-ray spectrometer for *in situ* analysis has been given recently.[2,4] The accuracy of the *in situ* analysis depends on the radioelement and its concentration.[4] For the results presented here an accuracy of $\pm$ 10% or better can be claimed for any uranium analysis greater than 8 ppm. Calibration constants are a function of the sample-detector geometry, crystal size and counting efficiency of the instrument, as well as channel width. If any of these variables is changed, the instrument must be recalibrated.

Field procedure

Since the instrument was calibrated for quantitative *in situ* analysis, it was important to maintain a constant geometry between the detecting crystal and the outcrop. The easiest way to accomplish this was to support the probe (2 in × 2 in NaI(Tl) crystal) about 3 in from the outcrop surface by a piece of wood; and a constant 180° geometry was obtained if the outcrop was relatively flat for a distance of about 7 ft from the probe. The proximity of the probe to the surface meant that small inhomogeneities in radioelement content in the outcrop would change the results significantly. To prevent taking a measurement at a location not representative of average concentrations on the outcrop, the spectrometer was switched to 'rate meter' and used as an ordinary scintillometer.

An area of the outcrop was scanned to locate a spot which was fairly representative of the average concentrations for the gamma-spectrometric measurement. This preliminary monitoring usually took only a few seconds. The cumulative count was then measured simultaneously in the three channels for a period of 5 min (timed by a stop-watch). The accumulated counts were then recorded by reading out each channel. Present commercial instruments are available with built-in timers, and subsequent field work showed that the counting time could be reduced to 2 min by increasing the size of the sodium iodide crystal from 2 in × 2 in to 3 in × 3 in. The resultant increase in weight is negligible and a greater number of measurements can be made in a day before the batteries require recharging. The counting time actually depends on the desired accuracy, and can be reduced when the activity of the rocks increases.

Periodically throughout the day, the energy calibration must be checked to ensure that the three channels are centred on the proper energies. This was accomplished by utilizing the natural potassium in rocks as an energy standard. It is also possible to carry a calibration source such as ^{88}Yt. In addition, throughout this survey a base station was monitored each day before any new readings were taken. This maintained a check on instrumental variations such as would be caused by changes in the characteristics of an electronic component with age, or by other unforeseen factors. About once a month a background reading was taken from a canoe on a lake, over a water depth exceeding 10 ft and at a location at least 1000 ft from shore.

An attempt was made to obtain measurements at 100-ft intervals up-section. Since the average dip of the strata in the Quirke Lake syncline is about 20°, this could be achieved by taking measurements at approximately 300-ft intervals along the surface. Where outcrops were sparse, availability of exposure determined the location of measurements.

The whole procedure of an *in situ* analysis can be completed in less than 10 min, but the number of measurements that can be completed in a day is limited by the life of the rechargeable batteries in the instrument. It was found that about 30 sets of U, Th and K analyses could be completed in a typical working day under average field conditions.

Geology and survey locations

The Quirke Lake syncline (Fig. 1) consists of gently folded Huronian sedimen-

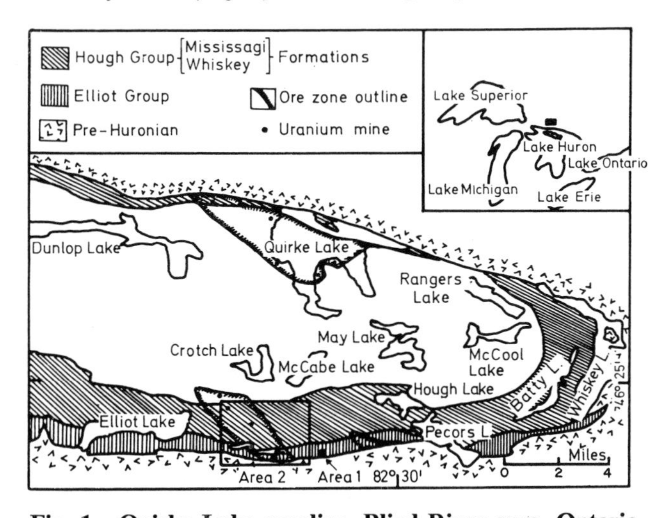

Fig. 1 Quirke Lake syncline, Blind River area, Ontario

tary rocks of the Penokean fold belt—a sub-province of the Southern Province of the Canadian Shield.

The Huronian sedimentary rocks consist of feldspathic quartzites, greywackes, argillites, limestones and various types of conglomerates. These are only slightly metamorphosed and folded into a broad syncline plunging gently west. Dips are about 30° on the north limb and 15° on the south, with thicknesses of up to 5000 ft in the central part of the syncline.

The Huronian rock formations of concern to this investigation are, from oldest to youngest, the Matinenda Formation and Nordic Formation of the Elliot Group, and the Whiskey Formation and Mississagi Formation of the Hough Group. Griffith and Roscoe[3] stated: 'The conglomeratic [uranium] deposits in Huronian strata near Elliot Lake underlie two separate areas about 6 miles apart [in the Matinenda Formation of the Elliot Group]. One, about 9 square miles in extent, is on the north flank of the Quirke Lake syncline; the other, about 3 square miles in area is on the south flank of the syncline. Both zones outcrop and extend down dip to depths of as much as 3,600 feet below surface'. The ore occurs in conglomeratic units up to 100 ft thick in the southern zone.

Survey locations

Two separate areas, referred to as areas 1 and 2, were investigated. Area 1 consists of a large outcrop of the Matinenda Formation on the south limb of the syncline approximately $\frac{1}{2}$ mile east of area 2. Area 2 comprises about six square miles on the south limb of the Quirke Lake syncline and includes the southern conglomeratic deposit zone described above, containing several uranium mines. The locations of both areas are shown in Fig. 1.

Results for area 1

A grid was marked on the outcrop of area 1, the length of a grid unit being set at 10 ft. In this manner an area of 11 by 9 units was set up for a possible 99 determinations. The outcrop was not perfectly flat but consisted of a series of beds dipping approximately 15° north and striking in an east–west direction. There was no difficulty in obtaining the 180° geometry for the determinations, rises between beds being only between 1 and 8 in.

The grid measurements represent an attempt to assess the variability of the concentrations and ratios of the radioelements over a small area. Since the Matinenda is a relatively highly radioactive formation (and the formation containing the uranium ores that are presently mined), it was of particular interest and suitability for this investigation.

Uranium concentrations (Fig. 2) range from less than 5 to more than 25 ppm, the greater part of the outcrop having a uranium concentration of about 6 ppm. An east–west trend can be seen, although it is not prominent. Two high areas stand out in the central and northeast section of the outcrop and two smaller highs are located in the south-central and southeast sections of the area.

Other determinations in area 1 indicate that thorium highs are associated with most of the uranium highs. The thorium concentrations range from less than 20 to more than 110 ppm, with an overall average of about 25 ppm. Potassium concentrations do not reflect the east–west trend of the geology and are more randomly distributed, averaging 3·5% K over area 1. For this reason the Th/K and U/K ratios simply reflect the variations in Th and U, respectively. Since Th and U over this area behave similarly, the Th/U ratio is relatively uniform, showing no trends, with an average of about 4.

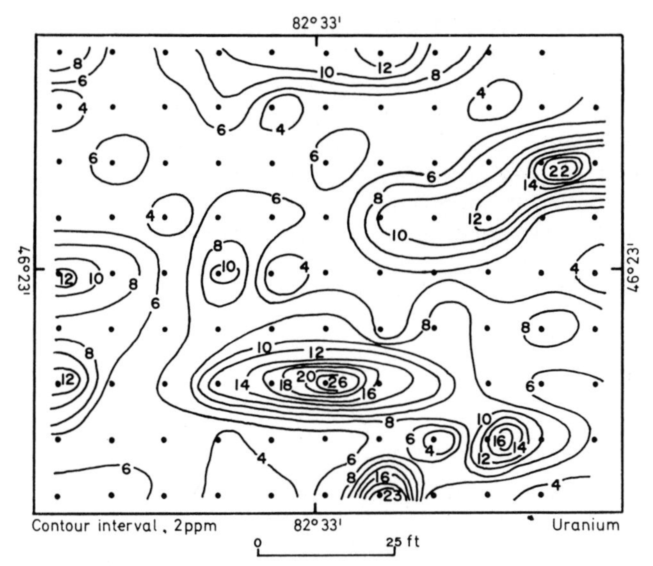

Fig. 2 Contoured uranium concentrations for area 1 (the grid)

The observed variability of radioelement content over a short distance will determine the field procedure for a large-scale survey and the field procedure discussed earlier was evolved from the knowledge of this variability, and represents an attempt to allow for it.

Results for area 2

Readings on seven traverses across the strike in area 2, on the south limb of the syncline, were compiled into contour maps of the radioelements and their ratios. The contoured uranium concentrations, ranging from less than 2 to more than 175 ppm, are shown in Fig. 3. The Mississagi Formation averages about

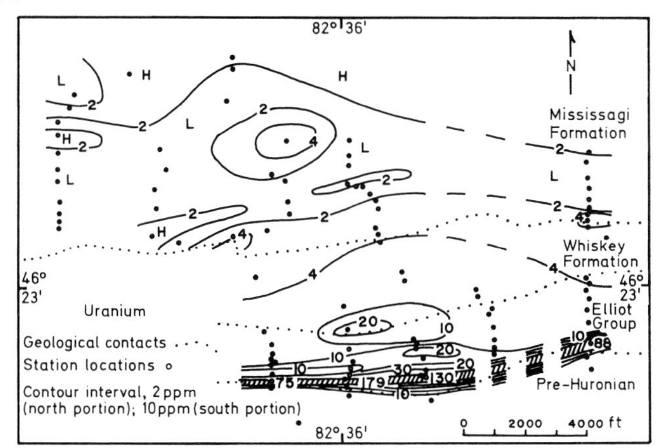

Fig. 3 Contoured uranium concentrations for area 2

2 ppm, concentrations increasing to more than 4 ppm in the Whiskey Formation. Uranium contents in the Elliot Group are more than 10 ppm on average, maximum concentrations falling in the basal Matinenda Formation. The highest concentration is coincident with the surface expression of the conglomeratic deposits containing the uranium mines of the southern ore zone of the Quirke Lake syncline.

The uranium contours show a linear trend parallel to the formational strike, and concentrations decrease both east and west of the central high.

The thorium distribution over area 2 is very similar to that of uranium, trending parallel to the formational strike and ranging from 4 to more than 390 ppm. Thorium in the Mississagi Formation averages about 6 ppm, increasing to in excess of 10 ppm in the Whiskey Formation. Thorium contents in the Elliot Group exceed 30 ppm, increasing to maximum concentrations in the basal Matinenda Formation.

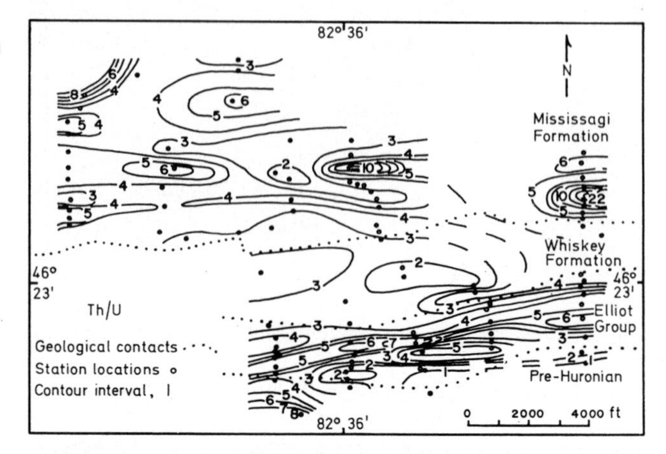

Fig. 4 Contoured values of Th/U ratios for area 2

The contoured values of the Th/U ratios are shown in Fig. 4. The trend is parallel to formation strike, ratios ranging from less than 1 to greater than 10. The Mississagi Formation has an average Th/U ratio of about 4, with several scattered spot highs. The Whiskey Formation contains ratios closer to 3, increasing rapidly to values greater than 4 in the Elliot Group. A relatively consistent band containing Th/U ratio values greater than 5 parallels the trend of the geology at a position about the middle of the Elliot Group. Values decrease rapidly to a ratio of 2 in the basal Matinenda Formation near the pre-Huronian. Ratios decrease to less than 1 at the location where the highest uranium concentrations were measured.

Potassium values were found also to parallel the trend of the geology, with average values of approximately 2% for the Mississagi Formation, 3% for the Whiskey Formation and 4% for the Elliot Group. Contoured ratios of Th/K and U/K primarily reflect the trends and anomaly patterns of the Th and U distributions, respectively, and they provide no particular advantage over direct observation of the Th or U distributions alone.

Conclusions

The uranium distribution adjacent to the mines on the south limb of the Quirke Lake syncline indicates that the surface expression of the 'ore valley structures' in the pre-Huronian[5] (which contains the Nordic, Lake Nordic, Milliken and Stanleigh deposits) can be mapped with the gamma-ray spectrometer. Extremely high thorium and uranium values have been recorded. The spectrometer readings of up to 180 ppm uranium in area 2 represent the detection of potential ore-bearing conglomerate. The distribution of high uranium concentrations determined with the spectrometer conforms with the results obtained from many thousands of feet of drilling carried out in the area to delineate the ore boundaries. A detailed preliminary survey with the gamma-ray spectrometer could therefore represent a considerable economic advantage in planning a drilling programme to outline ore of this nature.

The conclusions can be summarized as follows.

(1) The general trend of the distributions of the radioelements and their ratios is parallel to the strike of the geology. Trends can be followed by conducting traverses across formation strike.

(2) The radioelement content generally increases as the pre-Huronian rocks are approached from formations higher in the geologic section. Highest concentrations are in the Matinenda Formation at the base of the Elliot Group.

(3) The spectrometer can delineate areas of potential ore grade and sub-marginal ore-grade conglomerates where favourable geologic conditions, such as in the Quirke Lake syncline, exist.

(4) The variability of radioelement content over short distances must be taken into consideration for representative analyses of an area to be obtained.

(5) Material of highest radioelement concentration is characterized by low Th/U ratios.

(6) Studies of Th/K and U/K ratios do not seem to present any advantage to uranium exploration in this geologic situation, but they may be useful indicators of the presence of increased Th and U concentrations for areas where radioactivity is decreased by overburden cover.

References

1. Doig R. The natural gamma-ray flux: in situ analysis. *Geophysics*, **33**, 1968, 311–28.
2. Grasty R. L. and Darnley A. G. The calibration of gamma-ray spectrometers for ground and airborne use. *Pap. geol. Surv. Can.* 71–17, 1971, 27 p.
3. Griffith J. W. and Roscoe S. M. Canadian resources of U and Th. *Miner. Inf. Bull. Can. Dep. Mines* MR 77, 1964, 12 p.
4. Killeen P. G. and Carmichael C. M. Gamma-ray spectrometer calibration for field analysis of thorium, uranium and potassium. *Can. J. Earth Sci.*, **7**, 1970, 1093–8.
5. Roscoe S. M. Geology and uranium deposits, Quirke Lake–Elliot Lake, Blind River area, Ontario. *Pap. geol. Surv. Can.* 56–7, 1957, 21 p.

DISCUSSION

J. A. Robertson said that the authors and Dr. A. G. Darnley had used different stratigraphic nomenclatures: that was because their work was carried out at the same time as regional mapping was undertaken. Recently, the GSC and ODM formed a committee to resolve that matter. A nomenclature closer to that used by the authors than that used by Darnley had been agreed upon and had been used in recent ODM and GSC publications.* Only the new nomenclature would be used at the International Geological Congress in 1972. Nomenclature correlation tables could be obtained from the speaker.†

*Robertson J. A., Frarey M. J. and Card K. D. The Federal–Provincial Committee on Huronian stratigraphy: progress report. *Can. J. Earth Sci.*, **6**, 1969, 335–6.
Robertson J. A., Card K. D. and Frarey M. J. The Federal–Provincial Committee on Huronian stratigraphy: progress report. *Misc. Pap. Ont. Dep. Mines* 31, 1969, 26 p.
Robertson J. A. Geology and uranium deposits of the Blind River area, Ontario. *Trans. Can. Inst. Min. Metall.*, **72**, 1969, 156–71.
Robertson J. A. A review of recently acquired geological data, Blind River–Elliot Lake area. *Misc. Pap. Ont. Dep. Mines* 45, 1971, 35 p.

†Geological Branch, Ontario Division of Mines, Ministry of Natural Resources, Whitney Block, Queen's Park, Toronto 182, Canada.

Dr. S. H. U. Bowie said that Blind River, with its mixed uranium–thorium ores, was an excellent area in which to make gamma-spectrometric surveys. It was known that in that district the uranium:thorium ratio varied from 4:1 to 1:4: what significance could be placed on that variation in the ratio?

Dr. A. G. Darnley said that the work demonstrated the value of uranium–potassium ratios, which were clearly anomalous at Blind River and Agnew Lake. The Agnew Lake deposit showed a U:Th anomaly, although it was not as prominent as that around Elliot Lake. It also possessed a strong U:K anomaly, like all the other deposits in the region.

Dr. Killeen said that the variation in the Th:U ratio appeared to be a good indicator of potential uranium ore in the Blind River Area. Although uranium and thorium concentrations in rocks were known to vary together, in the anomalous situation of mixed uranium–thorium ores, a better indicator might be the U:K ratio, as Dr. Darnley suggested. The survey indicated that potassium was much less variable and represented a more stable base for comparison. Therefore the U:K ratio might be a much more generally applicable ore indicator than the Th:U ratio.

Exploring for uranium in the Northern Highlands of Scotland: a case history

M. J. Gallagher Ph.D., M.I.M.M.

Institute of Geological Sciences, London, England

550.8:553.495.2(411)

In 1968 a small field party of the Institute of Geological Sciences commenced the exploration of northern Scotland beyond the Great Glen as part of a regional reconnaissance for uranium in the United Kingdom on behalf of the United Kingdom Atomic Energy Authority. In this region of some 18000 km^2 known as the Northern Highlands uranium had previously been recorded from thin beds within Middle Old Red Sandstone (Devonian) sediments and from minor structures in a Caledonian (*ca* 400 m.y.) granitic pluton at Helmsdale.[7] Thorium minerals were also known to be associated with the Caledonian syenite mass of Ben Loyal (Fig. 1).

During two years of reconnaissance numerous new uranium occurrences were located in the northeastern district of the Northern Highlands (area *B* of inset to Fig. 1). With the exception of minor localities associated with Caledonian granitic rocks, no significant evidence of uranium mineralization emerged from the rapid surveys conducted over the remainder of the region. Subsequent work was therefore concentrated in this favourable district of some 2000 km^2, and the main results obtained have recently been published.[3] Because of the pronounced association of uranium with Old Red Sandstone sediments, the reconnaissance has been extended northwards to the Orkney Islands, which are largely formed of such rocks. Further uranium prospects have recently been located in the Stromness district of Orkney (Fig. 1 and Michie[4]), but discussion is restricted here to the 1968–70 work on the mainland.

General features of northern Scotland

The Northern Highlands of Scotland is a sparsely populated region (about five inhabitants per square kilometre) largely given over to sporting estates and forestry; it has a cool-temperate climate. Most of the population is found in the eastern parts of the region, which are underlain by the relatively soft rocks of the Old Red Sandstone (ORS). In the west, Cambrian rocks rest on late Precambrian unmetamorphosed sediments (Torridonian), which, in turn, overlie Archaean gneisses (Lewisian). All of these rocks are involved in a great zone of thrusting dominated by the Moine thrust above which was carried the Moine Schists, which form the spine of the Northern Highlands. The Moine thrust is regarded as a late event of the Caledonian orogenic period, as is the intrusion of numerous granite and syenite masses. Other granites of migmatitic character are probably older or originated at a deeper structural level and were affected by the regional metamorphism of the sediments now known as the Moine Schists. The late Caledonian granite of Helmsdale is of particular interest in that it was unroofed prior to the local formation of Lower ORS sediments. Subsequently, great thicknesses of Middle ORS sediments were deposited in the Orcadian cuvette. Thus, some 6000 m of calcareous siltstones, known as the Caithness Flags, are believed to be present in the central part of the ORS basin.

There are numerous faults in the region, the dominant one being the Great Glen fault in the southeast (see Fig. 1). This fault is believed to have had a long history of movement and has been correlated with the northeast-trending Helmsdale fault, which downthrows Jurassic sediments to the east. Evidence of soda metasomatism recognized very recently in the Inverness district is probably related to deep-seated carbonatite intruded in the plane of the Great Glen fault.[2]

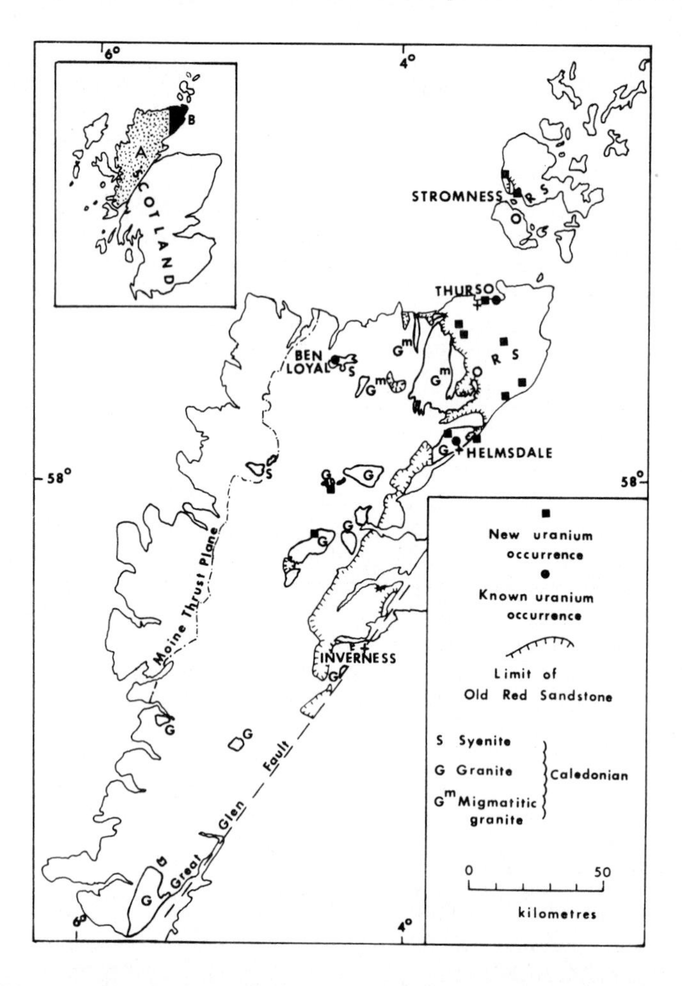

Fig. 1 New uranium occurrences found in recent exploration of Northern Highlands of Scotland. Generalized geology of region is also shown. Northeastern district (*B*) of region of exploration (*A*) indicated in inset map

There is no metalliferous mining at present in the Northern Highlands, but several vein deposits have previously been worked for lead—mainly in the last century. The most notable of these deposits is in the southwest of the region at Strontian, where non-uraniferous lead–zinc veins occur in Moine Schists at the edge of a Caledonian granite. The veins are, nevertheless, very much younger than their country rocks as they cut Permian dykes, so similar veins scattered through the region may also be appreciably younger than the enclosing ORS or Precambrian rocks. The distribution of mineral occurrences in the Highland regions of Scotland has been illustrated elsewhere.[3]

Radiometric methods of exploration

Evaluation of the relative success of different methods of mineral exploration is inevitably subject to qualification—perhaps nowhere more so than in the uranium field where aeroradiometric as well as ground radiometric and geochemical techniques can be employed. In the Northern Highlands of Scotland aeroradiometric work was limited to short total-count traverses along the mainly inaccessible coastal cliffs of the northeastern part of the region. This work confirmed the known presence of a fairly large ground anomaly (J. M. Miller, personal communication) and together with a boat-mounted survey over the same coastal section indicated the absence of comparable anomalies elsewhere.

Vehicle-mounted radiometric surveys were conducted over about 4000 km of roads and tracks in the Northern Highlands by use of a NaI(Tl) detector mounted parallel to and some 3 m above ground level either above or at the rear nearside of a Land Rover. Because many sections of road follow the sides of valleys, it proved necessary to travel 1500 km of road in both directions in order to achieve effective radiometric coverage. An optimum travelling speed of 30 miles/h was determined by trials across known anomalies. Two significant new anomalies were located in poorly exposed districts underlain by ORS sediments, and four minor anomalies were found elsewhere. In addition, known associations of anomalous radioactivity with Caledonian granitic and syenitic rocks and with potassium-rich Cambrian argillite were confirmed.

Thus, the vehicle-mounted scintillation counter surveys achieved appreciable success, but because of the low road density inland from the relatively populous east coast they did not provide adequate radiometric coverage. Moreover, many inland routes in the Highland regions of Scotland follow the bottoms of gravel-filled valleys or cross plateaux covered thickly by peat and glacial moraine—environments where the effectiveness of any method of radiometric reconnaissance is restricted.

Fig. 2 Close-interval radon probe (centre) and gamma-probe (left) survey of a radiometric anomaly. Probe measurements are made in shallow holes into overburden by use of a sliding-hammer steel bar (right)

Although patently unsuitable for blanket reconnaissance, on-foot radiometric traversing in selected districts was by far the most successful method in terms of number of anomalies detected. Selection was based essentially on geological criteria in the early stages of the work; but, as the investigations gathered momentum, it was guided increasingly by the results of the vehicle-mounted radiometric surveys and the geochemical sampling operations described later. Many thousands of scintillation counter readings were recorded on the large-scale geological maps (1 to 10 560 scale) available in manuscript for most of the region and also on large-scale plans of certain anomalies investigated by close-interval radiometric gridding. The standard instruments for this work were the commercially available Rank Nucleonics scintillation counter NE148 (AERE type 1597A), which incorporates a 25-mm diameter by 38 mm long NaI(Tl) detector and weighs 1·8 kg, together with the older AERE instrument (type 1413A), which is based on a similar detector and weighs 4·2 kg.

In the Northern Highlands, and particularly in the northeastern district where work was eventually concentrated, the rocks are generally buried beneath several metres of 'head' (locally derived materials), glacial moraine or fluvio-glacial material covered by peat or soil. Peat is especially widespread in upland areas of metamorphic and granitic rocks where, in consequence, radiometric traversing has been largely restricted to stream sections which afford some degree of rock exposure. In recognizing anomalies account had to be taken not only of the nature of the overburden but also of the surface and near-surface drainage, most notably the presence of springs and seepages. Small variations in scintillation counter readings (10–30 μR/h) characteristic of 'background' levels of gamma-activity over non-mineralized outcrops frequently provided the first decisive step in the location of significant anomalies in ground obscured by thick overburden.

The limitations of surface radiometry in districts where rock outcrop is poor or absent have been partly overcome by the use of gamma and radon probes[6] to delineate ill-defined surface gamma anomalies and the extensions of uranium mineralization recognized from scarce outcrops or transported 'float' material. The application of gamma and radon probes to regional reconnaissance has not been attempted because of the requirement for close-interval measurements (normally about 10 m with radon probes and about 2 m with gamma probes) based on detailed topographic control. A full account of the detection of concealed uranium mineralization by use of gamma and radon probes has been given elsewhere,[5] but two general points can be made here: (1) comparison of the two techniques shows that in relatively thick waterlogged peat best results are obtained with gamma probes, whereas both are effective when the cover is more permeable; and (2) in flat-lying ground the gamma and radon anomalies are generally coincident, but on hill slopes the displacement of radon anomalies by tens of metres downslope from gamma-probe anomalies has been observed.

Geochemical sampling methods of exploration

Drainage sampling was initiated at the outset of the investigation in the north-eastern district of the Northern Highlands, and has been extended gradually southwestwards as part of a geochemical reconnaissance of the Northern Highlands (covering nearly 8000 km^2 by the end of 1971). Scintillation counter measurements formed an integral part of this reconnaissance.

The drainage surveys involve collection of stream sediment and water at a mean site density of about 2 km^2,[9] and heavy mineral concentrates at most sites. A standard sediment sample consists of 1 kg of $-2{\cdot}5$-mm product, obtained by wet-screening large volumes of sediment through close-fitting nylon-mesh sieves mounted over a 45-cm diameter wooden pan. A standard heavy mineral concentrate is obtained by panning 2–3 kg of the same material. Uranium and several other metals of economic interest are markedly concentrated at fine-particle sizes in stream sediments characteristic of the main rock types of the region, and this has led to selection of the -100-mesh (B.S.I.) (150-μm) fraction of sediment for analysis. Sizing of the field sample is carried out locally after drying, and the fine sediment is further reduced in agate before taking a standard 1-g aliquot for delayed neutron measurement as described by Ostle, Coleman and Ball.[8] Table 1 gives analytical methods and detection limits for the more important of up to 20 other metals routinely determined on further aliquots of stream sediment in the laboratories of the Geochemical Division of the Institute of Geological Sciences.

Table 1 Abundances, methods of analysis and detectable limits for some metals of economic interest in Northern Highlands of Scotland

Metal	Abundance, ppm Caledonian granite*	Caithness Flags†	Limit of detection, ppm‡	Method of analysis§
Ni	5	50	3(10)	OES(AAS)
Cu	10	25	10	AAS
Zn	70	80	20	AAS
Mo	*3·5*	*3*	2	SWC
Ag	<1	<1	1	AAS
Sn	<3	<3	1(3)	SWC(OES)
Ba	1500	1000	60	OES
Pb	*40*	*75*	20	AAS
U	*9*	*5*	0·1	DNM

Metal abundance figures in italic type are regarded as values enriched relative to normal crustal values in comparable rock types.

*Based on analyses of 11 rock specimens from the Grudie granite, Sutherland (Gallagher *et al.*,[3] Table 2).

†Approximate average metal abundances based on analyses of rock samples from widespread localities in Middle ORS Caithness Flags (U. McL. Michie, personal communication).

‡P. J. Moore and P. B. Greenwood (personal communication).

§OES, optical emission spectrography; AAS, atomic absorption spectrophotometry; SWC, streamlined 'wet-chemical' techniques; DNM, delayed neutron measurement.

Stream-water sampling proved particularly effective in the environment of the ORS, where the waters are somewhat enriched in bicarbonate and possess a clarke uranium concentration significantly higher than that of stream waters derived from metamorphic and migmatitic rocks forming the western margin of the ORS basin. One significant uranium prospect (Gallagher *et al.*,[3] Fig. 20) and several lesser occurrences were located in the ORS as a direct consequence of an early water sampling survey by Dr. T. K. Ball. The distribution of uranium in stream waters of the northeastern district of the Northern Highlands, as determined in later high-density sampling by U. McL. Michie and Dr. L. Haynes (see Davis *et al.*,[1] Fig. 5), illustrates a fairly close correlation with that in −100-mesh stream sediment (unpublished work by Jane Plant), although locally there are differences. In upland districts of crystalline rocks the incidence of known uranium occurrences appears to be more clearly illustrated by the pattern of uranium in stream sediment. On the other hand, in low-lying ground underlain by the ORS Caithness Flags the pattern of uranium in stream water is more comprehensive, at least in part, because of the abundance of agricultural drainage suitable for water sampling but not for stream-sediment collection.

The uranium anomaly patterns in stream sediment and water related to the Caledonian granitic mass at Helmsdale have already been published (Gallagher *et al.*,[3], Fig. 8). The mineralogy of panning concentrates taken at many sampling sites shows that a significant proportion of the uranium in stream sediment is held in resistate minerals, such as thorite, monazite and secondary iron oxides. There is also a concentration of uranium in fine silt and clay fractions of stream sediments, especially in drainage related to districts of deeply weathered granite in the north of the pluton. As might be expected, uranium values are high (> 10 µg U/l) in stream waters derived from the same districts. Thus, the interpretation of uranium patterns shown by large-area drainage surveys becomes meaningful only when data are available from several sample types.

The average uranium content of a small number of samples of the Helmsdale granite is approximately 8 ppm U—similar to that of the Grudie granite (Table 1) and significantly higher than the average crustal abundance of approximately 4 ppm U in granite.[10] Thorium is also enriched in this pluton and the average Th/U ratio is about 5. In the districts of deeply weathered granite uranium is two to three times higher than usual and in the associated drainage there are concomitant increases in the levels of uranium in −100-mesh stream sediment and in stream water (normally about 20 ppm U and 3–4 µg U/l, respectively).

In the ORS Caithness Flags, where many new uranium occurrences have been found, the average uranium content (about 5 ppm) is accompanied by some degree of enrichment in lead, molybdenum and other metals. The geochemical information presently available does not permit the calculation of a metal balance between the ORS sediments and the older crystalline rocks from which they were presumably derived, but research along these lines is in hand.

Conclusions

By the application of relatively inexpensive ground radiometric and geochemical sampling surveys numerous new uranium occurrences have been discovered in the Northern Highlands of Scotland. The preliminary assessment of the gamma anomalies and uranium enrichments noted in water, stream-sediment and overburden sampling has been carried out by a combination of in-field radon-probe and gamma-probe measurements and by soil sampling.

The uranium mineralization forms weak disseminations and small veins in Caledonian granites, but it occurs more importantly as tabular enrichments, in phosphatic horizons and along fractures in Old Red Sandstone sediments.

References

1. Davis M. *et al.* United Kingdom research and development work in aid of future uranium resources. Paper presented at 4th International Conference on Peaceful Uses of Atomic Energy, Geneva, 1971.
2. Deans T., Garson M. S. and Coats J. S. Fenite type soda metasomatism in the Great Glen, Scotland. *Nature, Lond.* (*phys. Sci.*), **234**, 1971, 145–7.
3. Gallagher M. J. *et al.* New evidence of uranium and other mineralization in Scotland. *Trans. Instn Min. Metall.* (*Sect. B: Appl. earth sci.*), **80**, 1971, B150–73.
4. Michie U. McL. Further evidence of uranium mineralization in Orkney. *Trans. Instn Min. Metall.* (*Sect. B: Appl. earth sci.*), **81**, 1972, B53–4.
5. Michie U. McL., Gallagher M. J. and Simpson A. Detection of concealed mineralizations in northern Scotland. In *Geochemical exploration 1972* (London: Institution of Mining and Metallurgy, 1973), in press. (*Proc. 4th intn. geochem. Symp., London, 1972*)
6. Miller J. M. and Loosemore W. R. Instrumental techniques for uranium prospecting. This volume, 135–46.
7. Ostle D. Criteria for the selection of exploration areas in the United Kingdom. In *Uranium exploration geology* (Vienna: IAEA, 1970), 345–53.
8. Ostle D., Coleman R. F. and Ball T. K. Neutron activation analysis as an aid to geochemical prospecting for uranium. This volume, 95–107.
9. Plant J. Orientation studies on stream-sediment sampling for a regional geochemical survey in northern Scotland. *Trans. Instn Min. Metall.* (*Sect. B: Appl. earth sci.*), **80**, 1971, B324–45.
10. Rogers J. J. W. and Adams J. A. S. Uranium. In *Handbook of geochemistry* Wedepohl K. H. ed. (Berlin, etc.: Springer, 1970), II–2, pt 92.

DISCUSSION

R. J. Chico said that the evidence of Old Red Sandstone uranium occurrences was indicative of the existence of a uranium district in northern Scotland. He regarded that district as extending beyond Britain into the eastern Appalachians, U.S.A., and eastern Newfoundland, Canada, where, since early 1968, he had been involved in following up the results of regional reconnaissance work. No less than 200 uranium occurrences similar to those described by Dr. Gallagher were now known in eastern North America, but studies of those occurrences were not so far advanced.

The author said that it was interesting to learn of the recent evidence of widespread uranium distribution in eastern North America. It should be emphasized, however, that the Old Red Sandstone sediments of northern Scotland were an unmetamorphosed assemblage, whereas he understood that, in the Appalachian area of North America, the Middle Devonian was a period of orogeny, granite emplacement and regional metamorphism. On the basis of information published in 1965,* uranium occurrences in the Appalachian area were restricted to the immediately post-orogenic sediments of Upper Devonian to Lower Carboniferous age. No uranium had been found so far in the Upper Old Red Sandstone of northern Scotland.

S. A. Anzalone asked, with regard to the use of the term 'altered granite' in referring to the post-orogenic granite in the southern portion of the area of investigation, if that 'alteration' were some supergene effect or some form of hydrothermal alteration.

The author replied that the mode of origin of the extensive districts of weakly uraniferous and deeply altered granite in the Caledonian Helmsdale pluton was not clearly understood and was certainly deserving of much closer investigation. Although deep weathering had undoubtedly taken place, the available petrographical evidence did not preclude the possibility that the supergene processes were accelerated as a result of prior hydrothermal alteration. Such hydrothermal alteration could have been brought about by uraniferous fluids, most probably of granitic origin. There was limited field evidence of fracturing of the granite in the districts of deepest alteration and there was indirect evidence from the distribution of inclusions of schist country rock that the districts of altered granite were close to the roof of the pluton.

Although it was not known at what period or periods in the geological history of eastern Sutherland deep weathering had occurred, geochemical sampling had demonstrated that uranium was being liberated from districts of deeply altered and weakly uraniferous granite into the present-day drainage. The easily leached character of the altered granite suggested that it was improbable that it was enriched in uranium as a direct result of deep weathering beneath an ancient erosion surface.

Dr. G. Lawrence enquired, with regard to the radon monitoring, the depth to which the probe was sunk, and the sampling intervals used along traverses. In reply, the author and R. D. Beckinsale said that measurements of alpha activity in the field with the instruments developed were normally at depths of 2 ft, taken on grid patterns with centres of 20–50 ft. In the case of linear anomalies, measurement intervals along widely spaced traverses had been as little as 10 ft. Their application of that technique up to the present time had been entirely in the 'follow-up' phase of work on radiometric or geochemical anomalies. Alpha and gamma probes had been used concurrently with appreciable success in the surface assessment of anomalies, especially where those occurred in peat-covered ground. Alpha activity was measured immediately after withdrawing the hole-making implement (a steel bar), despite problems that might arise due to lack of equilibrium, in order best to standardize readings. In conditions of water-saturated peat the use of an alpha probe was inhibited by the risk of contaminating the detector.

They had attempted to measure thoron by counting over long periods of time rather than over the 1-min periods normal in their radon measurements; but they had not detected thoron in work to date. In northern Scotland emanometric studies had so far been made only during follow-up work at localities identified by surface radiometric work and geochemical drainage sampling. The encouraging results obtained led to the conclusion that radon measurements were desirable during the primary prospecting phase, especially in geologically favourable areas which had proved unresponsive to surface radiometric traversing.

*McCartney W. D. Metallogeny of post-Precambrian geosynclines. In Some guides to mineral exploration Neale E. R. W. ed. *Pap. geol. Surv. Can.* 65–6, 1965, 33–42.

FINAL DISCUSSION SESSION

Dr. S. H. U. Bowie, in opening the final session, said that Dr. Dodson did not commit himself on whether the Alligator Rivers deposits were syngenetic or epigenetic. But he did indicate that they ranged in age from 550 to 815 m.y. and that the host rocks were approximately twice as old. He also spoke of replacement orebodies in carbonate rocks as being a distinct possibility. The speaker would agree with him. Furthermore, he would suggest that galena age dating, used in conjunction with uranium–lead ages on uranium minerals, would be a valuable guide as to whether the mineralization was epigenetic or whether there had been repeated introduction of mineralization or reworking. That was also the answer to the question raised by Mr. Beckinsale concerning fresh introduction of uranium or minor recrystallization. If galena and concordant uraninite ages agreed in different parts of a deposit, there was little doubt that repeat mineralization had taken place, since a galena age did not change on recrystallization like that of uraninite. He would also suggest to Dr. Dodson that a study of wallrock alteration would provide valuable information on the mode of origin of the uranium mineralization.

Mr. Grutt concentrated on deposits in the U.S.A., particularly those of the Colorado Plateau and the Wyoming Basins. He suggested five possible modes of origin, and the speaker would agree that all five were possible, though he preferred 'deep-seated hydrothermal' to 'magmatic hydrothermal'. He wished, however, to raise some queries. (*a*) The chemistry that allowed uranium deposits to form in the absence of deposits of other elements of greater abundance in acid igneous rocks, such as iron, titanium or manganese, and the mechanism that permitted selenium and vanadium to form in relatively small amounts in association with some of the uranium deposits of the U.S.A. and copper and vanadium in ore grades with others. (*b*) The leaching hypothesis, which seemed very doubtful to him, especially as the tuffs, supposed to have been leached of uranium, were still existent and contained as much uranium when they were weathered down to bentonitic clays as they did when fresh. All the evidence was that uranium was not leached out of tuffs; they were still radioactive and contained approximately three times as much thorium as uranium. What was supposed to have happened to the thorium as they were leached? The same approximate U:Th ratio applied in the case of granites. It also seemed that M. Gangloff's model did not fit factual geochemical data, as in an enclosed basin a detrital concentration should have three times as much thorium as uranium.

He was anxious to see any analytical data on possible source rocks that were supposed to be unaltered, partly altered and completely leached of their uranium.

Dr. Barnes gave valuable information, essentially on costs per lb uranium at the exploration stage. However, he did not compare costs of geochemical or biogeochemical exploration with those of geophysical exploration. Monsieur Grimbert indicated a cost of $\approx$50F/km^2 for geochemical exploration, which compared well with U.N. figures of £4/km^2 for one sample/km^2; and a value of £20–200/km^2 for detailed soil surveys. That compared with about £30/km^2 for aeroradiometric coverage with a large detector volume gamma spectrometer.

The question of whether geochemical data should be obtained in the field or perhaps more accurately at a headquarters laboratory was largely a matter of personal preference and of environmental conditions, though there was much to be said for being able to follow indications immediately in the field.

As far as scintillometers versus G–M counters were concerned, it was generally agreed that the former were most suited for primary surveying and the latter for detailed studies of pits, trenches and underground workings. Instruments with β/γ facilities were specially useful for semi-quantitative and disequilibrium measurements.

It seemed to be generally accepted that total-count gamma was most suitable for initial reconnaissance surveying, with gamma spectrometry for the subsequent examination of anomalies. That particularly applied in areas where mixed U–Th ores were expected and the surface conditions were suitable for flying.

Aeroradiometric surveying was generally to be recommended when the time factor in discovery was important—which, however, was scarcely the case at present—or where general information on uranium distribution was considered valuable. There was new evidence that reconnaissance geochemistry, which was as inexpensive as any technique, could draw attention to a favourable region in which a uranium deposit was likely to occur, in the same way as regional aeroradiometric surveying could draw attention to a favourable area. He considered that more attention should be given to geochemical techniques such as had been described by M. Grimbert and Mr. Ostle. It was not realized widely enough that geochemistry could be used during reconnaissance and follow-up stages of exploration—M. Grimbert's 'strategic', 'detailed' and 'tactical' phases.

He wished to give a word of caution about accepting, as had been suggested earlier, that an area of higher than normal radioactivity necessarily contained ore deposits. The albite–riebeckite granites of Nigeria and the Conway granite of New Hampshire both had high activity over many square miles, but there were no known associated uranium deposits.

On the subject of costs in general, he said that he would not accept the axiom that the more money one spent the more likely would material be found to fill the ore bins. If prospecting were based on geological knowledge, and the most appropriate prospecting method were used, ore could be found at low cost. Moreover, he would not accept the suggestion that the larger the land area the more chance there was of finding uranium. That would only apply if the land area were favourable to uranium mineralization. He could think of thousands of square miles of country on which he would not recommend the expenditure of one penny. It was important to think more of cost effectiveness in prospecting. That was of greater significance than comparing costs as between different techniques on an area basis.

In general, it was wise not to extrapolate experience gained in one type of geological environment to another. Mr. Grutt made that point very clearly. Those concerned with prospecting through soil and other overburden should bear in mind that conditions in the Shield of Canada were widely different from those of the Northern Territory, Australia, or Equatorial Africa. In many conditions of tropical weathering radium, as well as ^{214}Bi, was leached out of the surface layers altogether and there was no anomaly at surface that could be attributed to uranium. That was where geochemistry and radon in ground air measurement became especially important; for he was sure that more deposits with no surface radioactivity as well as those that outcropped and could be detected by radiometric methods had to be found.

He had followed the detailed investigations of Mr. Dodd and Dr. Darnley over recent years with great interest, but felt that it would have been more significant if the latter had not chosen areas that had been opened up, for example, Bancroft, Blind River and Beaverlodge. He did not believe that it was now possible to tell what could have been detected by aeroradiometric techniques had they been used at those localities prior to the commencement of mining operations.

There was a danger in attempting to quantify aeroradiometric signals as the stripped count rate in the ^{214}Bi channel was rarely statistically meaningful. Furthermore, the count rate—total or spectral—was of little value in deciding whether or not an occurrence was of economic significance. That point had been made by Mr. Chico. As long ago as 1958 he and his co-workers, in describing the use, in Cornwall, of a total-count instrument with a 850-cm^3 NaI(Tl) detector had also made that observation.* He saw no reason arising from subsequent investigations to change that view.

On the historical aspects of Dr. Darnley's lecture he felt that participants would be specially interested in the work of Davis and Reinhardt† and Sakakura‡ in the U.S.A. Also, they might not be aware of the fact that gamma spectrometry was utilized by the CEA, France, as long ago as 1955–57.§ It was not new 13 years ago to suppose—despite criticism at that time—that total count, potassium, uranium and thorium could be measured semi-quantitatively from an aircraft.

The speaker, in concluding, said that it was now his pleasure to act as Chairman of

*Bowie S. H. U. *et al.* Airborne radiometric survey of Cornwall. In *Proc. 2nd Int. U.N. Conf. peaceful Uses atom. Energy, Geneva 1958* (Geneva: UN, 1958), vol. 2, 787–98.

†Davis F. J. and Reinhardt P. W. Instrumentation in aircraft for radiation measurements. *Nucl. Sci. Engng*, **2**, 1957, 713–27.

‡Sakakura A. Y. Scattered gamma rays from thick uranium sources. *Bull. U.S. geol. Surv.* 1052-A, 1957, 50 p.

§Berbezier J. *et al.* Methods of car-borne and air-borne prospecting: the technique of radiation prospecting by energy discrimination. In *Proc. 2nd U.N. Int. Conf. peaceful Uses atom. Energy, Geneva 1958* (Geneva: UN, 1958), vol. 2, 799–813.

the session, and he hoped that as many as possible would participate in the discussion. He had thoroughly enjoyed listening to the papers presented and to the case histories.

R. D. Beckinsale said that there appeared to be conflict between certain statements and a film depicting the Blind River discoveries, which indicated that originally there was no surface uranium anomaly, and data given by Dr. Killeen and Dr. Darnley, which showed clearly that there was a well-marked surface uranium anomaly. He asked the solution to that dilemma. Referring to the Colorado Plateau environment, Mr. Beckinsale pointed out that Dr. Bowie had doubted that there was evidence for surface leaching of uranium from basement rocks. The U–Th–Pb data of Rosholt and Bartel* showed clearly that the Granite Mountains in central Wyoming had lost about 10^{11} kg of uranium, i.e. at least one thousand times the ore reserves plus production in producing Wyoming Basins.

Referring to Mr. Beckinsale's latter comment, Mr. Grutt stated that Rosholt and Bartel were very fortunate in being able to obtain some very deep core and by sampling down the core for a distance of 700–800 ft they were able to show that there was considerable surface depletion of uranium.

The granites were highly jointed and fractured and provided ample opportunity for the circulation of groundwater accompanied by the removal of any intergranular uranium.

In the Wyoming Basin, at the close of Pliocene times, there was a plateau about 10 000 ft above sea level with Precambrian Monadnocks standing above that plane. Since then some 5000 ft of sediments had been removed. Those rocks were mainly tuffaceous and ample uranium was therefore available during the weathering to be re-deposited elsewhere. Volcanic rims remaining around the margin of the basin indicated that millions of tons of uranium were available for such re-deposition. That, of course, did not dispose of the question of the process of deposition, but certainly ample uranium had been available to account for the uranium deposits.

Dr. Bowie said that he was aware that uranium was available in the granite and that it was removed during weathering processes, but he questioned the mechanism and the chemistry by which the uranium became concentrated in deposits and was separated from other elements which occurred in much greater concentrations. Much of the uranium and thorium in unweathered granite was in resistate minerals and if those were broken down, it was necessary to explain what had happened to the latter element. Dr. L. W. Koch proposed that the reason there was little knowledge of other leachable rock constituents being reconstituted in sedimentary basins similar to the sandstone-type uranium deposits was because nobody had bothered to sample and analyse sedimentary rocks if they did not show traces of radioactivity. He indicated that systematic sampling of boreholes in sedimentary basins suggested that many of the rock constituent elements were present in measurable, although uneconomic, concentrations. They would not, however, necessarily coincide in location with the uranium concentrations.

Referring to Mr. Beckinsale's first point, Dr. S. M. Roscoe was invited by Dr. Bowie to speak on the history of the Blind River deposit. Dr. Roscoe stated that there was never any question about surface leaching of uranium in the Blind River deposits apart from that in the first account. When the deposit was first looked at in 1948, during a uranium 'rush', surface pits were found to be radioactive within the area, but samples taken from those pits were found to be of sub-ore grade. The values obtained were up to 0·2% uranium equivalent and also up to 0·15% actual uranium. Those pits, however, were not in the main ore shoot and further drilling located the ore shoot at a greater depth. Subsequent investigations showed that there was no evidence of surface leaching, but the story of the leaching was never refuted.

Dr. A. K. Ghosh enquired as to what were the conclusive geological criteria for regarding a deposit as of true deep-seated hydrothermal origin rather than of magmatic or metamorphic hydrothermal origin. Dr. Bowie replied that it was his belief that few, if any, uranium deposits were actually derived from granitic rocks. Uranium deposits throughout the world, ignoring some of the syngenetic types for the moment, were spatially—and probably genetically—associated with acid igneous rocks, either extru-

*Rosholt J. N. and Bartel A. J. Uranium, thorium, and lead systematics in Granite Mountains, Wyoming. *Earth planet. Sci. Lett.*, 7, 1969, 141–7.

sive or intrusive. In the case of mineralization associated with granites, in particular, the age of the mineralization post-dated the granite and the uranium was almost certainly not a differentiation of the granite itself, but the two had a common origin, i.e. they were both of the same parentage. Separation apparently took place at a considerable depth—either at the base of the crust or in the upper mantle. For that reason it was perhaps better not to talk about 'magmatic hydrothermal' because he did not think that the uranium deposits resulted from differentiation of a granitic magma. Deep-seated hydrothermal—as he had indicated in his opening remarks—would be a better phrase.

Dr. Bowie went on to point out that it was important to know whether the introduction of uranium had been a single or repeat event, or whether uranium had been subsequently remobilized. It was, therefore, very important to determine the ages of any associated galena in determining the origin of a uranium deposit.

Dr. Roscoe agreed that the role of time in the genesis of deposits was important. He pointed out that there was a considerable amount of uncertainty in the exact location of the dates shown, depending on which method of age determination one used; but, in general, they were relatively accurate. Bearing in mind that uncertainty, he estimated that the Archaean rocks were greater than 2500 m.y. old and that the Proterozoic was younger than that date. The conglomeratic uranium deposits were older than 2300 m.y. The division between the Archaean and the overlying Huronian was a very important one historically, Elliot Lake being the classic division point. Most Archaean areas were low in potassium, uranium and thorium, but there were certain high spots among those areas which were possibly the progenitors of uraniferous provinces of later ages. The rocks containing uranium conglomerates were generally of fairly restricted age. The Witwatersrand conglomerates were equally old and so were other occurrences throughout the world. The other great concentration of uranium in sedimentary rocks took place in the Phanerozoic. Dr. Roscoe argued that there was therefore a vast spread in the ages of sedimentary uranium deposits from 2300 to 300 m.y. He also pointed out that the pegmatitic uranium deposits, in North America at least, were very restricted in geological age. Vein deposits in Canada were virtually restricted to late Proterozoic rocks. They were also found in older rocks, but very close to regions containing older Middle Proterozoic rocks. That applied from Great Bear Lake to Beaverlodge and down to the Lake Superior Basin. Dr. Roscoe said that a similar analogy could be made to Australia, especially with regard to those ages, which were quite comparable. In Australia also the search for uraniferous conglomerates was being directed especially to Lower Proterozoic rocks. The Proterozoic rocks in Australia generally ranged from 1700 to about 2100 m.y. old, but the Huronian rocks were generally older than that. From about 2300 m.y. onwards there was a marked change in the composition of the atmosphere—from reducing prior to that to oxidizing at a later date. That was shown by, among other things, the presence of red beds. There was no evidence prior to that of an oxidizing atmosphere, especially with regard to the presence of red beds. The major observation was that there was a marked restriction of the types of uranium deposits with time. Dr. Roscoe also pointed out that the conglomeratic deposits and sandstone-type deposits did not seem to show any intergradation. Neither was there any gradation between the pyritic conglomeratic deposits and black sands, black sands not being found, as far as he was aware, in rocks older than 2300 m.y.

Mr. G. R. Ryan thought that Dr. Roscoe's comment relating to the oxidizing atmosphere change at 2300 m.y. was very important. The conglomeratic uranium ores were not the only types of ore to be confined virtually to rocks older than 2300 m.y. There was a family of massive sulphide ores, such as the taconite and jaspilite ores, which were also restricted to rocks of that age. The supposition could also be made that a conglomeratic uranium ore deposit was not likely to be found of later age. Mr. Ryan suggested that they were perhaps wasting their time in looking for uranium conglomeratic ores in anything younger than the Lower Proterozoic. Referring to Dr. Roscoe's comment on the age of the Lower Proterozoic in Australia, Mr. Ryan said that it certainly could be as old as 2500 m.y., but the dating had not been very certain. He had himself worked on those rocks in Australia and there was no evidence of glaciation, whereas the Huronian rocks in the type area contained ample evidence of that. He speculated as to whether the reason for the occurrence of conglomeratic ores was simply that there was glaciation present and wondered if there were any evidence of Witwatersrand conglomerates containing glacial deposits.

Dr. J. W. von Backström replied that in some conglomerates at the base of the Witwatersrand in the Dominion Reef, dates of about 2800 m.y. had been obtained, which seemed to be slightly older than the Lower Proterozoic of Canada. At the base of the Transvaal System there were conglomerates, and some of those had been mined for their gold content. Pebbles in the conglomerates were highly uraniferous, but the matrix did not contain appreciable quantities of uranium. According to the mineralogists, the uranium in the pebbles seemed to be syngenetic. Their origin, according to the Geological Survey, was in much older formations (about 3300 m.y. old). That was the Moody conglomerate, the Moody Formation being regarded as Archaean. Dr. W. Herzberg amplified some of Dr. von Backström's comments and stated that the Moody conglomerate appeared to contain pebbles of older sedimentary uranium deposits, suggesting that there were sedimentary uranium deposits in existence prior to 3300 m.y.

Mr. R. J. Chico supported Dr. Roscoe's statement of the importance of glaciation in the Blind River deposits and pointed out that the suggestion that the Blind River deposits were the result of channel infilling was not true. Referring to an earlier statement that the greatest concentrations of uranium were not generally related to derivation from an acid magma, he noted that the lithology of the Blind River deposit was not an arkosic type. Dr. Bowie asked Mr. Chico whether he was suggesting that the oligomictic conglomerates of the Blind River deposit were not, in fact, derived from acid igneous rocks. Mr. Chico replied that the composition of greywackes, sub-greywackes and arkoses must be taken into consideration. It depended upon which classification was used—Pettijohn's or Twenhofel's. Following sedimentological investigations of the Blind River deposits, it was found that those lithological types associated with the ore deposits were, in fact, related to sub-greywackes rather than to arkoses.

Mr. Dodd said it was very interesting that it was only after this 2300 m.y. event that limestones began to appear in the geological succession. Garrels had been doing work with partial pressures, solubilities of carbonate and bicarbonate, etc., and he speculated that, if indeed there was a change in the atmosphere, large quantities of the very substances that were being put forward as transport media for uranium at the present day were suddenly liberated to the weathering environment. That presumably took place over a comparatively short period of time; therefore, prior to that event there might have been considerable concentrations of uranium at the near surface which suddenly became capable of being moved.

Dr. Barnes said that he had worked for some time on the origin of uranium deposits and he would like to make a few comments in respect to age relationships. He did not think they should be bound by the laws of uniformitarianism; however, there were a few aspects about that problem which he thought it important to remember. First of all, if the atmosphere was not oxidizing prior to 2300 m.y., why was it that the uranium/thorium ratios in the Blind River conglomerates were so high? He also pointed out that uraniferous conglomerates with detrital uranium minerals, and he did not think that there could be any doubt that at least some of the uranium in the Blind River deposits was detrital, did not only occur prior to 2300 m.y. but were also found at the present time, even though the uranium minerals did not survive. In the Russian literature there were descriptions of a deposit which, in all respects, was very similar to the Blind River type, and it was of Cambrian age. He agreed with Dr. Roscoe's argument about the evolution of the types of sedimentation. The size of geosyncline, for example, increased from the Archaean to the present, and the types of geosynclines and the character of the rocks had all changed. He wondered if that related to the availability of the granitic crust at some time in the Archaean or to the nature of the mechanism which built plates or mountain chains such that that material could be mobilized. Once it had been introduced, it could be remobilized in successive cycles, through granitization and the building of acidic material with concentration and evolution of that geochemical cycle. He elaborated on the features of some of the vein-type deposits, for example, the true vein type (the five-element type), which was quite distinctive. Those were what Dr. Dodson had described and were typical of most of the important uranium 'vein types' of deposit. Those true 'vein types' did not represent a very important contribution to world ore reserves at present and would obviously not do so in the future. He did not see why the five-element vein type should not form, not only in late Proterozoic time, as in the Great Bear Lake area, but all the way through to the later Tertiary, as had happened in Western Canada. Of the other types of deposit which might also be called 'vein types' and 'replacement deposits' or 'chemical types', those might be formed allogenically, as opposed to orthogenically as in sandstones. Those types were not only known from the early Precambrian but were present by age up until 80 m.y. ago. He thought that when a large deposit was found it tended to be compartmentalized in a special place in that age. That occurred partly because of the evolution of geological processes, but at the same time that could occur throughout all geological times and it was dependent on the conditions which prevailed at various times in geological history.

Dr. T. F. Schwarzer said that she had not heard any discussion on the role of sulphur-reducing bacteria in the formation of uranium deposits. It was now generally recognized that the bacteria acted as a catalyst in many geological processes. They produced hydrogen sulphide which should result in the precipitation of uranium. Under those circumstances there was a biological factor involved. That process would have to be confined to the later geological times, certainly to the Phanerozoic. That might explain why certain types of deposit, especially the sandstone types, were confined to later geological periods and were almost non-existent in Precambrian times.

Mr. Grutt said that he did not know when bacteria were formed or developed on earth. He was unsure whether the sulphides associated with the Blind River deposits were a result of chemical precipitation aided by bacteria or not, but it was perfectly true that the bacteria produced hydrogen sulphide in their body cycle and fed on carbon as a nutrient. Those were indeed very powerful factors in the formation of the sandstone deposits. It had been demonstrated that they had a very important part to play in the production of uranium roll-type deposits, and they were also important in the fractionation of sulphur isotopes. Depletion of the heavier isotope, ^{34}S, was very marked in the interior of the ore rolls. The counterparts of the sulphur-reducing bacteria, *Thiobacillus* and *Ferrobacillus*, also played a very important part in the formation of roll deposits and might well be one of the factors in producing a condition for the formation of those deposits in the first place.

Dr. E. E. N. Smith mentioned the fairly ubiquitous occurrence of the hydrocarbon, thucholite, with pitchblende in fracture-type uranium occurrences, especially in northern Saskatchewan. He agreed largely with Dr. Roscoe regarding the relationship of uranium with geological time, but pegmatites were common in the Lower Proterozoic and also in the Archaean, and it was towards the end, during the Hudsonian orogeny, that pitchblende began to appear. From then on it had undergone remobilization over at least six major periods, the most recent dating being in the 25 m.y. range. It was the speaker's opinion that uranium was still being moved around in those zones in northern Saskatchewan.

M. Grimbert wished to amplify the remarks made by M. Gangloff, and mentioned by Dr. Bowie during his general comments. It was not the resistates which accounted for the accumulation of uranium in sedimentary basins. Following earth movements, all the accumulated uranium on the continent was removed and deposited in the basins.

Mr. W. Ching Chao enquired the minimum grade of ore which could be economically mined at the present time. Mr. Grutt replied that costs would very much depend on whether the deposit was being worked by open-pit or underground methods. As a rough guide, he could give some working cutoff levels in current operations in the United States. The cutoff for open-pit mining was of the order of 0·08% U_3O_8. That, of course, depended on the degree of stripping of overburden and other costs. Therefore, the ore grade would probably be not less than 0·15% U_3O_8. For an underground property one would think in terms of 0·25% U_3O_8. Cutoff grades in large stopes, however, would be down to about 0·08%. Average grades had been reduced since the start of mining from about 0·25% to grades being mined today of about 0·18% U_3O_8. That was largely a result of increases in the efficiency of mining operations as they moved out of the teething stages. Those costs applied to very large properties: the smaller the property, the higher was the grade required to give an economic return.

Mr. R. B. Sprague referred to earlier comments on the activity and role of bacteria in the formation of sandstone-type uranium deposits. There was an almost universal presence of carbonaceous material in the conglomerates of the Lower Proterozoic deposits, both in Canada and the Witwatersrand. He asked if anyone familiar with those would care to comment on the possible source of the carbon.

Dr. Dodson made another observation that the conglomerates of Rum Jungle fell into the 2300 m.y. age group very well. They looked remarkably like the Witwatersrand, but they lacked at least two features which were essential: first, they did not contain much carbon and, secondly, they did not contain much in the way of sulphides. Although that was negative information, it probably had a bearing on Mr. Sprague's question.

Dr. Bowie said he agreed in general with what had been said. The association with hydrocarbon was worldwide, but the origin of the hydrocarbon material was not always known. There were two likely modes of origin—one from various forms of plant life and the other from a deep-seated source. Very similar hydrocarbon occurred in pegmatites and the first known thucholite was described from a pegmatite in Ontario, Canada. Considering the material in the Blind River field, which actually replaced quartz pebbles and some of the earlier minerals like zircon, brannerite and so on, Dr. Bowie said that that had certainly been remobilized since the time of precipitation. He thought the same was the case in the Witwatersrand where there was an extensive development of hydrocarbon, for example, in the 'Carbon Leader'. That could have been derived by the polymerization of methane, which was still present in many underground workings in the Witwatersrand.

Discussion of geochemical and geophysical aspects of exploration

Mr. P. L. Kitto asked the current consensus of opinion on radon loss from a sample prepared for gamma spectrometry, especially within the initial period of build-up. Mr. Dodd replied that the radon was driven off during sampling and grinding and that the sample did not achieve equilibration unless it was hermetically sealed. If results were needed urgently, it was not necessary to wait for 5 or 6 half-lives, since corrections could be applied from the knowledge of the growth characteristics of the activity. All that would be required would be 2 counts separated by a reasonable time-interval. He noted that there was a very small loss due to thoron.

Mr. Kitto observed that a 5% variance was commonly obtained when preparing samples for gamma spectrometry, and Mr. Dodd replied that it would depend on how the comparison was made, what minerals were being dealt with, the emanation coefficient and whether, in fact, the sample was efficiently sealed, which was extremely difficult to achieve. The New Brunswick standards were 9–14% emanators in general, but if one started doing gamma spectrometry of other minerals from rocks, particularly secondary uranium minerals like liebegite and shrockingerite, then one might encounter 35–40% emanation.

Mr. Chico asked for further comments on the subject of costs for their various methods of exploration. Dr. A. G. Darnley, dealing with the costs of the high-sensitivity gamma aerial spectrometer, gave a breakdown of the way the costs were apportioned. On the basis of 50000 line miles a season, a 10-year amortization of the aircraft and a 5-year amortization of the spectrometer equipment, the costs were \$20/km^2 for high-sensitivity gamma spectrometry. That was on the basis of quarter-mile spacing and worked out at \$15/line mile. That was in the range being quoted commercially at that time and included a 100% factor for overheads. As far as other smaller systems were concerned, he had little information. The *Canadian Mining Journal*, he thought, gave, annually, average costs over the previous year for various types of airborne geophysical work. He believed that low-sensitivity airborne work costs were in the region of \$8 or \$9 per line mile. There was, therefore, a difference between what might be called minimal standard equipment and the most sophisticated standard of equipment that was commercially available.

Mr. Ryan said it was not possible to set costs on a large number of programmes, because they depended on a host of variables, such as the number of moves, instruments to be used, the size of the programme, the climate, available working time, etc. Usually, other instruments were also run in conjunction with airborne radiometric instruments.

Mr. Dodd was of the opinion that the large aeroradiometric system was not for the average individual mining company and he thought that the trend would be to obtain that type of work from a service company. It was therefore really the commercial price that was setting the pace, and one could get everything that Dr. Darnley had been talking about, with perhaps even more statistical evaluation of the data, for less than \$15 for 10 000-line mile bunches in North America and for 20 000-line mile bunches in some of the other continents. At the same time, for the addition of \$3/line mile, one could get a magnetometer survey and the magnetic data all plotted and worked up. Those were all published prices.

Dr. Barnes said that he would like to make a comment about flying other systems with aeroradiometric techniques. A magnetometer was very simple to install in an aircraft system, but the electromagnetic system, if carried in a 'bird', for example, would encounter problems because of the height flown, which would depend on the type of radiometric equipment being used. There might not be a great deal of advantage, and one might be losing the benefits of one's radiometric equipment by going higher. Another factor was that electromagnetic survey provided much geological information, but the speaker thought that in looking for uranium deposits other types of deposit were not necessarily going to be discovered. In Canada the areas of uranium search did not generally correspond with attractive areas for base-metal search.

Dr. Dodson said that in Australia there was a marked association of uranium with graphitic schist or shales, and one might well have an application for electromagnetic methods. Dr. Barnes thought that EM had an important geological as well as economic significance. It had been used to find massive sulphide deposits, but, basically, it was a useful geological tool. Dr. Dodson commented on the use of EM surveys of carbonaceous shales. In his experience very poor results were obtained by simply flying over carbonaceous shales. The character of the carbonaceous shale was such that it simply gave a large continuous anomaly. EM surveying did, however, give an indication of a change in lithology where the rock types were hidden by superficial deposits.

Dr. Davis made another point with regard to general costs. In the United Kingdom uranium reconnaissance search of selected parts of the country, i.e. those regarded as most favourable, had involved car-borne surveys, foot follow-up, geochemistry and, where appropriate, radon surveys and gamma spectrometry. The overall cost of that worked out at about \$64/square mile (\$25/km^2) of coverage, which represented a fairly economical survey.

Dr. Koch said that he would like to solicit comment on the use of colour photography. In his own experience the emphasis on aeroradiometric surveys was being reduced in favour of aerial colour photography, both real and filtered. Coloured photos helped in structural and morphological interpretation and were absolutely essential in locating geochemical anomalies.

Mr. Chico supported Dr. Koch and pointed out that colour photography was very useful in reducing the need for extensive geophysical coverage.

Mr. Grutt said the U.S.A. had spent something of the order of \$25 000 in the past few years to re-evaluate colour photography in order to see what it would show in five of their major uranium-producing areas. He stated that, broadly speaking, colour photography showed twice as much as black and white aerial photography would have shown. American companies were increasingly turning to colour photography in that it gave them an increasing range of forward planning. The cost factor was roughly double that of black and white photography.

Dr. D. A. O. Morgan asked whether anybody was working in infrared colour photography. Dr. Darnley said that, although infrared was used in Canada, it was not being used specifically in the search of uranium. Infrared scanning was used and another project being carried out was in multi-spectral photography. Dr. Koch added that false-colour photography was also of very great value.

Mr. Chico said that he would like to give some information about the costs of logging boreholes in the Wyoming Basin, using independent contractors. With a typical 700-ft hole, the setting-up charges amounted to \$25, mobilization to about \$0·15/mile. Self-potential, gamma and resistivity surveys for that hole worked out at about \$0·07/ft.

Dr. P. Johnson asked whether colour photography was as effective in cultivated areas as in uncultivated areas, and Dr. Koch said that in his experience it was probably the only means of aerophotography that could be used in cultivated areas, since with black and white one could hardly see anything. Mr. Grutt supported that and said that in many cases colour photography was especially useful when the ground had been newly ploughed. Changes in colour were seen here to a better degree than when the ground lay fallow or was covered with crops. It was often worth while to wait for the ground to be ploughed before taking aerial photographs.

Dr. Schwartzer referred to a statement by Dr. Koch that he was not terribly interested in radiometric aerial prospecting any longer, and that he was generally able to buy in the required information. She submitted that, if a total-gamma radiometric survey had been done at some time in the past, by use of a low-sensitivity system, in view of Dr. Darnley's remarks on the advantages of gamma spectrometry it might be worth while to refly some areas that had already been flown by the total-count method. Dr. Koch, in reply, said that it was largely a question of opinion, but in his experience nothing new had been discovered by reflying areas with more sophisticated equipment. Areas that had been indicated by previous surveys were continually being located, however.

Mr. Grutt said that he agreed with both the previous speakers, but pointed out that in the areas which had been extensively prospected for uranium in the past, gamma-ray spectrometry could be of considerable use, but more generally in the field of geological mapping rather than in finding specific uranium deposits. Dr. Darnley said that he was inclined to go along with that. He thought that he would want to know the geology of the area that he was dealing with, but, in terms of what he knew, rather distantly, about the Colorado Plateau area his opinion would be that it would not be worth reflying with gamma-spectrometric equipment.

Mr. Dodd visualized airborne spectrometry as being used for different purposes from the old anomaly-hunting with a gross-count system. That involved looking at uranium/thorium/potash ratios, to do geochemistry from the air, in effect, not only in the search for uranium but for other minerals as well. That, therefore, was a different type of flying with different types of information being sought.

Mr. Chico emphasized that it was important not to underestimate the value of the prospector for the location of target areas for uranium. After all, a significant number of uranium discoveries were made by prospectors. Dr. E. E. N. Smith said that relatively few discoveries were now being made by prospectors. There were very few prospectors remaining and those were probably employed by exploration companies who directed where they operated. The old-time prospector was a dying breed. Dr. Barnes said that it was still possible to find prospectors in parts of Canada and that their activity was largely a question of the type of metal that was being sought; if a new mineral was in demand, then probably the prospectors would come forward to look for it. Mr. J. L. Montgomery stated that prospecting associations in Canada had many geophysicist/geologist members who, by implication, felt that a good exploration scientist should be basically a prospector for economic mineral deposits. He personally considered himself a prospector who was using the scientific method to locate ore deposits. Prospecting with a light aircraft by a prospector acquainted with local conditions was a low-cost way of utilizing colour photography advantages.

Mr. B. L. Nielsen queried the mechanisms of disequilibrium in uranium ores and asked whether there was a possibility of measuring too much or too little uranium when estimating ore grades by use of gamma spectrometry. Mr. Dodd replied that disequilibrium was particularly important when dealing with gross-counting techniques. The primary problem was the opposing geochemistry of uranium and radium, which behaved almost as opposites in the earth's environment: sulphate, for example, making uranyl ions travel easily while at the same time fixing radium. On some occasions there was a problem of moving radium. A small amount of iron and a few ppm oxygen would cause almost total precipitation of the radium. Chloride waters depleted in oxygen, when introduced to a new environment, could certainly result in a concentration of radium with no uranium in the near vicinity. Radium could also 'plate out' on iron pipes, such as oil-well casing, which had been sitting in the ground for a few years. Mr. Dodd explained that there was another type of disequilibrium, due to uranium migrating away from the radium and leaving a high radiometric anomaly behind. He had yet to see a case where there was some uranium that did not have an accompanying gamma count. The system might not be in equilibrium, but there was always enough radium concentration to produce a really good anomaly. Dr. Bowie stated that in practice it was not generally very important to correct for disequilibrium in early assessment stages and that, in any case, measurements could be made quite effectively in the field β/γ ratio measurements. Mr. Dodd drew attention, however, to an effect which had been seen recently of rapidly moving groundwater flushing radon out of the local uranium concentration, giving up to 50% discrepancy between the actual uranium content and that obtained by gamma logging. On sealing the material it was found that the uranium and radium were in equilibrium and the disequilibrium was due to the radon getting away. That was a very rare occurrence and the speaker had only seen perhaps two instances of it in the last 15 years.

Mr. Nielsen asked why Geiger counters were better tools for underground investigations. Dr. Bowie replied that the Geiger–Müller counter in underground working had a number of assets. First, it was possible to make the tube size so small that it could be used in a fairly high background. It could be used for the measurement of ore-grade material, which was another great advantage as most scintillation counters were completely swamped in an underground working in ore grade material. In deep underground conditions there was little or no contribution from cosmic radiation to which G–M counters were particularly susceptible. Thus, the latter could be used for the precise delineation of grades, e.g. during development. Dr. Smith generally concurred and said that in his case the most important reason for using Geiger–Müller counters, rather than scintillation counters, for ore-grade control was that they could be cheaply constructed. At Eldorado he had 40–50 G–M counters produced which were compact and lightweight, for use by geologists and mine supervisory staff. The units cost about $50 each and they were thus considered to be almost expendable. Dr. Bowie agreed that low cost was important, but also that it was very easy to shield the probe so that ore faces could be graded accurately.

Dr. Killeen stated that he would like some clarification of the mechanisms involved in producing the roll-front features found in Wyoming, especially with respect to where the radioactivity was located. What were the typical dimensions of a roll front—were they of the order of a few feet or hundreds of feet in width? Also, how fast was the uranium moving across the front and was there a significant radiometric anomaly in the area containing the youngest uranium? Mr. Grutt said that the ore front was generally contained within impervious beds above and below. The ore fronts varied from 5 to about 100 ft in width. It was thought that the front was related to a solution front with a mildly oxidizing solution on one side and a strongly reducing solution on the other. Bacteria were also involved, including *Ferrooxidans* or *Thiobacillus* on the oxidizing side of the front and *Desulphovibrio*, sulphur-reducing bacteria, on the other. There was also a structureless humate present which might have represented the membrane between the oxidizing and reducing solution. Bicarbonate-bearing or sulphate-bearing waters also played a part. Mr. Grutt said there were two schools of thought about the stability of those roll fronts. One believed that once the roll front had been formed it was static and therefore never moved again; the other that there was a continuous evolutionary movement of uranium from one point of the roll front to the other. In old deposits there was a very good correspondence between radioactivity and the uranium content. In younger roll-front deposits, however, there was considerable variation, parts of them higher in uranium content than indicated by radioactivity. In Wyoming the roll fronts were generally young, whereas ore deposits such as the Uravan deposit were very old indeed.

In the course of discussion on geochemical methods, Mr. Chico suggested that more use might be made of paper chromatography, since a kit, costing about $300, was available commercially. The speaker had used that in the field, but still preferred the quantitative chemical methods. Nevertheless, the method should not be rejected, and he noted that quantitative measurements for uranium could be obtained with special standards. Mr. Grimbert said that it was very important to use chromatography to separate uranium from quenching agents when carrying out fluorimetric determinations for uranium. The quenching was largely due to iron and manganese, but other elements could also have a considerable effect. Paper chromatography was used to separate the uranium from the other elements in the sample. The paper was then cut, ashed and the uranium composition of the ash was determined by standard fluorimetric methods. M. Coulomb described details of the procedure included in M. Grimbert's paper.

Mr. Nielsen enquired as to participants' experience of geochemical soil sampling in Arctic or sub-Arctic areas, where the average annual temperature was very low and, consequently, the chemical weathering of the rock surface at a minimum. Dr. Darnley said that although geochemical soil survey studies had not been done specifically for uranium, certain pilot studies had been carried out, the results of which were very interesting.

Dr. H. W. Little said that he was interested to know that neutron activation analysis cost about 30% less than fluorimetric analysis as estimated by the Geological Survey of Canada. Neutron activation analysis would be fine for waters, but in the case of stream sediments it seemed to him that the uranium in refractory minerals would also be determined. If stream sediments were collected to trace the origin of a uranium deposit, only the uranium soluble in nitric acid would be useful, and that in the refractory minerals would be misleading. He suggested, therefore, that perhaps it would be better to use fluorimetric analysis on stream sediments. Mr. Ostle replied that he had shown in his presentation of data from northern Scotland that a study of the distribution patterns of total uranium in stream sediments and waters could provide an indication of the mode of occurrence of the uranium in a region, whether in refractory or soluble form, and therefore assist in the further selection of areas for follow-up. Uranium in stream sediments, in whatever association, had a much shorter dispersion train than that in water and, when sediment anomalies were detected, it involved the use of only a little mineralogy to identify the uranium association in the sample and, if that proved favourable, the source was likely to be easily located in the field. The advantages of the delayed neutron analytical method compensated for the possibility of having to introduce an additional step in the interpretation and follow-up procedure. M. Grimbert said that normally his group used only nitric acid dissolution and that uranium determination would therefore be made on the nitric acid soluble fraction of any stream sediment. Mr. R. F. Coleman pointed out that there was no reason why exactly the same procedure should not be followed and the neutron activation method used to analyse the solution for uranium, if it was required to determine acid extractable uranium. The neutron activation method therefore offered the options of analysing directly for total uranium or of determining acid-soluble uranium. He emphasized that the additional leaching steps put up the overall cost of the analysis.

Mr. H. Müller asked the panel what methods were available to distinguish the border between oxidizing and reducing zones. Mr. Grutt said there was no quick way—usually it depended on applying the geologist's experience. The most practical method was to use visual logging. He thought the solution would be a Redox log, but that was not yet available, and Mr. Dodd said that there were currently two commercial companies about to start using an Eh/pH log in the U.S.A. Dr. Koch said that the Redox normally changed so quickly that he had doubts as to the applicability of borehole logging. Mr. Dodd added that that was one of the objectives of the magnetic susceptibility log. The alterations in the rock destroyed a small percentage of the remaining magnetite in that section. It should be possible to see in the borehole the same effect as had been observed on actual samples in the laboratory, namely a lower magnetic susceptibility on the oxidized or altered size of the roll front.

Mr. I. R. Wilson asked whether Mr. Ostle could give any information on the uranium content of granites in the United Kingdom, particularly the differences, if any, between mineralized and unmineralized granites. Mr. Ostle replied that there were considerable differences in the uranium content of British granites. The Helmsdale granite had already been discussed by Dr. Gallagher. That was probably the highest in uranium of any in the country. The southwest of England granites were high, something of the order of 4–6 ppm (8–10 ppm for the Helmsdale granite). M. Grimbert added that in France the values of mineralized granites were considerably higher, but they varied according to the oxidation state of the granite. Below the water-table the value fell to about 10 ppm. That particularly applied to the granites in the Limousin area of France.

Mrs. Jane Plant asked M. Grimbert to what extent the use of an acid-soluble uranium technique increased the number of anomalies detected and how it eliminated the effect of organic concentration or secondary oxide concentration of the uranium. M. Grimbert agreed it was likely that there was absorption of uranium on iron and manganese oxides and that when the nitric acid attack was made it did take off part of that uranium. Professor Dall'Aglio said that in Italy uranium analysis on a routine basis had been carried out by use of nitric acid leach with chromatography and a fluorimetric finish. The tests had shown that approximately 50% of the uranium from the iron and manganese oxide had passed into solution. He stated that a practice had been made of ashing stream-sediment samples in order to eliminate the effect of carbonaceous matter on the analysis. That was necessary because determinations other than for uranium were frequently carried out on the same sample.

Mr. H. Kunzendorf asked whether anyone had considered the use of non-dispersive X-ray fluorescence spectrometry in uranium analysis. Dr. Bowie replied that XRF techniques were used extensively for the study of stream-sediment samples in the United Kingdom and that it was quite a useful method. He looked on it as an adjunct to the normal methods used, for example, neutron activation analysis. Commenting on the question of analysing for total or acid extractable uranium, Dr. Bowie referred to Mr. Coleman's observation that neutron activation analysis could be employed for either approach. He tended to favour analysis for total uranium.

Mr. G. A. Kingston asked whether, in geochemical analysis for uranium in soil and stream-sediment samples during a reconnaissance survey, account was taken of the iron and manganese content, since enhancement of the uranium would occur due to coprecipitation, and that information, together with Eh and pH, would be required in interpretation. M. Grimbert replied that it was very difficult and time-consuming to do all of those determinations directly on samples as a matter of course. Determinations of iron and manganese, pH, etc., should, however, be carried out on samples which were shown to be anomalously high in uranium. Commenting on the role of iron, Mr. Dodd said that was why so many anomalies were detected on the laterites in tropical weathered earth. Very seldom was any uranium found, even though the anomalies were considerable. Mr. Dodd suggested tentatively that at the time of the change from ferrous to ferric hydroxide and iron to limonitic material any uranium which had been held in the material at an earlier period had generated thorium daughter products. Uranium was released at the time of limonite and hematite formation and removed in the groundwater, the thorium remaining behind to fit readily into the hematitic lattice and continue its 10^{10} year half-life to generate ultimately thallium and lead. It was those which accumulated to give another kind of disequilibrium anomaly in the tropics. Mr. Ostle said that the process outlined by Mr. Dodd provided an explanation to observations he had made many years ago that the high radioactivity of red soils overlying some of the carbonatite complexes in Central Africa, Chilwa in particular, could be attributed entirely in thorium. Uranium occurred in pyrochlore in the calcareous rocks, but it never reported in the *terra rossa*, yet the latter were radioactive due entirely to thorium, the uranium having completely gone.

Dr. Bowie said that he would like to make the observation that mineralogy was extremely important in the matter of solubility. Mrs. Plant had raised that point and he would like to emphasize it. For example, if thorium were present in the uraninite lattice, the whole mineral became more resistant to attrition and to solution. In the Indus Valley, for example, uraninite with about 25% thorium was just as resistant as monazite, but if the thorium were not present, it would go into solution very readily indeed. That was one of the topics on which the speaker would like to have seen much more work because, in the case of the Blind River and the Witwatersrand deposits, most of the grains that looked as though they were detrital in form were found to be high in thorium (15–25% ThO_2). He did not accept that there was any special significance to be attached to a 2300 m.y. age, and considered it worth while repeating that undoubted detrital minerals in the Dominion Reef had been dated at 3100 m.y., whereas the main mineralization of the Witwatersrand was about 2100 m.y. old. The same applied in the Blind River field, where monazite and zircon had been dated at 2500 m.y., brannerite at 2000 to 1800 m.y., and uraninite at 1740 to 1680 m.y. old. Other near-concordant age determinations had shown that much uraninite at Blind River was 1200 to 600 m.y. old. That had to be taken into account before a valid pronouncement could be made on the mode of origin of conglomerate-type ore deposits. More detailed mineralogy, chemistry and isotope ratio studies were needed to supplement field observations.

In conclusion, he thanked all who had contributed to a very successful symposium and a useful discussion.

Name Index

Adams J. A. S. 107, 146, 158, 164, 171, 172, 205, 318
Adler H. H. 11, 75, 247, 273
Aitken R. N. 108, 241
Amiel S. 97, 100, 107
Anzalone S. A. 15, 319
Archbold N. L. 76
Armstrong C. W. 148
Armstrong F. C. 75
Austin S. R. 62, 65, 66, 75, 155, 266, 274, 275

Backström J. W. von 11, 24, 31, 147, 287, 324
Ball T. K. 12, 95, 106, 109, 117, 239, 317, 318
Baranov V. I. 239
Barnes F. Q. 12, 31, 32, 45, 78, 79, 93, 172, 287, 295, 320, 324, 327, 328
Barretto P. M. C. 173
Bartel A. 322
Basset W. A. 15
Beckinsale R. D. 15, 32, 319, 320, 322
Berbezier J. 321
Bergendahl M. H. 76
Berkman D. A. 44
Berkoff E. W. 253, 274
Berthollet P. 117
Bisby H. 11
Blake D. A. W. 221, 239
Blake R. W. 267, 274
Bowie S. H. U. 1, 11, 12, 13, 117, 172, 210, 239, 312, 320, 321, 322, 323, 324, 325, 326, 328, 329, 331
Bowles C. G. 74, 78
Boxer L. W. 30
Boyle R. W. 121, 132
Bristow Q. 206
Broding R. A. 273
Brooks J. H. 44, 293, 294
Buffam B. S. W. 44
Bunker C. 107
Bush W. E. 173

Cameron J. 109
Cannon H. L. 8, 12
Card K. D. 311
Carmichael C. M. 166, 171, 306, 311
Carter E. K. 294
Casentini E. 133
Chandler T. R. D. 12
Charbonneau B. W. 205, 206
Chenoweth W. L. 74
Chico R. J. 12, 30, 45, 77, 94, 133, 275, 300, 318, 321, 324, 326, 327, 328, 329
Ching Chao W. 325
Christ C. L. 123, 133
Christie A. M. 221, 239
Clary T. A. 63, 75
Coats J. S. 318
Cockcroft *Sir* J. D. 174
Coleman R. F. 12, 95, 107, 108, 109, 317, 318, 330, 331
Cook, B. 205
Cooper J. A. 44
Coulomb R. 254, 274, 277, 329
Courbois Th. 159, 171
Craig L. C. 76
Crew M. E. 253, 274
Crouthamel C. E. 171
Curry D. L. 68
Czubek J. A. 248, 259, 261, 262, 264, 267, 268, 273, 274, 275

Dahill M. P. 76, 77
Dahlberg E. C. 240
Dall'Aglio M. 15, 108, 121, 123, 133, 241, 287, 330
Danes Z. F. 261, 274
Daniels J. L. 293
Darnley A. G. 11, 16, 77, 78, 93, 139, 146, 148, 174, 177, 185, 202, 205, 206, 242, 288, 311, 312, 321, 322, 326, 327, 328, 329
Dass A. S. 239
Davidson C. F. 2, 11, 33, 35, 43
Davis F. J. 146, 321
Davis J. F. 75
Davis M. 17, 18, 30, 93, 242, 317, 318, 327
Dean B. G. 76
Deans T. 318
de la Hunty L. E.: *see* Hunty L. E. de la
Denham G. M. 181, 206
Denson N. M. 55, 74
Derriks J. J. 44
Derry D. R. 35, 43
deVergie P. C. 274
De Voto R. H. 146, 240
DiGiovani H. J. 274
Dodd P. H. 133, 146, 147, 155, 172, 173, 242, 243, 244, 247, 248, 252, 253, 266, 273, 274, 275, 276, 321, 326, 328, 329, 330, 331
Dodson R. G. 33, 43, 44, 119, 295, 320, 324, 326, 327
Doe B. R. 107
Doig R. 311
Domzalski W. 14, 209, 305
Donhoffer D. K. 206
Dorn F. E. 212
Douglas R. F. 44
Dowie D. L. 77, 288
Droullard R. F. 155, 247, 248, 252, 253, 268, 273, 274, 275
Dunham *Sir* K. C. 13

Dunn P. R. 43
Durham C. C. 239
Duval J. 205
Dyck W. 6, 11, 109, 117, 133, 212, 239, 240

Eargle D. H. 69, 72, 75, 76
Ellis J. R. 155, 273, 275
Elliston J. 43
Elworthy R. T. 216, 240
England R. N. 35, 36, 43
Eschliman D. H. 146, 244
Evans R. D. 173
Everhart D. L. 294

Faulkner R. L. 19, 30
Finch W. I. 76
Fischer R. P. 44, 60, 75, 76
Flanagan F. J. 102, 107
Flawn T. P. 76
Fleet M. 206
Fleischer M. 102, 107
Frarey M. J. 311
Frondel C. 206
Fuchs H. D. 45, 107, 171, 242

Gabelman J. W. 12, 13
Gale N. H. 107, 108
Gallagher M. J. 107, 313, 317, 318, 330
Gangloff A. 14, 320, 325
Garrels R. M. 55, 74, 76, 123, 133, 294, 324
Garrett R. G. 121, 132, 228, 239, 240
Garson M. S. 318
Gasparini P. 158, 171, 205
Gautier R. 11
Ghosh A. K. 322
Gobel D. 68
Godby E. A. 206
Gordon G. E. 107
Gott G. B. 54, 65, 74, 75
Granger H. C. 61, 75, 76
Grasty R. L. 11, 185, 202, 205, 206, 311
Graveson R. T. 274
Greenwood P. B. 317
Gregory A. F. 206
Griffith J. W. 12, 308, 311
Grimbert A. 107, 110, 117, 240, 241, 320, 321, 325, 329, 330, 331
Grundy W. D. 274
Gruner J. W. 55, 74
Grutt E. W. Jr. 47, 72, 74, 75, 77, 93, 133, 320, 321, 322, 325, 327, 328, 329, 330
Guitton J. 262, 273

Hagmaier J. L. 54, 74, 133
Hallenburg J. K. 247, 273
Halligan R. 293
Hamaguchi H. 107
Hansen J. 171
Harshman E. N. 76
Harsveldt H. M. 109
Hart O. M. 65, 75
Haynes L. 133, 317
Heier K. S. 107
Heinrich E. W. 43, 206, 293
Herzberg W. 324
Hiemstra S. A. 43
Higgins F. B. Jr. 213, 240
Higgins L. J. 173
Hilpert L. S. 60, 75
Hinds G. W. 75
Hobbs J. D. 228, 239
Holen H. K. 62
Hood P. J. 182, 206
Horbert M. 240
Horwood J. L. 206
Hose R. K. 75
Huggett M. G. 12, 148, 208
Hughes F. E. 44
Hunty L. E. de la 293

Ibraev R. A. 133
Israël G. W. 240
Israël H. 215, 240
Iten K. W. B. 294

Johnson P. 156, 327
Joubin F. R. 10, 43

Kapkov Yu. H. 240
Kauranen P. 241
Keith M. L. 240
Keller W. D. 61, 75
Kelley V. C. 44, 76
Kerr P. F. 1, 11
Kidwell A. L. 76
Killeen P. G. 31, 166, 171, 306, 311, 312, 322, 329
King J. W. 65, 66, 75, 266, 274
Kingston G. A. 331
Kitto P. L. 147, 276, 326
Kleinhampl F. J. 12
Klepper M. R. 33, 43
Klinge U. 156, 275
Klohn M. L. 76
Koch L. W. 275, 288, 322, 327, 328, 330
Koeppel V. 11
Krcmar B. 215, 240
Krumbein W. C. 228, 240
Krusiewski S. V. 12
Kulp J. L. 11
Kunzendorf H. 171, 331
Kvendbo B. O. 156, 241

Lambert I. B. 107
Lambert R. 192
Langen R. E. 76
Lanphere M. A. 15
Lathan C. P. 274
Lawrence G. 209, 301, 304, 305, 319
Lawrence L. J. 294
Lebreton F. 274
Lenda A. 261, 264, 274
Leutz H. 171
Link J. M. 33, 43
Little H. W. 209, 330
Loosemore W. R. 11, 135, 147, 148, 318
Loriod R. 117, 240
Louis P. 282
Løvborg L. 108, 147, 157, 166, 171, 275
Love J. D. 67, 75
Low G. H. 294
Lowder W. M. 146, 158, 171, 172, 205
Lynch J. J. 239
Lyubavin Yu. P. 267, 274

McAndrew J. 44, 294
McCartney W. D. 319
McKelvey V. E. 294
MacLeod W. N. 294
McNail E. E. 107
Mabile J. 30
Macdonald J. A. 227, 234, 240
Madigan R. T. 294
Malan R. C. 57, 60, 75
Mapel W. J. 76
Meehan R. J. 274
Meilleur G. 239
Meschter D. Y. 54, 74
Meyer W. 15, 32, 78, 147
Meyer W. T. 45
Meyertons C. T. 148, 242
Michie U. McL. 313, 317, 318
Miller D. S. 11
Miller J. M. 11, 135, 147, 148, 242, 275, 315, 318
Montgomery J. L. 289, 328
Moore J. M. 15
Moore P. J. 317
Moore R. 107
Morblay C. M. 75
Morgan B. D. 294
Morgan D. A. O. 94, 119, 327
Morgan J. W. 107
Morlock J. S. 35, 43
Morse R. H. 12, 117, 219, 240, 243
Moulton G. F. Jr. 75
Mukashev F. A. 133
Müller H. 156, 243, 330
Munro D. L. 44

Neale E. R. W. 319
Nel L. T. 2, 11
Newson N. R. 77, 241
Nicholls G. D. 107
Nielsen B. L. 172, 276, 287, 328, 329
Nininger R. D. 1, 11, 30
Noldart A. J. 294
Norton D. L. 69, 75
Novikov G. F. 240

Obellianne J. M. 117
Onuma N. 107
Osterwald F. W. 76
Ostle D. 12, 95, 106, 107, 108, 109, 117, 119, 147, 239, 305, 317, 318, 321, 330, 331
Ovchinnikov A. K. 267, 274

Page L. R. 1, 11
Partington J. R. 240
Patterson J. A. 274
Peacock J. D. 11, 240
Pelchat J. C. 239
Pickens W. R. 76
Pirson S. J. 274
Plant J. 317, 318, 330, 331
Pleizier G. 107
Plumb K. A. 43
Pontecorvo B. 174
Prain *Sir* R. L. 94
Pratten R. D. 171
Prichard C. E. 44, 294
Prutkina M. I. 173
Puibaraud Y. 149

Rackley R. I. 76, 77
Ramdohr P. 35, 43
Raumer J. von 171
Reinhardt P. W. 146, 321
Renfro A. R. 65, 75
Ridge J. D. 75, 76
Roberts H. G. 43
Robertson D. S. 44
Robertson J. A. 240, 311
Robinson C. S. 76
Robinson S. C. 44
Rod E. 43
Rodger T. H. 109, 147
Rogers A. S. 213, 240
Rogers J. J. W. 318
Roscoe S. M. 11, 43, 308, 311, 322, 323, 324, 325
Rose A. W. 228, 240
Rosholt J. N. Jr. 133, 266, 274, 322
Rouse G. E. 146, 240
Roux A. J. A. 31
Rubin B. 77
Rummerfield B. F. 273
Russell R. T. 74
Rutherford *Lord* 216
Ruzicka V. 240

Ryan G. R. 13, 45, 94, 119, 155, 171, 172, 241, 296, 323, 326
Rybach L. 161, 164, 171

Sakakuva A. Y. 321
Santos E. S. 76
Satterly J. 216, 240
Saum N. M. 33, 43
Schiltz J. C. 108
Schmidt-Collerus J. J. 53, 74
Schnabel R. W. 65, 74
Schultz L. G. 76
Schwarzer T. F. 109, 148, 172, 209, 325, 328
Scott J. H. 173, 252, 253, 254, 255, 256, 266, 274
Sears J. D. 71, 75
Sharp B. J. 247, 273
Shashkin V. L. 173
Shatwell D. O. 44
Shawe D. R. 76
Shay R. S. 173
Shockey P. N. 76, 77
Simmons G. C. 76
Simpson A. 318
Smith A. R. 158, 171
Smith A. Y. 212, 240
Smith E. E. N. 31, 45, 93, 239, 325, 328, 329
Smith G. H. 12
Smith H. B. 1, 11
Soister P. E. 66, 75
Sørensen H. 168, 171
Sprague R. B. 14, 325, 326
Spratt R. N. 44, 294
Steacy H. R. 43
Stephens J. G. 55, 74
Stern T. W. 12
Stevens D. N. 146, 240
Stewart J. H. 76
Stewart J. R. 294
Stieff L. R. 12
Stocking H. E. 1, 11
Suhr N. H. 240
Sullivan C. J. 294
Syromyatnikov N. G. 133

Tanner A. B. 144, 146, 172, 213, 240, 274
Tatsumoto M. 107
Taylor J. 42, 44
Tipper D. B. 301, 305
Tittle C. W. 274
Tittman J. 273, 274
Tonani F. 123, 133
Tremblay L. P. 218, 221, 241
Tugarinov A. I. 8, 12

Vaes J. F. 44
Van Gelderen L. 171
Van Houten F. B. 76
van Wambeke L.: *see* Wambeke L. van
von Backström J. W. : *see* Backström J. W. von
von Raumer J.: *see* Raumer J. von

Wahl J. S. 273
Walker K. R. 294
Walpole B. P. 35, 43, 294
Wambeke L. van 2, 11
Waters A. C. 76
Wedepohl K. H. 318
Weeks A. D. 72, 75, 76
Weiss O. 9, 12
Weitz J. L. 75
Wennervita H. 241
Whalen J. F. 72, 75
White J. H. 38
Williams R. M. 30
Williamson R. 11, 240
Wilson I. R. 44, 171, 330
Wollenberg H. A. 107, 158, 170, 171
Wood H. B. 56, 60, 74
Wyant D. G. 33, 43
Wyatt J. D. 294

Yoli A. H. 274

Zeller H. D. 55, 67, 74, 75
Zimens K. E. 172

Subject Index

^{227}Ac 136
^{228}Ac 136, 137
Acid rocks 2
  igneous rocks 3, 4, 10, 12, 33, 122
Aeroradiometric surveys 1, 4, 5, 10, 13, 15, 82, 84, 110, 113, 138, 139, 150, 174–211, 289, 296, 304, 315
  aerial photography 5, 85, 113
    black and white 85
    colour 85
  total-count surveys 1, 4, 13, 138, 153, 177, 188
Ages and names of host rocks (sandstones) 48
  review of selected criteria 48
    depositional environment of host rock 49
    features of host rock 52
    physiographic province 51
    provenance of host rock 49
    special stratigraphic factors 50
    tectonic element 51
    type of host rocks 50
Agnew Lake Mines 88, 89, 311
Airborne gamma-ray survey techniques 174–211
Airborne scintillation counters and spectrometers available for purchase 182–3
^{28}Al 99, 108
Allanite 4, 42
Alligator Rivers deposits 33, 34, 39, 41, 42, 45, 289, 304, 320
Alpha probes 144, 319
Alpha radiation 149, 150, 154, 319
Altered basalt 39
^{241}Am 139, 143, 181
Ambrosia Lake deposit, New Mexico 15, 48, 53, 62
Analogue profiles 301
Analytical techniques for geochemical analysis of uranium 96
  delayed neutron measurement 96, 97, 112
    block diagram of neutron detection system 99
      experimental procedure 98
    irradiation of sample in reactor core 98
    sampling procedures 97
Anatase 62
Angola 4
Anomalous radioactivity 38, 40, 54, 73, 104, 116, 131, 289, 301, 302, 306, 315
Appalachians 13, 87, 319
Application of computers to the assessment of uranium deposits 277–88 (*see also* Computers)
  checking, correction and storage 280
  continuous recording of radiation logs from developed deposits 278
  data storage and retrieval 278
  'deconvolution' principle 283
  digitization 279
    automatic scanner digitization 279
    point by point on a plotter 279
  mining projects 287
    determination of most probable ore site 287
  non-continuous radiometric logging 278
  ore-reserve estimation 282
    count rate against distance 285
    gamma logs 286
    plots of flight lines superimposed on base map profiles 286
    punching machines and punch codes 278
    regression analysis 283
    type of data 278
Argillite 40, 293, 308, 315
Arlit deposits, Niger Republic 119
Arsenic 3, 53, 66
Assessment of uranium by gamma-ray spectrometry 157–73
  automatically operating gamma spectrometer 163
  core-scanning apparatus 164
  data recording and data reduction 159
  field gamma spectrometry 165
  laboratory spectrometry 158
    background suppression 158
Assessment of uranium occurrences 9
Astragalus (indicator plant) 7, 53
Athabasca surveys 192, 195
Athabasca to Soulier Lake survey 198
Atomic Energy Board of South Africa 31
Atomic Energy Research Establishment, Harwell, Didcot, Berkshire, England 135, 143
Atomic Weapons Research Establishment, Aldermaston, Berkshire, England 95, 96
Auger drill holes 38
Automatic processing of hydrogeochemical data 129
Autunite 38, 132

^{10}B 98
Background spectra 158, 159

Bancroft area, Ontario 4, 178, 209, 212, 213, 218
Bancroft surveys 192, 197, 202
Basic rocks 2
Beaverlodge area, Saskatchewan, Canada 2, 213, 218
  analyses of stream waters, lake waters and sediments 229–38
Bedford Downs area, Western Australia 292
Bentonite 60, 69, 72
Bentonitic mudstones 60
Beryllium 4
Betafite 4
Beta-probe attachment 9
Beta radiation 150
Beta-radiation surveys 6
^{210}Bi 8
^{214}Bi 5, 14, 136, 150, 178, 184, 185, 186, 207, 209, 242, 266, 267, 301, 306, 321
Bingham, Utah 15
Black Hills, South Dakota and Wyoming 48, 51, 54, 63, 64, 78
Blanket or trend deposits 51, 52, 64
Blind River 3, 10, 13, 14, 79, 289, 311, 312, 324
  sandstones 13
Borehole logging 145, 146, 154, 155, 172
Borehole logging techniques for uranium exploration and evaluation 244–76 (*see also* Logging programmes and applications)
  count rate (gamma intensity) 251, 252, 254, 255
  diagram of the natural decay series of uranium, thorium and potassium 249
  GAMLOG
    computer program 256
    data 256
  plot of identification and correlation of lithologic rocks and mineralized zones (gamma-ray, resistance and SP logs) 246
  techniques and correction factors (logging and analysis) 257
    correction for disequilibrium 265, 275
    correction for interstitial water 260
    correction for atomic number, Zeq, effect 261
    dead time 257
    effects of borehole diameter and fluid 258
    effects of casing 259
    non-standard borehole conditions 258
    non-standard conditions in the formation (ore zone) 260
  theory and practice of log analysis 253
    automated analyses by computers 253
Boron 100, 127
Boron trifluoride counters 98
^{87}Br 98
Brannerite 14
Breccia 37, 40, 43
  skarn 42
Breeder reactors 31, 32
  use of thorium 31, 32
Broken Hill, New South Wales 37
Bureau of Mineral Resources, Geology and Geophysics, Canberra, Australia 33, 35, 43, 290, 296

Cabinda 4
Caesium 108
Calcareous concretions 65, 74
Calcilutite 36
Calcite 54
Canadian Shield 2, 13, 78, 85, 86, 93, 175, 184, 208, 209, 212, 218, 239, 308
  Huronian rocks 2, 87, 88
  uraniferous conglomerates 2, 88
Carbonaceous materials 36, 45, 53, 55, 60, 61, 64, 66, 67, 69, 72, 73
Carbonatite 314
Catamarca and Mendoza Provinces, Argentina 36
Chalcocite 38
Chalcopyrite 38, 45
Chalk River Laboratories, Canada 175
Chelation (fixation of uranium) 53
Chlorite 62
Chromatography as aid to determination of uranium 96, 112, 118
Clastic particles 119
Clastic rocks 73
Climax deposit 15
^{57}Co 139, 181
^{60}Co 158
Coarse clastic formations 289
Cobalt 3, 38
Colorado Plateau 2, 3, 7, 15, 36, 47, 51, 289, 320
Commissariat à l'Energie Atomique, Fontenay-aux-Roses, France 110, 149, 277
Compton scattering 5, 6, 13, 137, 148, 165, 179, 185, 188, 189, 190, 193, 250, 261, 262, 263
Computers (Commissariat à l'Energie Atomique, France 277 (*see also* Application of computers to the assessment of uranium deposits)

main computer, IBM 360–91 (4000K) at Saclay 277
Concentration factor (in aeroradiometric surveys) 5
CON-COR technique 299
Conglomerate deposits 2, 33–6, 39, 45, 63, 66, 293, 295, 308
Copper 3, 53, 60
native 38, 40
Copper mineralization 45
Copper minerals 38
Cordillera, Canada 51, 87
Core drilling 300
Crystal–photomultiplier assembly 158, 159
^{137}Cs 139

Danish Atomic Energy Commission, Research Establishment, Risø, Denmark 157
Decay series diagram (U, Th and K) 249
Denison mines 89
Desulfovibrio bacteria 53
Detectors 137–48, 176, 273
collimated 166
cryogenic semiconductor detectors 135
germanium–lithium 136–50
^{3}He 275
inorganic 148
isotropic 166, 167
multi-crystal 210
NaI crystal 139, 154, 175, 262
NaI(Tl) crystal 5, 137, 138, 140–5, 150, 152, 155, 158, 165, 181, 262, 265, 267, 301, 306, 315
radon 212
semiconductor gamma detectors (Ge-Li or SiLi) 137, 146
Determination of uranium in natural water 118
Detrital minerals 2, 14, 62
Development of radon detectors 213
Diabase dykes 2, 12
Diamond drilling 9, 35, 38, 85–92, 299, 300
average costs in the Huronian, Ontario, Canada 85
rotary drilling costs in Tertiary sandstones, U.S.A. 90
Distribution map of main deposits of uranium in Australia 34
Dolerite 39, 40, 302, 304, 305
Dolomite 36, 37
Dominion Reef, South Africa 6, 324
Doppler control 296
Downhole assaying techniques 299
Drill core assay by gamma spectrometry 164

Eldorado mine property 203
total radioactivity survey map 203
Eldorado surveys 203, 204
Electrical conductance value 128, 129
Electromagnetic equipment 83, 93
Elements associated with uranium 53 (*see also* Indicator elements)
Elliot Lake–Agnew Lake areas 6, 87, 88, 107, 213, 218, 308, 312
Elliot Lake mines 35, 89, 91, 93
Elliot Lake surveys 192, 193, 194, 196
Environments of formation of uranium deposits 2–4, 33–46
Epigenetic uranium impregnations in sandstone 47
Equipment costs and airborne survey lease and contract rates 83
Eudialyte 169
European Nuclear Energy Agency 21
Experimental survey with a portable gamma-ray spectrometer, Blind River area, Ontario: a case history 306–12
field procedure 306
geology and survey locations 307, 308
Exploration costs of finding uranium 79
Exploring for uranium in the Northern Highlands of Scotland: a case history 313–9
abundances, methods of analysis and detectable limits for metals of economic interest 317
boat-mounted survey 315
Caledonian granite 314, 315, 317, 318
Caledonian syenite, Ben Loyal 313
granitic pluton, Helmsdale 313
Great Glen 313, 314
map of new uranium occurrences 314
methods of exploration 315
aeroradiometric 315
geochemical techniques 315
ground radiometric 315
Middle Old Red Sandstone (Devonian) sediments 313
Moine schists 313, 314
Moine thrust 313
Northern Highlands 313
Stromness district, Orkney, Scotland 313
vehicle-mounted radiometric survey 315

Fast reactors 27, 28, 32
Feldspars 4, 42, 45, 50, 77
Feldspathic quartzites 308

Ferghana Basin, U.S.S.R. 36
Fergusonite 4
Ferricrete 119
Finding costs (uranium) 81
Fluorimetric determination of uranium in soil 117
Fluorimetry (in uranium analyses) 96, 112, 117
Fluorine 125
  correlation between uranium and fluorine 131
Fluorite 125
Front Range, U.S.A. 78, 89
  vein-type deposits 89
Fulvic compounds 74
Fusion reactors 32

Galena 8, 40, 45
Gallium 108
Gamma anomalies 316, 318
Gamma logging 10–2, 299, 300
Gamma-ray emitting decay products of uranium 96, 135, 136
  in situ measurement of radiation from ^{214}Bi(RaC) 110
Gamma-ray probes 90, 316, 318, 319
Gamma spectrometry 4–9, 12–5, 79, 82, 110, 137, 157–70, 174–211, 301
  automatically operating gamma spectrometer 163, 170
  car-mounted scintillation counter 139, 140
  core-scanning apparatus 164
    continuous scanning 164, 169, 170, 171
    step scanning 165
  drill core assay 164
  field gamma spectrometry 165, 166
  from the air 5, 40, 89, 93, 138, 139, 174–211
    crystals 83
    influence of Th and K on count rates 5, 110
    linear radioaltimeter 5, 138
    scintillometers 14, 83, 92, 110, 137–46 (*see also* Detectors and Scintillation counter)
  gamma-ray digital spectrometer 82, 83
  laboratory spectrometry 158
    background suppression 158
  portable field spectrometer 110, 137, 141, 142, 157, 165–70, 177, 306
  three-channel gamma-ray counting 161, 163, 165
Garnet 42
Gas Hills, Wyoming Basins 48, 54, 65
Gatineau Hills, Quebec 213, 218
Geiger–Müller instruments 5, 6, 83, 84, 110, 137, 145, 149, 150, 152, 154, 155, 170, 282
Ge(Li) detector 136, 150, 179
Generation of nuclear power 17, 26
  assumed growth of capacity 19
Geochemical haloes 178
Geochemical mobility of elements 122, 123
Geochemical techniques in uranium prospecting 6, 110–20, 127
Geological Survey of Canada, Ottawa, Ontario, Canada 144, 174, 175, 208, 212
Geological Survey of Greenland 170
Geophysical methods 8
  electromagnetic methods 8, 298
  induced polarization 8
  resistivity 8, 298
Gold 45
Granitic rocks 2, 4, 10, 34, 41, 42, 49, 50, 55, 66, 67, 69, 73, 89, 153, 175, 227, 292, 296, 302, 304, 305, 313–8
Grants Mineral Belt, New Mexico 61, 77
Great Artesian Basin, Australia 295
  Mesozoic carbonaceous sediments 295
  phosphatic sediments 295
Great Glen, Scotland 313
Gulf Minerals Company 88
Gummite 38
Gypsum 129

^{4}He 10
Helmsdale granite 102, 104, 105, 106
Hematite 45, 67
Herald reactor 98, 100
Humic compounds 74
Huronian rocks 2, 87, 88, 306, 307, 308
Hydrogeochemical prospecting for uranium 6, 121–34
  automatic processing of hydrogeochemical data 129
    multiple-regression analysis 129
  climatic conditions 125
    desert climates 125
    tropical climates 125
  element content in brines during solar salt production 124
  elements (in stream waters) as a function of electrical conductance values 128
  geochemical mobility of the elements 123
  groundwaters 121, 125
  hydrogeochemical aureoles 125
  hydrosphere 122
  interpretation criteria 127

planning of hydrogeochemical surveys 126
Hydrothermal–mesothermal deposition 41, 42

^{131}I 143
^{137}I 98
Igneous rocks 32, 122, 163
Ilímaussaq, Greenland 4, 157, 158, 167–70, 172
Ilmenites 113
Indicator elements 7, 55
arsenic 7, 66
cobalt 7
copper 7, 53, 60
lead 7
manganese 53, 67
molybdenum 7, 57, 64, 66, 72, 73
phosphorus 53, 66
selenium 7, 62, 66, 67, 72, 73
vanadium 7, 57, 60, 61, 64, 65, 67
Indicator plant, *Astragalus* 7, 53
In-hole spectrometry 244, 262
evaluation of low-grade resources 244
evaluation of mixed U–Th ores 244
identification of the anomalous isotope 244
indication of geochemical environments 244
investigation of disequilibrium 244
resolution of low-intensity radium–uranium haloes 244
Institute of Geological Sciences, London, England 1, 11, 95, 96, 135, 143, 313, 317
Institute of Petrology, Copenhagen University 170
Instrumental techniques for uranium prospecting 135–48
Interior Plains, Canada 87
International Atomic Energy Agency 109
Ionization chamber 8, 150, 154, 212, 215
Ionizing radiation 149
Isomorphism of the uranyl ion with other elements 33

Johannite 38
Joint Congressional Committee on Atomic Energy 29

^{39}K 157, 161
^{40}K 110, 135, 150, 153, 158, 161, 179, 185, 186, 206, 207, 248, 301, 306
Kaolinite 62
Katherine–Darwin Province 296
Archaean granite 296
Nanambu granite 296
Kimberley Range, Australia 295

Laboratorio Geominerario del CNEN, Rome, Italy 121
Lac Forestier, Quebec Province, Canada 156
Laterite 119, 293, 304
Lawrence Radiation Laboratory, Berkeley 170
Lead 38, 45, 318
radiogenic in non-radioactive minerals 8, 10
Leucocratic rocks 3
^{9}Li 98
Light water reactors 20
Lignites 4, 54, 69
Limonite 67
Lisbon Valley, Utah 48, 55, 57
Lithium 100
Logging programmes and applications 245
gamma-ray log 9, 247, 269
resistance and resistivity log 245, 269
simultaneous gamma-ray and resistance logging 245
spontaneous-potential (SP) log 245, 269
Logs (supplementary) for uranium exploration 247
caliper log 247
experimental logs 248
Compton/photo ratio log 244, 248
induced polarization (IP) 248
KUT log 248
magnetic susceptibility 248
gamma–gamma density log 247, 270
magnetic susceptibility log 273
neutron–neutron log 247, 261, 269, 270, 271, 275
quantitative gamma-ray log 248
origin and characteristics of gamma rays 248
Löllingite ($FeAs_2$) 133

Magmatic differentiation 2, 171, 172
Magmatic uranium 4
Magnesite 36, 37
Magnetic susceptibility measurements 155, 244, 273
Magnetites 113
Magnetometer 14, 83, 93, 94, 156, 283, 296, 301, 305
Malachite 38
Manganese 53, 133
oxides 67
Mary Kathleen, Queensland 4, 34, 41, 42, 43, 289
Marysvale deposit, Utah 15
Mass spectrometer 8

Maybell and Baggs areas, Colorado and Wyoming 48, 71, 72, 77
Measurement of radiogenic heat 8, 15
Melanocratic rocks 3
Mercury 119
Metamorphic rocks 33, 34, 49, 66, 67, 69, 72, 73, 153, 295, 304, 316
Methods of collecting radon 216
  from natural waters 217
  from soil emanations 216
Molybdenum 53, 57, 62, 64, 66, 72, 73, 318
Monazite 169, 317
Montmorillonite 62
Montmorillonitic clay facies 61
Monument Valley–White Canyon area, Utah and Arizona 48, 57, 60, 77
Mount Isa area, Australia 42, 289, 292
Mount Painter, Australia 34, 37
Mudstone 66, 67, 73
Multiple migration–accretion concept 55
Multiple-regression analysis of data 129, 130, 224–8, 239
Multi-system surveying of large areas 301

^{17}N 98, 99, 100, 108
Nabarlek 39, 40, 41, 45
  Ranger deposit 40, 41, 42, 45
Nabarlek area, Arnhemland, Australia: a case history 301–5
  four-channel differential spectrometer 301
  gamma-ray spectrometric profiles 303
  geophysical and ancillary instrumentation 301
  geophysical results 302
  magnetometer 301
  Nabarlek anomaly 302
  NaI(Tl) detectors 301
Närke and Västergotland districts, Sweden 4
Neutron activation analysis as an aid to geochemical prospecting for uranium 1, 95–109, 112
  analysis of uranium by delayed neutron measurement 97
  block diagram of neutron detection system 99
Neutron breeder reactors 27
New Brunswick Laboratory of U.S. Atomic Energy Commission 160
Nickel minerals 38
Niger (Republic of) deposits 36, 46
  Arlit deposit 119
Niobium 4
Normalized gamma spectra
  of ^{40}K, assuming ^{39}K/^{40}K = 8500 161
  of ^{232}Th in secular equilibrium 160
  of ^{238}U in secular equilibrium 160
Nuclear power generation 17, 26
  assumed growth of capacity 19
Nuclides in equilibrium with ^{238}U or ^{235}U 136

^{17}O
Operating methods and cost in uranium search 81
Ore blocking 299, 300
  use of CON-COR technique 299
Orogenic processes 2

^{231}Pa 136
^{234}Pa 136
^{206}Pb 8
^{207}Pb 8
^{208}Pb 8
^{210}Pb 8, 158
^{214}Pb 147, 266
Pegmatite 3, 4, 6, 87, 93, 131, 175, 219, 241, 293, 304
Peneconcordant deposits 3, 61, 72
Peneplanation 45
Peralkaline rocks 4
Percussion drilling 298
Photoelectric absorption 250, 262, 263
Phosphorites 4
Phosphorus 53, 66
Phosphuranylite 38
Pitchblende (uraninite) 3, 4, 15, 16, 37–40, 42, 43, 45, 53, 295, 304
Planning and interpretation criteria in hydrogeochemical prospecting for uranium 121–34
Plots of radioactive emanations from soils (Gatineau Hills) 219
Plutonium 26
^{210}Po 8, 9
^{218}Po 147
Portable equipment for uranium prospecting 12, 137, 141, 142, 149–56, 157, 165–70, 212, 306
Potassium 5, 82, 150, 310, 311, 312
  radioactive isotope ^{40}K 110, 135, 150, 153, 158, 161, 179, 185, 186, 206, 207, 248, 301, 306
Powder River Basin, Wyoming 48, 54, 67, 68
Precambrian rocks 2, 78, 289, 313, 314
Precambrian Shield areas 2
  sediments 3, 4
Preferential concentration of uranium 177
Probe monitors 145
Production of uranium 24
Prospecting criteria for sandstone-type uranium deposits 47–78

Pulsed neutron activation 244
for direct determination of uranium 244
Pump monitors 144, 145
Pyrite 4, 8, 35, 36, 38, 40, 45, 52, 65, 72, 73, 156
Pyrite-bearing carbonaceous facies 3
influence on uranium mineralization 3
Pyrometasomatic deposits 4

Quartz 4, 40, 62
Quartz-pebble conglomerates 2, 10, 35
Quirke Lake syncline, Blind River area, Ontario 306–11

^{223}Ra 136
^{226}Ra 7, 136, 140, 243, 261, 265, 266
Radiation from uranium and thorium 135
Radiation spectrum at different heights 180
Radioactivity of uranium ores 150
elements ^{238}U, ^{235}U and ^{232}Th and their daughter products 150
Radioaltimeter 302
Radium 7, 114, 115, 119
measurement in water 114
Radium C (^{214}Bi) 5, 110, 150, 153, 178, 184, 209, 267
Radium Hill, Australia 34
Radium–uranium haloes 244, 266
Radon 6, 8, 9, 38, 135, 144, 147, 148, 172, 212–43, 319
alpha emission 154
anomalies 316
collection from natural waters 217
collection from soil emanations 216
counters 214, 215
detectors 213
measurement by radon gas emanometer apparatus 241
measurement in ground air 1, 6, 10, 212
measurement in overburden 144
measurement in soils 89, 115, 144, 156, 212
measurement in water 10, 84, 89, 114, 144, 212
portable counter 215
probes 316, 318
prospecting instruments 144
alpha probe 144, 145, 218
pump monitor 144, 145, 147
thorium-derived ^{220}Rn 144
Radon methods of prospecting in Canada 212–43
collecting radon from natural waters 217
collecting radon from soil emanations 216
radon, radium and uranium contents of stream waters, lake waters and sediments (Beaverlodge, Saskatchewan) 229–38
soil gas tests 218
surface water tests 219
Reconnaissance 82, 84, 86
application of geophysics and geochemistry, 82, 84
Requirements of uranium 27
estimated annual requirement 27
estimated cumulative requirement 28
Rhyolite 293
Rio Algom Quirke mine 219
^{219}Rn 135, 136
^{220}Rn 9, 135, 144
^{222}Rn 8, 10, 135, 136, 144, 242, 243, 266
Roll-type deposits 3, 15, 52, 54, 61, 65, 66, 67, 72, 73, 77, 133
Rössing, South West Africa 4
Rotary-percussion drilling 298
Rum Jungle, Northern Australia 34, 35, 37, 38, 41, 42, 45, 119, 289, 295
summary of main features of mines 38

Saleeite 38, 298
Sample specification for a high-sensitivity gamma-ray spectrometer survey 206
Sandstone areas 48, 55
Ambrosia Lake, New Mexico 48, 61
Black Hills, Wyoming and South Dakota 48, 63, 78
Gas Hills, Wyoming 48, 65
Lisbon Valley, Utah 48, 55
Maybell–Baggs, Colorado and Wyoming 48, 71, 72, 77
Monument Valley–White Canyon, Arizona and Utah 48, 60, 77
Powder River Basin, Wyoming 48, 67
Texas Gulf Coast 48, 68, 77
Uravan mineral belt, Colorado 48, 60
Sandstones 3, 13, 33–6, 45, 47–74, 155, 275, 293, 295, 304
aeolian sandstone 49
arkosic 3, 50, 55, 61, 65, 67, 73, 77
calcareous 36
feldspathic 50, 60, 72, 73, 77
ferruginous 63
fluvial 49, 53, 55, 60, 61, 63, 65, 67, 72, 73
fluviomarine 63, 64
hematitic 67

Jurassic 49, 73
marginal marine 49, 63, 69
Old Red Sandstone 105, 106, 313, 317, 318, 319
permeable feldspathic 50, 60
quartzose 2, 50, 60, 63, 69, 72, 73, 77
Tertiary 49, 73, 90
Triassic 49, 73, 81, 89
tuffaceous 51, 69, 77
Saskatchewan, Canada 4, 37, 88, 175
Beaverlodge area 213
Eldorado area surveys 203, 204
Saskatoon 289
Uranium City area 198
Scandium 108
Scapolite 42
Scintillation counter 14, 40, 83, 84, 89, 110, 137–48 (*see also* Detectors)
anthracene scintillation crystal assembly 175
car-mounted counter 140
hand-held 40, 298
organic scintillator 138
plastic or liquid scintillators 150
Rank Nucleonics scintillation counter 316
thallium-activated sodium iodide scintillation detector 5, 137, 138, 140, 141–5, 150, 152, 155, 158, 165, 179, 181, 262, 265, 301, 306, 315
Seal Lake, Labrador 4
Sedimentary rocks 33, 60, 65, 67, 68, 72, 87, 92, 153, 271, 273 (*see also* Sandstone areas and Sandstones)
Selenium 7, 53, 62, 66, 67, 72, 73, 133
Semiconductor gamma detectors 137
GeLi 136, 137
SiLi 137
Sericite–chlorite schist 40
Shales 4, 35, 49, 63, 73, 295
black 45
black pyritic 38
carbonaceous 39, 41, 49, 64, 66
chloritic 36, 38
graphitic 36
lignitic 54, 72
Siltstone 36, 39, 40, 60, 63, 66
bands 46
ferruginous 39
micaceous 39
pyritic carbonaceous 39
siliceous 39, 40
Sklodowskite 38
Slopoke reactor 109
Sodium metasomatism 314
South Terras, Cornwall, England 96
uranium–radium property 96
Spectrometers 179
differential (window) 181
four-channel differential 301
GSC Skyvan 186, 187, 188, 190, 192
integral (variable threshold) 181
multi-channel 179
portable three-channel 306
Scaler 299
Stack deposits 52, 61, 62
Stanleigh mine, Canada 89
Status of uranium prospecting 1–16
Stillwellite 42
Stratiform deposits 33
Stromness district, Orkney, Scotland 313
Supergene deposits 45
Survey grids 85
Syenite 313
Syenitic rocks 315
Syngenetic deposits 4, 41

Tantalum 108
Texas Gulf Coastal Plain 36, 47, 48, 51, 52, 68, 69, 77
^{227}Th 136
^{230}Th 136
^{231}Th 136
^{232}Th 100, 135, 136, 150, 160, 248
^{234}Th 136, 267
Thorite 169, 317
Thorium 4, 82, 93, 108, 150, 159, 170, 295, 308, 310, 318
thorium-derived ^{220}Rn 144
use of in breeder reactors 31, 32
Thorium C″ (^{208}Tl) 5, 110, 153, 179
Thorium minerals 313
Thoron 9, 135, 319
decay products 9, 135
Thucholite 35, 38
Tin 3
^{208}Tl 5, 136, 137, 179, 185, 186, 207, 301, 306
Torbernite 38
Total-count rates 137–43, 301
Total-count surveys 4, 13, 138, 140, 142, 148, 153
Tourmaline 40
Trend deposits 61, 62

^{234}U 136, 265, 266
^{235}U 97, 100, 135, 136, 146, 150, 248, 267, 275
^{238}U 135, 136, 144, 150, 157, 158, 160, 170, 248, 265, 266
Ultrabasic rocks 33
United Kingdom Atomic Energy Authority, London, England 17, 313
University of Western Ontario, London, Ontario, Canada 306
Uraninite (pitchblende) 3, 4, 15, 16, 37–40, 42, 43, 45, 53, 295, 304

Uranium
  assessment of occurrences 9
  association with Precambrian rocks and Phanerozoic rocks 2
  estimated resources 22, 23
  exploration costs 79
  finding costs 81
  from sea water 4, 10, 12
  genesis of deposits 1, 2
  ore minerals
    allanite 4
    betafite 4
    coffinite 53
    fergusonite 4
    torbernite 38
    uraninite (pitchblende) 3, 4, 15, 16, 37–40, 42, 43, 45, 53, 295, 304
    uranophane 4
    uranothorite 4
  production 24
  prospecting techniques 1
    aeroradiometric surveys 1, 4, 5, 10, 13, 15, 82, 84, 110, 113, 138, 139, 150, 174–211, 289, 296, 304, 315
    airborne geochemistry 18
    airborne magnetic surveys 301
    beta-radiation surveys 6
    boat-mounted survey 315
    colorimetric method (uranium in water) 7
    gamma-spectrometry surveys 4, 5, 6, 8, 9, 12–5, 40, 79, 82, 89, 93, 138, 139, 174–211
    geobotanical or biogeochemical methods 7
    geochemical surveys 6, 82, 92, 95, 96, 110–20, 127, 315
    geophysical methods 8, 82, 83
    hydrogeochemical prospecting 42, 113, 121–34
    hydrogeochemical, stream-sediment and soil sampling methods 6, 10, 106, 113, 121–34 (*see also* Hydrogeochemical prospecting for uranium)
    indicator-element techniques 7 (*see also* Indicator elements)
    instrumental techniques 135
    measurement of radiogenic heat 8, 15
    measurement of radon 6, 10, 84, 89, 114, 115, 144, 145
    neutron activation analysis 1, 7, 9, 95, 112, 146
    on-foot radiometric surveys 6, 153
    radiogenic lead in non-radioactive minerals 8, 10
    reconnaissance 82, 84, 86
    thermal infrared scanning 8
    vehicle infrared scanning 8
    vehicle-mounted radiometric survey 5, 146, 315
  reserves and production 21, 29
  resource regions, western United States 49
  supply and demand 17, 25, 26
  time parameter in deposition 2
  types of deposits and their environments 2
    magmatic uranium (Ilímaussaq, Greenland; Seal Lake, Labrador) 4
    oligomictic conglomerates (Blind River; Witwatersrand) 2, 3
    pegmatites (Bancroft area, Ontario; Rössing, South West Africa) 3
    pyrometasomatic deposits (Mary Kathleen, Queensland) 4
    sandstones (Colorado Plateau; Wyoming Basins) 3 (*see also* Sandstone areas and Sandstones)
    syngenetic origin 41
    syngenetic uranium in sediments (Västergotland and Närke, Sweden; North and South Dakota; Saskatchewan; Phosphoria and Bone Valley Formations, U.S.A.; Cabinda; Angola; West African localities) 4
    vein deposits 34, 37, 45
    vein-type deposits (France; Spain; Portugal; Czechoslovakia) 3, 15, 156
    vein-type occurrences in Australia 37
    vein-type occurrences in Canada 84, 87, 89
Uranium City area, Saskatchewan, Canada 198
Uranium exploration costs 79–94
Uranium search in Australia, 1968: a case history 289–95
Uranium supply and demand 17–32
Uranophane 4
Uranothorite 4
Uravan Mineral Belt, Colorado 48, 52, 53, 59, 60
U.S. Atomic Energy Commission, Grand Junction, Colorado 29, 47, 92, 93, 161, 244, 275
U.S. Geological Survey 138, 185
Use of geochemical techniques in uranium prospecting 110–20

Vanadium 3, 40, 53, 57, 61, 64, 65, 67, 73
Västergotland and Närke districts, Sweden 4

Vehicle-borne radiometric surveys 5, 146, 315
  Geiger–Müller counter 5
  scintillation equipment 5
Vein-type deposits 3, 15, 37, 45, 84, 87, 89, 92

Westmoreland, Australia 34, 41, 43
Witwatersrand System 2, 3, 14, 35, 289
Wyoming Basins 3, 15, 36, 47, 51, 52, 55, 77, 133, 320

X-ray fluorescence technique 146
X-ray (portable) generators 146

^{88}Y 207
^{88}Yt calibration source 307

Printed in England by Stephen Austin/Hertford